Herbal Formularies for Health Professionals

"Dr. Stansbury is a prescient, loving steward of the Standing Green Nation. Deeply loved by herbalists globally, her mentoring and clinical mastery powerfully manifest in this set of formularies. It is at once classic, complete, and enduring, offering herbal medicine savvy and deep research through articulate narratives and formulations for many conditions. Dr. Stansbury is always modest, grateful, and curious. Her friends, the sentient plants of a living earth, know this. Pachamama, too. The volumes themselves are beautiful objects whose rich content and elegant design are exemplary and respectful of this hard-earned, cumulative wisdom."
—DAVID J. SCHLEICH, PhD, president,
National University of Natural Medicine

"*Herbal Formularies For Health Professionals* is a gift to all healthcare practitioners seeking to heal their patients through a natural medicine approach. The expertise of Dr. Stansbury is extraordinary. This text is not a simple guide; it is a fascinating educational read and treatment formulary that is practical and evidence-based."
—MIMI GUARNERI, MD, president, Academy of Integrative Health and Medicine

"There are many herb books, but few address the fine points of clinical herbal prescribing. To remedy this deficiency, Dr. Jill Stansbury, ND, has written perhaps the best herbal formulary since the days of the renowned Eclectic physicians. . . . Each section and page clearly shows the skill, experience, and expertise of the author and will help even the seasoned naturopathic physician or herbalist improve their clinical practice and success in treating their patients."
—DAVID WINSTON, RH (AHG), dean of David Winston's Center for Herbal Studies; founder of Herbal Therapeutics Research Library

"*Herbal Formularies for Health Professionals, Volume I* is simply the best text available for the modern clinical practice of Western herbal medicine. It's likely to become an instant classic and essential book for any student of herbal medicine, as well as seasoned practitioners."
—CHRISTOPHER HOBBS, PhD, LAc, fourth-generation herbalist

"Fabulous! This book is an outstanding resource. Dr. Jill Stansbury, one of the most respected herbalists and naturopathic doctors of our time, has created an up-to-date formulary that combines the best of traditional formulas with modern research and her own extensive clinical experience."
—ROSEMARY GLADSTAR, herbalist; founder of Sage Mountain Herbal Retreat Center

"Esteemed clinical herbalist and naturopathic physician Jill Stansbury shares thirty years of experience with botanical medicines from all over the globe in this incredibly helpful text, which is part guidebook, part *materia medica*, and part recipe-based formulary, laced with an enthusiastic appreciation of all things herbal. It will no doubt be a frequently referenced text in the library of any herbal practitioner."
—TIMOTHY MILLER, ND, MAc, LAc, RA, founder of Naturopathic CE

"Accomplished naturopathic physician Dr. Jill Stansbury brings her vast clinical experience, deep knowledge, commitment, and passion for herbs to this exciting modern-day formulary. *Herbal Formularies for Health Professionals, Volume 1* is an extensive reference guide to the creation of classic herbal formulas using traditional herbal healing wisdom as well as modern science. This text should be a foundational part of herbal education. There is no other formulary as comprehensive or as accessible to the clinician; this book truly fills an empty space in the list of current herbal clinical education publications."

—DR. MARY BOVE, ND, author of *Encyclopedia of Natural Healing for Children and Infants*

"*Herbal Formularies for Health Professionals, Volume 1* delves deeply into herbal formulas for digestive, liver, urinary, and skin problems, along with descriptions and, where known, causes of these problems. In a chapter on the art of formulation, Jill Stansbury outlines example cases and provides appropriate formulas for each. Her considerations are fascinating."

—HENRIETTE KRESS, herbalist; founder of Henriette's Herbal Homepage

"Dr. Jill Stansbury's *Herbal Formularies for Health Professionals* is a formidable accomplishment and historic contribution to the rapidly developing field of botanical medicine. *Herbal Formularies* is practical and immediately usable for anyone interested in the applications of herbal medicine. Its depth will satisfy senior clinicians, academic researchers, and medical educators, while its accessibility will greatly further the education of students and the serious lay person."

—DAVID CROW, LAc, founder of Floracopeia Aromatic Treasures

"Reading Dr. Stansbury's formulary has quickly supported us in updating and filling in gaps in our dispensary stock to better serve our customers. I believe this book will similarly inform the dispensing practices of practitioners of all experience levels looking to glean insight from one of the most experienced herbalist physicians in the United States.

"Previous efforts have been predominately geared toward licensed acupuncturists or naturopathic physicians, but these volumes will speak to a broader, dynamic audience. This book offers mature templates for improvisation to treat a myriad of complex health conditions with botanical medicines."

—BENJAMIN ZAPPIN, LAc, herbalist; cofounder of Five Flavors Herbs

"Formulation is one of the key elements of a clinical herbalist's practice. It is also an aspect that intimidates many practitioners, especially those newer to herbal medicine. Jill Stansbury's book is a valuable guide to formulation that will lend a strong guiding hand for herbalists of all skill levels. Her language is clear, instructive, and easy to read. She gives multiple examples drawn from her extensive practice, and her explanations offer direction for working with individual patients. *Herbal Formularies for Health Professionals, Volume 1* is a comprehensive and practical book that I will recommend to my herbal students."

—7SONG, director of Northeast School of Botanical Medicine; Director of Holistic Medicine, Ithaca Free Clinic

HERBAL FORMULARIES FOR HEALTH PROFESSIONALS

VOLUME 3

HERBAL FORMULARIES FOR HEALTH PROFESSIONALS

VOLUME 3

ENDOCRINOLOGY

INCLUDING THE
ADRENAL AND THYROID SYSTEMS,
METABOLIC ENDOCRINOLOGY, AND
THE REPRODUCTIVE SYSTEMS

DR. JILL STANSBURY, ND

Chelsea Green Publishing
White River Junction, Vermont
London, UK

Project Manager: Sarah Kovach
Editor: Fern Marshall Bradley
Editorial Assistance: Patryk Madrid
Copy Editor: Nancy N. Bailey
Proofreader: Deb Heimann
Indexer: Shana Milkie
Designer: Melissa Jacobson

Printed in the United States of America.
First printing May 2019.
10 9 8 7 6 5 4   24 25 26 27 28

ISBN: 978-1-60358-855-3

Library of Congress Cataloging-in-Publication Data
Names: Stansbury, Jill, author.
Title: Herbal formularies for health professionals. Volume I, Digestion and elimination, including the gastrointestinal system,
    liver and gallbladder, urinary system, and the skin / Dr. Jill Stansbury.
Other titles: Digestion and elimination, including the gastrointestinal system, liver and gallbladder, urinary system, and the skin
    Description: White River Junction, Vermont : Chelsea Green Publishing, [2017] | Includes bibliographical references and index.
    Identifiers: LCCN 2017044410| ISBN 9781603587075 (hardcover) | ISBN 9781603587082 (ebook)
Subjects: | MESH: Formularies as Topic | Phytotherapy—methods | Digestive System Diseases—drug therapy | Urologic Diseases
    —drug therapy | Skin Diseases—drug therapy
Classification: LCC RM666.H33 | NLM QV 740.1 | DDC 615.3/21—dc23
LC record available at https://lccn.loc.gov/2017044410

Chelsea Green Publishing
White River Junction, Vermont, USA
London, UK
www.chelseagreen.com

To the Standing Green Nation—
all plants everywhere as living, breathing, sentient beings.
May we learn to treat them with respect and care.

And to all the traditional peoples, cultures, and nations
who have lovingly entered into sacred relationship with
the earth and the plants, and contributed to the art
and craft of herbal wisdom shared in these pages.

— CONTENTS —

Research in Herbal Medicine

In this set of herbal formularies, I draw from folkloric traditions and the wisdom of many indigenous plant-based healing practices of the world. I blend that traditional knowledge with molecular, animal, and clinical research on herbal medicines to help support, explain, and validate the long-standing practices of using specific herbal medicines in specific situations. Relying on modern research to validate herbal medicine is a double-edged sword. On the one hand, as the medical underdogs of the present era, many herbal clinicians strive to show the reasonably large volume of studies on plants published in medical journals to combat the oft-heard criticism that there is "not a shred of evidence" that herbal medicine works. Not only are there thousands of studies that demonstrate the efficacy of herbal medicine, but anyone who has ever put aloe on a painful sunburn or gulped down a large cup of prune juice might hold some truths to be self-evident. On the other hand, the type of herbal research being conducted, the claims being made on the basis of the research, and the public understanding of such data leaves a lot to be desired.

The problems with herbal research range from a lack of high-quality trials to unethical bias in trials funded by entities that have a vested interest in the results but a lack of understanding of the complex nature of whole plant medicines. Very little of the research on herbal medicine is double-blind placebo-controlled human clinical trials—the gold standard of research. And the majority of such high-quality clinical studies are conducted in China. These studies do not involve the herbs most commonly used in North American herbalism and typically employ complex herbal formulas, making it impossible to suss out the effects of any single herb.

Much of the research on herbal medicines is molecular studies or animal-based investigations. This is problematic because the finding that an herb has certain effects in mice, rats, dogs, rabbits, or monkeys does not necessarily mean that the herb would have the same effects in humans. Animal studies are used to create models of specific diseases. For example, plant extracts are tested on animals to assess their effectiveness at protecting pancreatic beta cells from diabetes-inducing toxins or the brain from ischemic stress. Or a study may be designed to show the effectiveness of a plant extract at supporting the recovery of hepatic stellate cells in animals in which cirrhosis has been induced. Performing ovariectomies on rats is a way to model menopause and allow researchers to investigate whether pre- or post-treatment with isoflavones or certain whole herb extracts can prevent the loss of bone density that occurs with a rapid decline in hormones. This type of animal research is commonly used to study pharmaceutical drugs as well; it is not unique to herbal research.

Some vendors, manufacturers, and commercial and educational websites, however, will use such findings to justify statements such as "builds bones in menopause" or "treats osteoporosis," when the only result the research demonstrated is a prevention of bone density decline in ovariectomized rats. Nonetheless, animal research is an enormous arm of medical research, and when a volume of animal and molecular research supports traditional ethnobotanical usage of the plant, perhaps it does add up to a meaningful conclusion. Some animal studies involve treatment that is ethically questionable, and that taints the research from an ethical standpoint. However, it would be silly and wasteful to ignore knowledge gleaned from this expenditure of life, and in these volumes, I cite many results of animal studies, for what they are worth. But I believe that there are other ways to mature our understanding of herbal medicine, and I hope to see animal research on herbal substances dispensed with altogether within my lifetime.

Cell cultures are another type of investigation used in herbal medicine research, but such research can also be misleading because cell cultures provide information on only one small sliver of the whole metabolic picture.

In cell cultures, tissues or cancer cell lines are cultivated to note whether various agents have antiproliferative effects on tumor cells, or whether a particular herbal constituent can inhibit the release of inflammatory cytokines from blood cells, for example. Cell culture studies might also focus on various compounds that can bind hormone receptors in prostate or endometrial cell cultures.

One example of the limitations of cell culture research is comparisons of studies of pharmaceutical serotonin reuptake inhibitors versus the action of St. Johnswort (*Hypericum perforatum*) on inhibition of serotonin uptake. Because St. Johnswort's effect was so much weaker than that of the pharmaceuticals, researchers concluded that the plant could not be a useful antidepressant. Next, St. Johnswort was found to promote dopamine activity, but researchers found it to be so much weaker than pharmaceutical dopaminergic agents that they doubted it could have clinical efficacy. And then St. Johnswort was noted to bind to GABA receptors, but because its effect was weaker than that of Valium, researchers doubted the plant could be an effective anxiolytic agent. Each of these investigations attempted to draw a conclusion based on a single molecular mechanism of action. Researchers failed to grasp that St. Johnswort's multiple mechanisms of action could yield a significant anxiolytic and antidepressant effect overall. When clinical studies were conducted administering St. Johnswort medications to human subjects suffering from depression, the herb demonstrated a significant mood-elevating effect: The action on the whole was greater than the sum of the individual molecular mechanisms. In hindsight, several factors contributed to the misjudgment of St. Johnswort. One problem was that scientists researching single mechanisms of action did not have cultural competence in the field of herbal medicine. In addition, no experienced herbalists were consulted about study design, and the results, therefore, were not meaningful to herbalists.

Another example of poor study design are the handful of clinical trials investigating *Echinacea* in the treatment of cold and flu. These expensive studies reported *Echinacea* to be no better than a placebo and became publicized as evidence that *Echinacea* "didn't work." However, study designers failed to understand that the traditional literature emphasizes *Echinacea* for decay, necrosis, and tissue breakdown rather than colds and flu. *Echinacea* is somewhat helpful for upper respiratory infection and influenza, but herbs such as *Sambucus*, *Thymus*, *Euphrasia*, or *Allium* are better indicated. When *Echinacea* is effectively used to treat colds or influenza, it is often one component of a synergistic herbal blend crafted by a skilled herbalist aimed at best addressing specific cases, individual constitutions, and underlying contributing factors. Thus, herbalists' response to the study conclusions was: "Of course *Echinacea* didn't work to reduce cold and flu symptoms; that's not how the plant is best used!"

A similar limitation of some current research is that herbalists choose herbs based on a symptom pattern, an energetic presentation, and a holistic understanding of the person's entire being—not just a *particular diagnosis*. In clinical practice, such discernment is crucial in choosing which herb for which person—a nuance that the pharmaceutical model misses and that study designs do not allow for. *Echinacea* may be effective for colds involving severe lymph congestion, with swollen, fetid tonsils and mucosal ulcerations, but may not be very helpful for simple rhinoviruses without such intense symptoms.

Another common type of herbal research involves the use of a single compound isolated from a plant. This could be a single flavonoid such as curcumin, a single alkaloid such as piperine, or a resin such as capsaicin. Research on isolated constituents may reveal some of the mechanisms of action of the whole plant from which it was extracted. However, any given plant contains hundreds if not thousands of constituents, and such research cannot represent the action of a plant in its entirety. For example, *Digitalis* contains dozens of cardiac glycosides, some of which are stimulating to the heart and some of which are suppressive. The sum total of *all* the alkaloids gives *Digitalis* a cardiotonic effect. Many researchers prefer to conduct studies on single molecular compounds because such work matches the pharmaceutical model—a single substance with a single primary mechanism of action. When I first started presenting lectures on herbal medicine to pharmacists, I frequently was asked, "What is the active compound?" When I responded that herbalists prefer to view plants as containing dozens of active compounds that act synergistically, some of my listeners became frustrated, as if they didn't like such a messy approach to medicine. Others were disbelieving, as if it were implausible that a plant could contain more than one active compound. But if one follows that logic, then an orange is nothing more than a handy wrapper for vitamin C. Still other pharmacists in my audiences actually developed an amusing deer-in-headlights appearance, as if their worldview had been so shaken that they were stunned, in need of time to recover and let the information sink in.

It is surprisingly difficult to abolish such mechanistic thinking among some groups of health professionals, and this, too, affects herbal research and how it is interpreted. The assumption that a drug or an active molecule will have only one precise mechanism of action is so ingrained that it is very difficult for some people to shift to vitalistic thinking. *Vitalism* embraces the notion that living beings have an innate healing wisdom that enables self-ordering of molecules, supports homeodynamic balance, and compels physiologic process to continually restore and repair the body. As living entities, most herbal medicines contain vitamins, minerals, and enzymes, in addition to active compounds, that nourish the tissues and support the vital force. Where pharmaceutical medicine generally employs strong substances with a single, precise action, herbal medicines are mostly nutritive and protective, supporting tissue repair and regeneration and organ tone and balance. *Allo-* means opposite, and allopathic medicine attempts to oppose symptoms and pathology. For example, antibiotics oppose microbes, anti-inflammatories oppose inflammation, and anodynes oppose painful sensations. It is difficult to demonstrate the vitalistic aspect of herbal medicine using a mechanistic research model that studies a single compound's single activity, in a single situation. Some of the core herbal medicines are the *alteratives* and the *adaptogens*, neither of which have equivalent counterparts in allopathic medicine. Both alteratives and adaptogens are whole plant agents that work to support organ function, enhance the elimination of wastes and detoxification pathways, and support multiple innate healing mechanisms at once. Many adaptogens have been found to *lower* cortisol level when it is excessively elevated by stress yet *increase* cortisol production when insufficient. This dichotomy can sound implausible to a mechanistic thinker, but indeed, many herbs are capable of such polar actions. If one were to read only a single study, one that happened to show an adaptogenic herb to increase cortisol output, it could lead a mechanistic thinker to assume that herb was contraindicated in states of cortisol excess, yet nothing could be further from the truth.

Another example may be found with regard to the research on herbs that contain phytosterols. Many such herbs are shown to competitively bind to estrogen receptors in hormone-sensitive tissues such as the breasts or endometrium. Due to being weaker agonists than endogenous estrogens, phytosterols lower the net estrogenic activity in the body when the endogenous estrogen load is high by crowding out estrogen, effectively reducing estrogen binding to the receptors. Yet phytosterols will also provide estrogenic activity in cases of very low endogenous estrogen, binding to the relatively vacant estrogen receptors and promoting an increase in estrogenic activity. In this way, phytosterols reduce hormonal stimulation when excessive yet provide estrogenic effects in menopausal women or those suffering from premature ovarian failure. Failing to grasp this amphoteric action of phytosterols has led some health professionals to insist that phytosterols be avoided by those at risk of hormonal cancers, erroneously equating phytosterols with the dangerously proliferative effects of Premarin and other synthetic hormones. This is the kind of misinterpretation of herbal research that has also led some professionals who have a poor grasp of herbalism and vitalistic thinking to purport that *Echinacea* and *Astragalus* are contraindicated in autoimmune disorders because these herbs promote immune function. This kind of unipolar, mechanistic thinking also leads to some of the misinformation regarding drug-herb interactions and presumed contraindications of herbs; assumptions are made based on the drug model. For example, some institutions will suggest that a plant found to be a monoamine oxidase (MAO) inhibitor can't be taken with tyramine-containing foods, simply because that recommendation holds true for pharmaceutical MAO inhibitors. Another example: I have heard it said that St. Johnswort cannot be taken with a selective serotonin reuptake inhibitor (SSRI) because it is contraindicated to take two pharmaceutical SSRIs in tandem. And yet, I have used St. Johnswort dozens of times to wean people from their Prozac or a similar prescription with no ill effects.

So how do these rumors get started? Such claims result from theoretical concerns that are stated as fact rather than theory, even in the absence of a single adverse event on record. Those making such claims are well meaning and aiming to avoid harm, but they have been influenced by modern pharmacology and attempt to apply those guidelines to herbs. The examples above are just a few of the many I could describe to argue that such logic does not necessarily equate. When theories are put forth by respectable authoritarian entities, it is difficult for a wee herbalist to refute them, and the theoretical concern is disseminated as if there have been thousands of cases of SSRI–St. Johnswort toxicity, or thousands of rheumatoid arthritis and multiple sclerosis patients harmed by *Echinacea*, which is simply not the case.

Another issue that affects herbal research is, of course, the profitability of the research, or lack thereof. Large numbers of institutions and individuals conduct research on plant-derived molecules in pursuit of the goal of creating a synthetic version or manipulating the molecule such that it can be patented. Conducting a dozen million-dollar trials on garlic would not pay off if the fruits of that research simply showed that eating whole garlic is beneficial to health but resulted in no drug patents or proprietary secrets. However, if the research were on curcumin (a compound in turmeric having robust antioxidant, anti-inflammatory, and anticancer effects but poor absorbability), and if that research led to the development of a manipulated molecule with better absorbability, then that molecule *could* be owned, patented, and highly profitable. And, in fact, there are patented curcumin compounds now on the market, such as Meriva and Theracumin.

Many researchers are examining individual plant compounds such as curcumin for novel mechanisms of action. Their motive, however, is not curiosity as to how plants work but is the wish to create new pharmaceuticals. Finding something promising, the scientists then elucidate the compound's molecular shape and strive to mimic that shape in a synthetic compound that can be manufactured. In contrast, only a small pool of underfunded scientists are interested in doing whole plant research with a goal of the betterment of herbal medicine.

Even when whole plant extracts *are* studied, the interpretation of the results is complicated. Unlike pharmaceutical drugs, where a set dosage of the substance should elicit the same result time and again, the quality and composition of botanical medicines can vary greatly depending on the precise product and preparation being tested. Fall-harvested versus spring-harvested plant material, fresh versus dried material, an alcohol versus an aqueous extract, and leafy versus root tissue can all have different constituent profiles. Water-soluble constituents may be absent in a tincture or present in a tincture, but absent in an encapsulated dry herb. The presence or absence of such constituents can influence the absorption and pharmacodynamics of the medicine, the effective dose, the efficacy in a condition being tested, or the incidence of side effects. All this helps to inform our understanding of the variable outcomes in some clinical trials.

As in so many aspects of contemporary life, another reality of the herbal research arena is *spin*. Elaborate marketing campaigns with glossy ads tout a too-good-to-be-true weight-loss herb, but the evidence for the claim consists of only a few animal studies, hardly worthy to support the claim of a new breakthrough. The spin on herbal research may include both exaggerated claims for one herb's actions and a dismissal of other herbs that have excellent safety profiles and demonstrated efficacy. In contrast to this sensationalism in some herbal advertising are clinical trials in which an herb is shown to be at least as effective as a pharmaceutical for pain or for allergy symptom relief or for improving sleep, but the studies fail to conclude by embracing the herb as a safer alternative to the drug. Instead, they conclude weakly, saying "more research is needed." When I read something like that, I suspect that the researchers didn't think that they could get the study published unless they adopted an underwhelmed attitude and certainly were careful to avoid saying that an herb could replace a synthetic drug. When allergy relief or sleep enhancement has been the traditional use of an herb for centuries, it is hard to understand why, exactly, more research would be needed before using or prescribing said herb, especially if it were safer or had fewer side effects than available drugs.

While spin may be likened to a white-lie interpretation of molecular research, there are more serious cases of fraud, deceit, and corruption within the medical research arena. There is often publication bias—journals that publish studies on herbs if the result is negative or shows a harmful side effect but choose not to publish research showing herbs as effective for anything. Far worse, there are also cases of falsifying results of pharmaceutical drug trials, suppressing evidence of toxicity or side effects of drugs, or hiding the fact that the entity funding a trial and paying the researchers and staff is the same company who is making the drug and is positioned to profit from a positive study. Pharmaceutical giants are also known to conduct research and never publish the results, presumably due to negative outcomes or concerning side effects, effectively depriving the scientific community of such information and skewing the perceived risks-to-benefit ratio of the drug. For example, GlaxoSmithKline failed to publish its research on SSRIs that suggested a possible increase in suicidal ideation in teens using Paxil, and Pfizer buried the studies on gabapentin showing negative outcomes for neuropathy.[1] There is also bias in what sort of research is deemed creditable, where studies done in the United States and Europe are often taken seriously, while those done in China, India, or Latin America are more likely to be taken with a degree of skepticism until someone in the United States repeats the research. The unfairness of this brings up the question of research racism. The bias against foreign research is couched within statements that the study was limited by its small sample size, the study medication was poorly defined, or that the research had design flaws. While many such criticisms can be true, the fact is that even in the face of the obstacle of insufficient resources, researchers around the world are contributing basic information that should be respected as preliminary findings—and not dismissed outright.

In summary, bias, snobbery, and spin are among the political issues affecting herbal research. Lack of vitalistic research models to conduct research that herbalists would find meaningful is another limitation in showcasing what herbal medicine is truly capable of. Lack of funding and facilities to conduct large-scale clinical trials on herbal medicines is another obstacle and is one that is difficult to overcome, in part because such research is not aimed at winning lucrative patents. One exception is the arena of novel molecules with important medical applications: immunotoxins that could perform molecular surgery for cancer patients, nootropic agents that can penetrate the blood brain barrier and offer new therapies for dementia or Parkinsonism, or liposomal nutraceuticals that could greatly increase the delivered dose of important plant-based compounds compared to what is possible with herbal teas, tinctures, and encapsulations.

I find molecular research fascinating, and I cite a great deal of molecular research in this set of herbal formularies. But as the science of herbalism comes of age, I hope to see the field of herbal medicine mature, with more large-scale clinical trials having the potential to demonstrate that herbs can have anticancer properties; may prolong life in those with heart, liver, or renal failure; may optimize cholesterol and blood sugar levels; and may successfully treat anxiety, depression, menopausal symptoms, allergies, infertility, and a myriad of other common health issues. Many of these conditions are managed allopathically with pharmaceutical medicines that have detrimental effects to health, such as harsh chemotherapeutic drugs for cancer, hepatotoxic statin drugs for elevated lipids, and vascular-damaging synthetic hormones for menopause. Many herbs have been found to boost the efficacy of chemotherapeutic agents, offer hepatoprotection, and relieve menopausal symptoms, and furthering the herbal research in these and many other arenas would be highly valuable to millions of people.

About This Book

This text is the third in a set of five comprehensive volumes aimed at sharing my own clinical experience and formulas to assist herbalists, physicians, nurses, and allied health professionals create effective herbal formulas. The information in this book is based on the folkloric indications of individual herbs, fused with modern research and my own clinical experience.

I have organized this set of volumes by organ systems. Volume 1 features the organs of elimination—the gastrointestinal system, the liver and biliary system, the urinary system, and the skin. Herbalists know these organs are foundational to the health of the entire body. The treatment of many inflammatory, infectious, hormonal, and other complaints will be improved by optimizing digestion and elimination. Volume 2 covers respiratory and vascular issues, including both cardiovascular and peripheral vascular complaints. In this volume, metabolic and reproductive endocrinology is covered, with a focus on adrenal and thyroid disorders, diabetes and metabolic dysfunction, and male and female reproductive disorders.

Each volume in this set offers specific herbal formulas for treating common health issues and diagnoses within the selected organ system, creating a text that serves as a user-friendly reference manual as well as a guide for budding herbalists in the high art of fine-tuning an herbal formula for the person, not just for the diagnosis. Each chapter includes a range of formulas to treat common conditions as well as formulas to address specific energetic or symptomatic presentations. I introduce each formula with brief notes that help to explain how the selected herbs address the specific condition. At the end of each chapter, I have provided a compendium of the herbs most commonly indicated for a specific niche, a concept from folklore simply referred to as *specific indications*. These sections include most herbs mentioned in the corresponding chapter and highlight unique, precise, or exacting symptoms for which they are most indicated. Please note that these listings do not encompass *all* the symptoms or indications covered by the various herbs, but rather only those symptoms that relate to that chapter—the indications for metabolic distress, male and female reproductive issues, and adrenal and thyroid disorders. You'll find certain herbs repeated in the specific indications section of all three system chapters in the book, but in each instance, the description will feature slightly different comments. Readers are encouraged to refer back and forth among the chapters to best compare and contrast the information offered.

The Goals of This Book

My first goal in offering such extensive and thorough listings of possible herbal therapies is to demonstrate and model how to craft herbal formulas that are precise for the patient, not for the diagnosis. It is my hope that after studying the formulas in this book and other volumes in the set and following my guidelines for crafting a formula, readers will assimilate this basic philosophic approach to devising a clinical formula. As readers gain experience and confidence, I believe they will find that they rely less and less on these volumes and more and more on their own knowledge and insight. That's what happened to me over the years as I read the research and folkloric herb books and familiarized myself with the specific niche-indication details of a wide range of healing plants. I now have this knowledge in my head, and devising an herbal formula for a patient's needs has become second nature and somewhat intuitive. But from talking with my herbal students over several decades of teaching, I have come to understand that creating herbal formulas is one of the most challenging leaps between simply absorbing information and using it to treat real, live patients. Students often feel inept as they try to sift through all their books, notes, and knowledge and struggle to use "information" to devise a single formula that best addresses a human being's complexities. Thus, I felt that it was high time that I created a user-friendly guide to help students refine their formulation skills and to help all readers develop their abilities to create sophisticated, well-thought-out formulas.

Another goal I aim to achieve through this set of herbal formularies is to create an easy-to-use reference that practitioners can rely on in the midst of a busy patient day. In this "information age," it is not hard to track down volumes of information about an herb, a medical condition, or even a single molecule isolated from a plant. The difficulty lies in remembering and synergizing it all. While this text doesn't pretend to synergize the "art" of medicine in one source, I believe it will help health professionals quickly recall and make use of herbal therapies they already know or have read about by organizing them in a fashion that is easy to access quickly.

Naturopathic physicians are a varied lot. Add in other physicians and allied health professionals, and the skill sets are varied indeed. I rely on my naturopathic colleagues to inform me about the latest lab tests, my allopathic colleagues to inform me about new pharmaceutical options, and my acupuncture colleagues to inform me on what conditions they are seeing good results in treating. This text allows me to share my own area of expertise. I have included a large number of sidebars that feature some of the more in-depth research on the herbs and individual molecular constituents, helping to provide an evidence-based foundation for the present era of medical herbalism.

I realize that not all clinicians, not even all naturopathic physicians, specialize in herbal medicine. I hope that this formulary will serve as a handy reference manual for those who can benefit from my personal experience, formulas, and supportive discussions.

Creating Energetically Fine-Tuned Formulas

Much like a homeopathic *materia medica*, this set of formularies aims to demonstrate to clinicians how to choose herbs based on *specific indications* and clinical *symptoms* and *presentations*, rather than on diagnoses alone. For example, I do not offer a one-size-fits-all menopause formula. Instead, I offer more precisely aimed formulas to address menopausal hot flashes, insomnia, mood disorders, and uterine bleeding irregularities. Similarly, this volume offers possible therapies for diabetes to address the many associated issues ranging from vascular inflammation to an increased risk of infections, to the overlap with hypertension, hyperlipidemia, and full metabolic syndrome. I include supportive research on herbs that helps to explain why a particular herb is chosen for a particular formula as well as endnote citations that provide details of specific studies for those interested. I also provide findings from research on individual herbs that are essential to the treatment of the various conditions featured in a chapter. To make the text as useful as possible for physicians and other clinicians, I also offer clinical pearls and special guidance from my own experience and that of my colleagues—the tips and techniques that grab attention at medical conferences year after year.

The Information Sourced in This Book

The sources of the information in these volumes are based on classic herbal folklore, the writings of the Eclectic physicians, modern research, and my own clinical experience. Because this book is designed as a guide for students and a quick reference for the busy clinician, the sources and research are not rigorously cited, but enough so as to make the case for evidence-based approaches. When I offer a formula based on my own experience, I say so. I also make note of formulas I've created that are more experimental, due to lack of research on herbs for that condition or my lack of clinical experience with it.

My emphasis is on Western herbs, but I also discuss and use some of the traditional Asian herbs that are readily available in the United States. In some cases, formulas based on Traditional Chinese Medicine (TCM) are featured due to a significant amount of research on the formula's usage in certain conditions. I readily admit that TCM creates formulas *not* for specific diagnoses, but rather for specific energetic and clinical situations. However, I have included such formulas, perhaps out of context but with the overall goal of including evidence-based formulas, with the expectation that readers and clinicians can seek out further guidance from TCM literature or experienced clinicians where possible. In reality, TCM is a sophisticated system that addresses specific presentations, and I have borrowed from this system where I thought such formulas might be of interest or an inspiration to readers. I admit that listing just one formula for a certain condition based on the fact there have been numerous studies on it is somewhat of a corruption of the integrity of the TCM system, which is aimed at precise patterns and energetic specificity. Nonetheless, I chose to do so with the goal of creating a textbook to help busy clinicians find information quickly, while still encouraging individualized formulas for specific presentations.

While I have endeavored to create herbal formulas to address as many different conditions and presentations as possible, this volume does not cover in-depth protocols for reproductive cancers. It does present some of the risk factors that may contribute and some of the research on the safe use of phytosterols in those with hormone-driven cancers or a history of such cancers. The herbal formulas presented for such situations, as well those for adrenal or thyroid cancers, are not meant to replace an expert's care, or other therapeutic options, but are simply a way to share my own experience using herbal and complementary medicine in a wide variety of endocrine disorders. While common reproductive infections from vaginal yeast and prostate infections to management of human papillomavirus are discussed, this text does not cover treatment of HIV infections or related acquired immunodeficiency syndrome (AIDS), as I have not had the clinical experience or delved into the research to do the topics justice.

How to Use This Book

Each chapter in this book details herbal remedies to consider for specific symptoms and common presentations of various diagnoses. Don't feel that you must be a slave to following the recipes exactly. When good cooks create a food recipe, they are always at liberty to alter the recipe to create the flavor that best suits the intended meal—the big picture. A formula listed should not be thought of as *the* formula to make, but rather as a guide and an example, inviting the clinician to tailor a formula for each individual patient.

To create an herbal formula unique to a specific person, the clinician should first generate a list of actions that the formula should perform (respiratory antimicrobial, expectorant, bronchodilator, mast cell stabilizer, and so on), and then generate a list of possible herbal *materia medica* choices that perform the desired actions. If these ideas are new to you, you may want to begin by reading chapter 1, "The Art of Herbal Formulation," before you start generating lists.

Look to the formulas in chapters 2 through 4 that address specific symptoms for guidance and inspiration.

Unity of Disease (Totality of Symptoms)

The concept that any given health issues a person may experience are actually one disease, as opposed to a number of disparate diagnoses to be treated individually, is a core tenet of naturopathic medicine and the philosophical underpinning of holistic medicine in general. Any one symptom does not provide the full story, and just because you can label the symptoms with a Western diagnosis and offer the established therapy for that diagnosis does not mean you are really helping a person to *heal*. A careful consideration of the sum totality of all symptoms is important to reveal underlying patterns of organ strength or weakness, excess or deficiency states, nervous origins versus nutritional origins, and, of course, a complex overlap of all such issues. The most effective therapies will address *all* issues in their entirety and involve an understanding of the entire energetic, mental, emotional, nutritional, hereditary, situational, and other processes creating a complex web of cause and effect—the unity of any given individual's "dis-ease."

(These formulas are grouped within the chapter by a general diagnosis, such as "Formulas for Diabetes" or "Formulas for Pregnancy and Related Issues.") Regard the lists and formulas I have provided as starting points and build from there. In my commentary on the individual formulas and in sidebars that focus on specific herbs, I offer further guidance as to whether the formula or individual herbs are safe in all people, possibly toxic in large doses, intended for topical use only, or indicated only in certain cases of that particular symptom. Once herb and formula possibilities have been identified, the reader should then review "Specific Indications" at the end of the chapter to narrow in on choices of which herbs would be *most* appropriate to select and to learn more about how those herbs might be used. Herbalists can narrow down long lists of herbal possibilities to just a few *materia medica* choices that will best serve the individual. In many cases, the reader/clinician will be drawing upon herbal possibilities from a number of chapters and organ systems as the clinical presentation of the patient dictates. Thus, you are not making a formula by throwing together all the herbs listed as covering that symptom or symptoms, but you are studying further and narrowing down the list of possibilities to consider, based on the sum totality of all the symptoms. In some cases, you will rule out herbs on the list for a particular symptom after reading the specific description of that herb at the end of the chapter. In some cases, you might decide to put one herb in a tea and another in a tincture due to flavor considerations. In other cases, you might decide that you will prepare only a topical remedy. And in other urgent situations, you might come up with a topical, a pill, an herbal tea, *and* a tincture to address the situation as aggressively as possible. Aim to select the best choices and avoid using too many herbs in one formula. Larger doses of just a few herbs tend to work better than smaller doses of many herbs, which can confuse the body with a myriad of compounds all at once. The use of three, four, or five herbs in a formula is a good place to start; this approach also makes it simpler

to evaluate what works when the formula is effective as well as what is poorly tolerated, should a formula cause digestive upset or other side effects.

Learning from the Formulas in This Book

In reviewing the formulas in this book, notice how specific herbs are combined with foundational herbs to create different formulas that address a variety of energetic presentations. There are a handful of all-purpose immune modulators, all-purpose alterative herbs, and all-purpose anti-inflammatories that can be foundational herbs in many kinds of formulas. Such foundational herbs can be made more specific for various situations by combining with complementary herbs that are energetically precise. Notice how the herbs are formulated to be somewhat exacting to address specific symptoms and make a formula be warming, drying, cooling, or moistening and so on. Also, note how acute formulas may have aggressive dosages and include some strong herbs intended for short-term use, while formulas attempting to shift chronic tendencies are dosed two or three times a day and typically include nourishing and restorative herbs intended for long-term use. Also notice how some potentially toxic herbs are used as just a few milliliters or even a few drops in the entire 2-ounce (60 milliliter) formula. These dosages should not be exceeded, and if this is a clinician's first introduction to potentially toxic herbs, further study and due diligence are required to fully understand the medicines and how they are safely used. Don't go down the poison path without a good deal of education and preparation. I am able to prepare all of the formulas in these texts upon request, but I can only offer those containing the "toxic" or cardiac glycoside-containing herbs (*Aconitum, Atropa belladonna, Convallaria, Digitalis*, and so on) to licensed physicians.

It is my sincere hope that this book helps you in your clinical work and efforts to heal people.

DR. JILL STANSBURY

CHAPTER ONE

The Art of Herbal Formulation

Creating an effective and sophisticated herbal formula is somewhat of an art, and like all art, it is difficult to put into step-by-step directions; however, this is my attempt to do just that. This book aims to explain how to create specific formulas for *presentations* rather than *diagnoses*—rather than offering a single formula or two for a general condition such as hypertension or irritable bowel syndrome, this text offers exacting formulations to best address the precise presentation of the person. I have personally seen many *different kinds* of insomnia, prostatic inflammation, chronic vaginitis, and other conditions that conventional medicine tends to treat with one across-the-board medication or therapy. My aim is to coach readers on how to create numerous finely tuned herbal options for treating the person and not the diagnosis—a core tenet of natural medicine.

Creating effective herbal formulas and treatment plans requires many skills: knowledge of the herbs; herbal combinations best for a particular situation; the proportions to use in a formula; the starting dose, frequency of dose, and how long to dose; what form of medicine is best, such as a tincture, a tea, a pill, or a topical application; broader protocol options that may include diet, exercise, nutritional supplements, or referral to allied health professionals; and the follow-up plan.

Hippocrates said, "It is more important to know what kind of person has a disease than to know what kind of disease a person has." To know what sort of person has a disease requires careful listening and skilled and nuanced questioning in a safe and comfortable setting (see "Asking the Right Questions" on page 10). Listen for underlying causes, for overarching emotional tone, for what a person is able to do for themselves (sleep more, exercise more, eat better, brew a daily tea), and for what they are resistant to (sleep more, exercise more, eat better, brew a daily tea). Address underlying causes and start where people show some interest and capacity. Cheerlead to instill enthusiasm if required, so that people

become better educated and thereby better motivated to make important changes or adopt valuable healing practices. Giving the right medicine is only one aspect of doing healing work with a person; creating a sacred space that invites truth and sharing is key to getting to the point where you know what sort of person has a disease.

Healing also stems from understanding pain, suffering, challenges, and unique situations, from nonjudgmental listening to the stories so often linked to our physical ailments. It comes from giving encouraging words and sympathy, congratulating and complementing people's efforts and accomplishments, and giving people the tools, resources, and support to succeed. This kind of true caring plus an earnest effort to provide real support are among our best medicines.

The Importance of Symptoms

Naturopathic medicine has a different philosophical stance on physical symptoms than allopathic medicine, especially infectious and eruptive symptoms, and that view is worth briefly describing here for those who may be unfamiliar.

The Biochemical Terrain

The terrain of the human body invites microbes specific to the chemical composition of the ecosystem, and when infectious microorganisms have consumed those specific chemical substrates, the disease-producing microbes are no longer supported. The microbes themselves change the chemical composition of the tissue by consuming the "food" that invited them, and when those nutritional resources are exhausted, the biochemical makeup of the tissues is changed for the better. Antibiotics are not the best way to treat chronic infections. Instead, optimize the ecosystem to support the desired beneficial flora. As when bacteria start to stink up the compost bucket on the kitchen counter, we don't spray the compost with a germicide—we clean the bucket! For example, some diabetic patients are prone to opportunistic infections because the constantly elevated blood sugar creates a hospitable environment for such pathogens. It would not serve such a person to use repetitive courses of antibiotics because of the likelihood of developing antibiotic-resistant strains. Instead, it would be worth the more lengthy and arduous approach of using diet and exercise to get better control of blood sugar while employing gentle herbal antimicrobial and alterative agents that reduce the propensity to infections over time. Even though toenail fungus and chronic yeast vaginitis may seem far removed from the gut, optimizing intestinal flora can actually reduce such infections by helping to eradicate opportunistic fungal infections. And for those where every splinter in the skin becomes infected, or for those whose every respiratory virus turns into a sinus infection or pneumonia, systemic immune support, blood sugar control, and vitality-building therapies may reduce the chronicity and support better health in the long run than treating every new infection with an antibiotic time and again.

The Role of a "Healing Crisis"

An acute infection, such as a simple cold, is a classic example of a "healing crisis"—meaning the acute symptoms are actually a part of the body's attempt to heal itself, allowing the infectious organisms to consume the "morbid matter" and thereby restore a healthier ecosystem. The symptoms of the cold—runny nose, fever, loss of appetite—are part of the body's process to heal itself.

Naturopathic philosophy embraces the symptoms of a healing crisis as a triumph of the vital force. Our symptoms serve us and call attention to the imbalances requiring changes. Be thankful when the body has the vitality to manifest a healing crisis. Be concerned when infections, eruptions, and discharges stop, and allergies, autoimmunity, joint pain, ulcers, blood pathology, and so on emerge. These are not healing crises, but rather signs that vitality is being damaged and pathology is becoming deeper and more serious.

Because the symptoms of a healing crisis are the way in which the body can heal itself, such symptoms should not be suppressed, but rather supported and made as tolerable as possible. Only when such symptoms are so severe as to threaten damage to the body should they be suppressed. Use herbs to help reduce blood sugar to treat chronic infections in diabetics, support the liver with alterative herbs to improve conditions associated with hyperestrogenism, and use adaptogens and nervine herbs to help mitigate and recover from the effects of acute and chronic stress to support adrenal health and the adrenals' role in immune and endocrine function. Herbs can be used to support core vitality and hormone balance, to improve the clearance of normal metabolic wastes as well as environmental toxins and exogenous estrogen, and to optimize metabolic function to reduce oxidative stress in a manner that protects the vasculature and organ function in those with diabetes and metabolic syndrome.

The Harm in Suppressing Symptoms

Habitual suppression of symptoms over time can force the body to give up its struggle for health. For example, laxatives may relieve constipation temporarily, but ultimately damage normal peristalsis. Stimulants may initially provide energy, but ultimately exhaust the adrenals and nervous system. For example, a pot of coffee may jolt an exhausted person awake, but it will not improve core energy status in the long run and will only further exhaust it. Antibiotics can kill infectious pathogens, but unless the underlying ecosystem that supports such microbes is significantly improved, the same pathogens will be supported a second, third, and fourth time and become resistant to antibiotics if they are given repeatedly, such as for chronic otitis, cystitis, or sinusitis.

Asking the Right Questions

Learning how to ask questions that will elicit relevant information is as much an art form as creating an herbal formula. Aim your questions to gather information in several categories. What follows here is a broad, but not exhaustive, list of questions that can yield helpful information.

Etiology

How long has this been happening?

How did it begin?

How has it evolved over time?

What else was going on in your life at the time this began?

Have any other symptoms, complaints, problems accompanied this complaint?

What is your health history?

Any previous episodes? Related pathologies? Family members with this complaint?

What is the predominant emotion associated with this complaint? Did the complaint begin during a period of grief? Anxiety? Ambition?

Did the complaint begin following hard labor or an injury, stress, eating a new food, traveling out of the country, starting a new medication (and so on)?

Quality and Occurrence of the Complaint: "PQRST"

Provocation. Does anything seem to bring on the complaint? Does anything alleviate it? What makes the complaint better or worse?

Quality. What is the character of the complaint: burning, throbbing, dull, sharp, shooting, aching, and so on?

Radiation. Does the pain travel? Does this symptom affect any other organ? Is this complaint associated with any other symptom?

Severity. On a scale of 1 to 10, how severe is this? Does it interfere with sleep, activities, work, sex, relationships, child rearing, creativity, and hobbies?

Time. Timing throughout the day, throughout the month (hormonal fluctuations?), throughout the year (seasonal allergies? seasonal affective disorder? Asian or Native American concepts of seasons?). Timing may involve an association with eating food or going without eating, an association with anxiety/relaxation, an association with menstrual cycle, or an association to a certain environment or allergen exposure. Is the complaint any better or worse with sleep?

Concomitant Symptoms

Do you feel hot or cold when this occurs?

Is there lethargy or anxiety, heart palpitations, or weakness?

Does it occur during times of stress and activity, during sleep, or after prolonged sitting?

Is there fear or fatigue, mania or depression, weakness or restlessness (and so on)?

Is there a desire to be consoled or to be left alone?

Constitution and Energetic Considerations

Note the constitution by asking the right questions and observe the person to get additional clues as follows:

Complexion. Pale? Yellow? Cyanotic? Flushed? Haggard? Dry? Damp? Inflamed? Quick or slow to perspire? Oily or dry skin?

Pulse. Strong or weak? Fast or slow? The quality? The variability?

Tongue. Large or small?

Muscles. Well developed, overdeveloped, underdeveloped? Spastic, atrophic? Soothed by pressure and massage or aggravated by pressure or massage?

Senses. Hyperacute includes sensitivity to odors, noises, bright light; pain; racing thoughts. Hypoacute includes loss of taste, hearing, sensation; poor ability to concentrate, remember, respond.

Diet and appetite. Hungry, anorexic? Can eat large quantities or tolerate only small amounts? Hypoglycemic symptoms? Unusual cravings for a particular flavor or food? Unusual aversion or aggravation by particular foods or flavors? Are the symptoms better following meals or between meals when the stomach is empty? What is the general diet, nutrient intake, bowel habits, and quality of the stool?

Thirst. Large thirst versus small or no thirst? Thirst due to dry mouth? Thirst due to compulsion? Thirst for large gulps or thirst for small sips? Thirst for warm fluids or thirst for cold fluids?

Sex drive. High, low? Markedly cyclical?

Sleep. Requires more than 8 hours of sleep to feel rested, able to function with little sleep? Sleeps soundly all night, or wakes many times? Takes a long time to fall asleep but then sleeps well, or falls asleep readily but wakes at a particular time unable to sleep further? Eating before bed disturbs or improves sleep? Restless during sleep or wakes with a jolt? Has nightmares or difficult dreams?

The "Triangle" Exercise

Once you have listened thoughtfully and well to a description of symptoms, I recommend starting with a traditional and simple method of thinking through the choice of the components of a formula: *the triangle*. Visualize a triangle with a horizontal base and two slanting sides that meet at a point at the top of the triangle. Like the triangle, your formula needs a *base* on which to rest—that's where you'll start—and it also needs two axes that "point" the formula in the right direction.

In general, the herb or herbs chosen as the base should be nourishing and nontoxic—something tonifying, restorative, alterative, adaptogenic, or nutritive to the main organ system, tissue, or issue of concern. For example, cardiovascular formulas might have *Crataegus* as a base herb on which to rest. A formula for insomnia with exhaustion might rest on *Withania*, and a skin condition formula might rest on *Calendula* or *Centella*. There is an old Wise Woman saying that all healing begins with nourishing. Everyone needs nourishment and tonification, so such herbs are always appropriate. Simple, right?

From there, herbs for the other two axes of the triangle are selected to point the formula in a more specific direction. The base herbs tend to be nonspecific and indicated in a very wide number of clinical situations, so to make your formula more specific for an individual patient, you will need to drive it in the right direction.

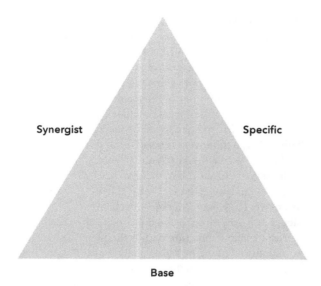

Figure 1.1. The Triangle—A Pragmatic Tool for Crafting Herbal Formulas

Through the ages, herbalists have used varied vernacular to refer to such herbs as *synergists*, *specifics*, *energetics*, *kickers*, *directors*, and so on. The choice of names does not matter, but for the sake of discussion, I use the words *synergist* and *specific* to apply to the other two axes of the herbal formula triangle. In Traditional Chinese Medicine (TCM), royal terms such *Emperor*, *Minister*, and *Assistants* may be used.

The choice of a *specific* requires a detailed understanding of *materia medica*, such that one or two energetically and symptomatically precise herbs can be selected. The lists of herbs offered in "Specific Indications" at the end of each chapter help provide exacting details of symptoms for which individual herbs are most indicated. If there is more than one specifically indicated herb, use them both! If one tastes good and one tastes bad, put the tasty one in a tea and put the less palatable option in tincture or pill. The point is to identify one or two specifics that best address all symptoms possible as well as the constitution, energy, and the entire body's strengths and weaknesses.

As an example, if your case mainly involves hypertension, you may choose to rest the formula on *Crataegus*. The unique identifying symptoms are that the woman's blood pressure began going up during menopause and she is also experiencing episodes of tachycardia and heart palpitations. After reviewing the specific indications for various herbs, you might choose *Leonurus* as your specific to point or drive the formula in the right direction and to the right place. If, on the other hand, the patient is a long-term smoker with high cholesterol and vascular inflammation, your specific may be *Allium*. And if the hypertension formula is for a highly stressed Type A personality who also suffers from alarming hypertension accompanied by a throbbing headache, an appropriate choice of specific might be *Rauvolfia*.

A *synergist* is less specific but takes into account underlying contributors such as other organ systems involved and other energetic considerations that contribute to the overall case. To follow up on the hypertension examples above, the menopausal hypertension formula might include *Actaea* as a synergist to also offer hormonal and nervine support. The long-term smoker may benefit from *Ginkgo* as a synergist to combat circulatory stress, and the Type A severe hypertensive might benefit from *Piscidia* as a vasodilating synergist.

The "Specific Indications" in chapters 2, 3, and 4 offer details about individual *materia medica* options to help choose specifics and synergists to best address the sum total of the symptoms. A synergist may be very "specific"—it's all semantics. The point is to take into consideration various individual factors and unique presentations that contribute to the case.

Returning to the example of the menopausal woman with hypertension and episodic heart palpitations, if she also was suffering from stress and poor sleep as underlying contributors, you might select *Withania* as a synergist. If, on the other hand, she was sleeping fine and did not feel particularly stressed but was having some difficulty with constipation and hemorrhoids, you might select *Aesculus* as the synergist. Or if she was having constipation, hemorrhoids, and episodes of heavy menses as part of the menopausal transition, you might select *Hamamelis* as the synergist. Review "Specific Indications" at the end of each chapter to start developing a familiarity with each individual herb and its "personality" or the specific symptoms that it best addresses.

Whatever the details you are presented with, you will use the model of the triangle and select at least three herbs: a nourishing base (tonic), a synergist, and a specific. There is no reason not to select more than one nourishing base or more than one specific or synergist if something very appropriate pops out. If, for example, the menopausal woman with hyptertension had a long-standing history of anxiety, stress, mood swings, and emotional lability well before the onset of the recent hypertension, then her nervous system may need nourishment and tonification as much as, or more

More about Formula Components

Base. Also called the lead herb or the Director, the base should be a tonic that has a nourishing and restorative effect on the main organ system affected. This herb does not require a great amount of skill to choose as it is often among the herbs best known for having an affinity to a particular organ system.

Synergist. Also called the Adjuvant, the Balancer, or the Assistant, this component helps correct or complement the action of the lead herb and helps drive it to the desired tissues. The selection of these herbs requires an in-depth understanding of the case and person being treated in order to address underlying causes and give the formula the needed energetic specificity.

Specific. Also called the Kicker or the Energetic Specific, this component is selected not just for a specific condition or diagnosis, but for a specific quality, essence, or expression of any given disease or disorder. Such qualities include the pulse, the tongue, the affect, the pathology, the etiology, and the person themselves to guide the selection of the most specific medicine. The practice of learning and using specific medications is gleaned from careful study, observation, and clinical practice.

so than, her cardiovascular system. In that case, you might choose two nourishing herbs upon which to rest the formula—one such as *Crataegus* for the vascular system and one such as *Avena* for the nervous system. Or, if she had a history of these stress-related symptoms and they also tended to cause irritable bowel and diarrhea on many occasions, then you might choose *Crataegus* and *Matricaria*, which provide a nourishing base for the vascular, nervous, and digestive systems.

Therefore, when you reflect on all the details of a case, other symptoms often emerge that are not the chief complaint but are highly important accompanying considerations that should not be overlooked. A simple presentation of perimenopausal hypertension with episodic heart palpitations and no other symptom details may lead to the formula as described above: *Crataegus*, *Leonurus*, and *Actaea*. However, if you learn that this woman has had gallstones, a history of fat intolerance, and many digestive symptoms and that the heart palpitations are worse after meals, you might choose *Curcuma* or *Silybum* as synergists because biliary congestion may contribute to hypertension and, for that matter, may deter the processing of hormones and postprandial lipids. This is another example of an important underlying factor that should not be overlooked. Or, if you find that she is a highly allergic person who sometimes has wheezing with chemical exposure, occasional hives, and chronic low-grade eczema on her hands that flares up after a day of heavy cleaning with exposure to a lot of water and cleaning products, you might choose *Angelica* as a second specific because it is specific for hypertension, asthma, eczema, and hives.

Mastering the Actions of Herbs

Herbal clinicians should have an excellent grasp of primary actions while gaining a solid knowledge of *materia medica* and specific indications. Clinicians should know which herbs are the best antispasmodics for a variety of situations, the best anti-inflammatories, the best vulneraries, the best nervines, the best antimicrobials, and so on. Such actions of herbs are also foundational considerations when creating an herbal formula or when considering which herbs to select as a tonic base, synergist, and specific of a formula triangle.

Another exercise that I often encourage my students to undertake is learning basic categories of actions of herbs. Actions include antispasmodics, antimicrobials, carminitives, alteratives, adaptogens, demulcents, vulneraries, and so on. I encourage my students to type up pages or create a "little black book" that helps to remind them of the hundreds of herbs they are learning, organized as to their categories of action. And from there, they can go deeper. For example, individual antispasmodic herbs might be categorized as having an affinity for a certain organ system or being best suited for a particular quality of spasm. Consider the following antispasmodics: *Lobelia* is especially indicated for respiratory smooth muscle and cardiac muscle spasms, *Dioscorea* is specific for twisting and boring muscle spasms about the umbilicus, *Piper methysticum* can allay

Supporting Vitality Instead of Opposing Disease

Western medicine has its "differential diagnosis" where the presenting symptoms can generate a list of possible (differential) causes, ultimately leading to the diagnosis. Although this approach has some value, herbal medicine is less concerned with the formal diagnosis. Instead, herbal medicine aims to carefully consider organ system strengths and weaknesses, underlying causes (stress, toxicity, poor nutrition, poor sleep, circulatory weakness, inflammatory process, allergic hypersensitivity, and so on), and how to support and nourish basic organ function and systemic vitality. Medicines employed by herbalists are typically aimed at restoring function and supporting the innate recuperative powers of the body. The intelligent wisdom of the body to heal itself is sometimes referred to as the "vital force," and herbalists aim to support the vital force more so than to oppose the symptoms of disease. Almost all herbs offer at least some nutrition, being more like foods than drugs, and by nourishing the body and stimulating the vital force, the body is supported in healing itself.

acute musculoskeletal pain and urinary spasms, and *Viburnum opulus* or *V. prunifolium* is especially effective for spasms of the uterine muscle. Sidebars throughout this book offer quick reference lists that summarize various actions.

It is best to avoid a "What herbs are good for prostatitis?" or "What herbs are good for heavy menstrual bleeding?" style of creating herbal formulas. When treating heavy menses, for example, the condition may be associated with a high stress level causing pregnenolone to be shunted into producing cortisol rather than progesterone. Or there may be problematic exposure to xenoestrogens or an issue with tone of the uterine muscle

itself. I encourage clinicians to ask themselves, "What *actions* do I need this formula to perform?" In this case, we need to tailor the formula to the individual and select herbs with specific desired actions. We may want to emphasize herbs having an adaptogenic *action* in a formula for heavy menses, such as *Glycyrrhiza* or *Withania*, to support the adrenal gland role in synthesizing progesterone. Or we may want our formula to have a strong alterative *action* to help the liver to efficiently conjugate estrogen and eliminate toxins and exogenous estrogens, choosing *Taraxacum*, *Silybum*, and/or *Curcuma*. Or we may want our formula to have a uterine tonifying and trophorestorative *action* and choose *Viburnum* as the base herb to help support optimal uterine muscle tone.

When thinking through an herbal formula, especially for difficult or complex cases, it is useful to write down what actions you wish the formula to perform and then list several herbs that perform this action. The next step is to consider any specific indications for the listed herbs to help narrow in on the best choices. And finally, the most nourishing herbs may be chosen as supporting the foundation of the triangle, a base on which the formula may rest. Other herbs from the action lists may then be chosen as specifics or synergists to best offer all the needed actions, while creating energetically specific, finely tuned formulas.

Energetic Fine-Tuning

Using the triangle method and an awareness of the actions of herbs, as detailed above, will assist you in selecting three or more herbs that address a case in its entirety, and as such, the formulas are likely to be effective and successful. To fine-tune your formulas even further, an added tier of specificity is the energetic state of your patient. TCM philosophy often depicts the energetic state as a mixture of polar opposites in keeping with the Taoist philosophy of yin and yang polarities. For example, is your patient hot or cold? Tight and constricted or loose and atrophic? Excessively damp in the tissues or excessively dry? Tired and lethargic or energized and manic—and so on. Ayurveda, the traditional medical system of ancient India, sets up a three-pronged system of doshas—vata, pitta, and kapha—rather than the two-pronged polar opposites of Taoism but is similarly aimed at addressing differing constitutions and energetic presentations. The four-elements theory of ancient Western herbalism looks for symptoms or presentations categorized into earth-, air-, fire-, and water-related symptoms, with herbal therapies being chosen accordingly. Again,

Pharmacologic versus Physiologic Therapy

Pharmacologic Therapy. Pharmacologic prescribing is the use of a potent medicine to force a rapid pharmacological response. The energy to catalyze changes, movement, and homeodynamic balance seems to come from the medicine itself. The chemical constituents have strong actions in the body and act as cardiosedatives, emetics, antibiotics, vasoconstrictors, antispasmodics, diuretics, and so on. Pharmacologic medicine can be used heroically and can save lives, but it doesn't build the vital force and restore organ function, plus its use may be needed repeatedly. Most pharmaceutical drugs are pharmacologic in nature and are often suppressive to the body's vital force—for example, acetaminophen to suppress a fever, antibiotics to kill pathogens, or bisphosphonate drugs to halt bone cell turnover rates.

Some of the more toxic herbs can have pharmacologic activity, but in general, herbs are more like foods, offering nutrition and physiologic support. Pharmacologic medications do very little to nourish, tone, or deeply "cure" anything; the symptoms return as soon as the medication is removed. For example, steroids may suppress wheezing in the lungs or suppress eczematous skin lesions, but because they do not alter the underlying condition, wheezing and itching will recur as soon as the medications are stopped. At times, for serious and acute situations, an herb with a pharmacologic action may be chosen as a synergist or a specific in an herbal formula, but never as a lead herb upon which to base a formula.

Physiologic Therapy. Physiologic prescribing involves the use of gentle medicines over an extended period of time to nourish organs and restore normal tone and function. Physiologic medications do not have rapid or strong pharmacological actions and even if prescribed inappropriately would be unlikely to push the limits of homeostasis or cause undesirable side effects. Physiological medications balance, nourish, and tone, and they help restore optimal physiology through gentle support of assimilation, elimination, detoxification, metabolism, perfusion, and nerve function. Most herbs are physiologic in nature, capable of restoring normal functioning, organ tone, and homeostatic balance. In contrast to the above pharmacologic examples, herbs may be used to support a fever when needed to allow the body to fight infections, or to encourage the body to build new bone cells rather than halt all bone cell turnover. Because physiologic medications repair and restore tissue and organ function, they can usually eventually be stopped as the body becomes capable of maintaining balance without them. Herbs with nutritive physiologic actions should be those tonics upon which all formulas are based.

the precise system, vernacular, and approach do not matter as long as you are aware of some sort of energetic presentation. Whether or not you take it upon yourself to learn the doshas, TCM, or four-elements thinking and prescribing, you can still begin to notice whether a patient is hot or cold, damp or dry, for example, and choose herbs based on the specific clues or symptoms that you discern through thorough questioning and from simple and obvious physical exam findings.

For example, a man with erectile dysfunction may present with chronic urinary symptoms including dribbling of urine, cloudy appearance to the urine, and difficulty initiating urination, all symptoms of an enlarged prostate gland interfering with normal urine flow. The patient is elderly and in generally good health but is often cold, especially in the hands and feet, and has mild chronic arthritis, and the erectile dysfunction involves no difficulty achieving an erection, but difficulty maintaining an erection—all symptoms of a cold, deficient energetic state. The herbal formula in this situation might feature fluid-moving herbs such as *Angelica* and *Ginkgo* as the base and combine them with general genitourinary tonic herbs such as *Serenoa* or *Urtica* root as adjuvants and with warming chi tonic herbs such as *Panax ginseng* as a specific. Another formula for erectile dysfunction, however, might be appropriate for a younger man who has difficulty both achieving and maintaining an erection, which is associated with the diabetic state where elevated glucose and lipids have impaired renal and pelvic circulation. A formula for this patient might feature *Ginkgo, Angelica,* or *Salvia miltiorrhiza* as a base to enhance circulation, and combine them with *Commiphora mukul* as an adjuvant to help control hyperglycemia and *Epimedium* as a specific for erectile dysfunction due to aging, in this case premature due to metabolic distress. And a last example is a formula for erectile dysfunction in a long-term smoker who now has COPD and poor circulation to the limbs, as well as impaired microcirculation in the penile vasculature. A formula for this person might feature *Ginkgo, Salvia miltiorrhiza,* and *Angelica* once again as general circulatory tonics as a base, *Curcuma* and/or *Vaccinium* as adjuvants to reduce oxidative stress in the vasculature, and *Pausinystalia yohimbe* as a specific because none of the nourishing herbs may be strong enough to achieve an erection in a case with this degree of vascular damage.

When I teach, I often present sample cases and lead discussions with my students to explore how to think through the choices of herbs for the base, synergists, and specifics, based on details of the particular patient. It can be helpful to work through two examples of people with the same "diagnosis," such as insomnia or acne, and discover how differently a formula evolves based on the patient's unique constitution, energetics, and symptoms. In the first sample case, I offer an extensive discussion that arrives at a specific formula. Following that, I present a more condensed sample case, with several options for base, synergist, and specific herbs. See how you do at finishing the selection process for the sample cases described below. There is no single right or wrong answer.

SAMPLE CASE: VARIATIONS OF INSOMNIA

Patient 1: This insomnia patient has been exhausted for years, with an accompanying history of chronic hay fever and occasional respiratory infections requiring medical attention. The insomnia is of recent onset and involves not being able to fall asleep for many hours, lying in bed very tired and exhausted, but awake. The person will finally fall asleep and wake in the morning groggy and still exhausted. She is cool with cold hands and feet in bed at night and even during the day. She experiences gas and bloating if she eats raw broccoli and onions. For such a patient, you might consider basing a formula on "energy or chi" tonics, such as *Panax, Eleutherococcus, Astragalus, Rhodiola, Ganoderma, Cordyceps,* or

Oplopanax. You might settle on a base of *Panax* because it is more warming than some of the others. For the specific, you might select *Astragalus* because it is also a chi and immune tonic, and it is also specific for allergies and respiratory infections that linger and exhaust. The synergist might be *Zingiber* because it enhances circulation in the cold extremities and improves digestion by warming and stimulating core organs. It could help drive the other ingredients in the formula where they need to go by being a heating, stimulating, and moving herb.

Patient 2: This second patient has suffered from restless sleep for many years, but it is presently worse. The pattern has been to fall asleep readily but to wake after a few hours and then fall back to restless sleep for the remainder of the night, waking briefly every half hour or so, tossing and turning with stiff muscles, and going back to sleep for short stretches only. She has occasional episodes of feeling too hot in bed and rolling away from her partner, feeling uncomfortably warm, and taking a while to fall back asleep. The person has come for a consultation because recently, instead of frequently waking and rolling over and going back to sleep, she now lies awake for half an hour, sometimes several hours, before falling back to sleep. She also experiences back and joint stiffness at night in bed and for a short while in the morning upon waking. She reports having only a few minor colds per year, recovers quickly without needing treatment, and

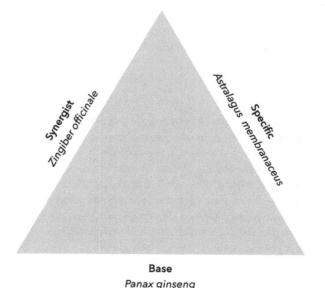

Figure 1.2. Formula for Insomnia Patient with Exhaustion

doesn't even feel terribly ill with them. She reports having oily skin, still prone to breakouts on the central face, especially premenstrually, and having some minor PMS symptoms, primarily emotional lability and crankiness.

There are many relevant details here. For this restless, stiff insomnia patient, who is excessive in movements, muscle tension, nervous tension, and heat, you want to rest a formula on something cooling and relaxing,

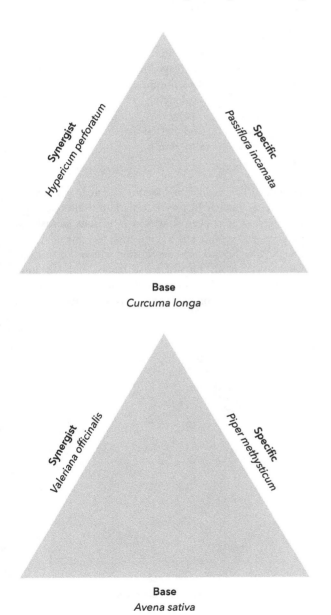

Base
Curcuma longa

Base
Avena sativa

Figure 1.3. Two Formulas for Restless Insomnia

particularly to the nervous and musculoskeletal systems, such as *Avena*, *Passiflora*, or *Scutellaria*. While *Valeriana* and *Piper methysticum* are two of the most powerful herbs specific for both sleep and muscle relaxation, you might choose not to "rest" the formula on them because they are rather hot and could be a problem long term or if put in the formula in too great a proportion. Not only are they heating, but they are not particularly nourishing and never listed in the folkloric literature as nervous system restoratives or daily long-term-use herbs. *Valeriana*, *Actaea*, and *Piper methysticum* could certainly be used in the formula, as long as care is taken to use them as synergists or specifics in smaller proportion than the other ingredients in the formula, while basing the formula on something more restorative to the nervous and musculoskeletal system and cooling to the body, such as *Avena* or *Scutellaria*. Another approach might be to consider adding *Valeriana* or *Piper methysticum* to a formula for short-term use or to be taken only before bed to reap the benefits of the more powerful muscle-relaxing effects of these herbs. In addition, you would create a separate formula to be used during the day and over the long term, attempting to restore the nervous and muscular tone so that the more powerful muscle relaxers and before-bed formula are no longer needed. *Scutellaria*, *Matricaria*, *Eschscholzia*, *Hypericum*, *Passiflora*, *Avena*, and *Melissa* are cooling, soothing, nourishing, and restorative to the nervous system, with *Scutellaria* and *Passiflora* having the greatest effects on muscle tension as well. Many people with insomnia, who are not particularly exhausted like the first insomnia patient, benefit from sedating relaxing herbs like *Valeriana*, while patients with insomnia and long-term exhaustion might only become further tired or even lethargic and depressed with strong sedatives like *Valeriana* or *Piper methysticum*.

Other considerations or clues for the second insomnia patient are her oily skin, skin eruptions, aggravation by elevated premenstrual hormones, and hormone-related emotional tension. Although these are all very minor and very common in the general population, they are all heat symptoms. These symptoms also suggest that the liver and hormonal metabolism may be contributing underlying factors in this person's overall constitution and balance. The liver not only processes hormones and removes waste products from the blood that can otherwise contribute to acne and skin eruptions, but, in TCM, the liver is said to "rule" the joints and tendons. In many traditions, including TCM, liver herbs are often said to be specific for vague muscle stiffness that does not represent tendonitis, arthritis, fibromyalgia, or other condition. Therefore, a good synergist for this case might be a cooling liver herb noted to improve skin and hormonal balance. Some choices here might be a simple alterative such as *Taraxacum*, *Arctium*, *Silybum*, or *Curcuma*. Thus, for the second insomnia patient, you might end up with a formula such as *Hypericum*, *Passiflora*, and *Curcuma*, to be taken multiple times per day for many months, with a before-bed formula of *Avena*, *Piper*, and *Valeriana*. Yes, there is some arbitrariness in the selection of herbs, but only to a degree. The arbitrariness stems from a somewhat capricious choice between herbs that have very similar energies and specific indications, or mixing and matching formulas in such a way that the overall base and specific energy is the same. It is possible to create several variations of a formula having nearly identical action. For example, both *Taraxacum* and *Arctium* roots may be somewhat interchangeable as an alterative base in a formula. Either fennel or caraway seeds may offer interchangeable carminative effects in an IBS formula. *Valeriana* or *Piper methysticum* might both be effective in the above insomnia case. Or a synergist of *Curcuma* or *Silybum* might be logical to help the liver process hormones.

SAMPLE CASE: VARIATIONS OF ACNE

Patient 1: The first case is a 25-year-old woman with acne. History reveals frequent heavy menses, PMS, including menstrual headaches, breast tenderness, and significant anxiety, along with chronic constipation as concomitant complaints. She rarely suffers from infections but when she does, they will come on quickly with painful sore throats, high fevers, and acute illnesses. She will recover quickly in a day or two. She also has frequent episodes of insomnia and muscle pain in the neck and upper back, most often premenstrually.

Figure 1.4 shows possible herbal combinations based on these considerations of the patient's constitution, action of herbs, and energetics.

4 Elements	Actions of Herbs	Energetics
Seek water to balance fiery symptoms	Seek alteratives, cholagogues	Seek cooling
Seek earthy therapies and herbs to ground the volatile tendencies	Seek hormonal-balancing agents	Seek sedating
	Seek nervines	Seek yin tonics
	Seek detox	

Patient 2: The second case is also a 25-year-old woman with acne, but her history reveals minor allergies, hay fever, frequent upper respiratory infections, occasional yeast vaginitis, and tetracycline use for several years. She is chilly, has a damp constitution with much mucous production, tends toward loose stools, and has low-grade infections that linger a long time. Figure 1.5 shows possible formula combinations appropriate for this case.

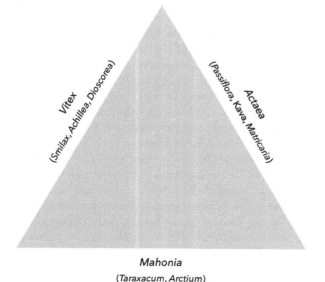

Mahonia
(*Taraxacum, Arctium*)

Figure 1.4. Formula Possibilities for Acne with Fiery Symptoms

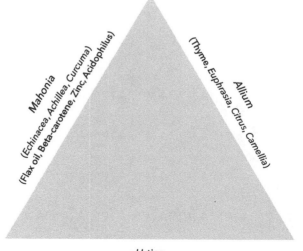

Urtica
(*Taraxacum, Calendula*)

Figure 1.5. Formula Possibilities for Acne with Dampness and Infections

4 Elements	Actions of Herbs	Energetics
Seek warming herbs to balance cold	Seek immune stimulants	Seek chi tonics
Seek drying herbs to balance damp	Seek antimicrobials	Seek yang tonic
Seek fiery herbs	Seek intestinal, vaginal, and respiratory astringents	Seek warm and drying tonics
	Seek antiallergy herbs	Seek to purge fluid and dampness
		Seek to move medicine to head, skin, upper respiratory tract

SAMPLE CASE:
VARIATIONS OF RHEUMATOID ARTHRITIS

Patient 1: Our first example is a 50-year-old woman with a chief complaint of rheumatoid arthritis, primarily in the hands, wrists, neck, and shoulders. Onset was associated with conflicts and issues with children, loss of control in influencing children's lives, and disappointments. She reports minor anxiety and frequent bouts of insomnia. She also suffers from frequent constipation, a chronic cough due to a dry scratchy throat, and occasional brief episodes of cystitis. Her hands become red and swollen in acute episodes that are experienced as aching, burning, sore, and tender and then settle down over a month's time. Her neck becomes tight and spastic, with a throbbing headache and burning sensation on the skin. Symptoms wax and wane independently of diet or activity, but perhaps correlate to stress. She has a trim build, is often warm, is often thirsty, has a big appetite and a fast metabolism, and is very active. See figure 1.6 for formula possibilities for this case.

4 Elements	Actions of Herbs	Energetics
Seek to cool fire	Seek nervines	Seek to cool and moisten
Improve dryness, heat with watery, cooling herbs	Seek anti-inflammatories	Aim to soften, lubricate, and smooth
Improve insomnia, restlessness, and worry with grounding, earthy herbs	Seek antispasmodics	Seek to quiet and calm energy
Aim to ground with moist earthy herbs	Seek tissue lubricants	Aim to move medicines to nerves
	Seek nerve and connective tissue tonics	

Patient 2: The second example is a 50-year-old woman with a chief complaint of rheumatoid arthritis—the arthritic pain is in multiple joints including the low back, hands, wrists, shoulders, and hips. There is a family history of rheumatoid arthritis and osteoarthritis. Constant mild to moderate stiffness is worst after sleeping or prolonged inactivity and also is aggravated with exertion.

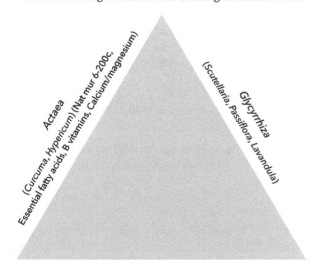

Figure 1.6. Formula Possibilities for Rheumatoid Arthritis with Stress

Actaea (*Curcuma*, *Hypericum*) (Nat mur 6-200c, Essential fatty acids, B vitamins, Calcium/magnesium)

Glycyrrhiza (*Scutellaria*, *Passiflora*, *Lavandula*)

Piper methysticum (*Avena*, Gotu kola, *Althaea*, *Symphytum*)

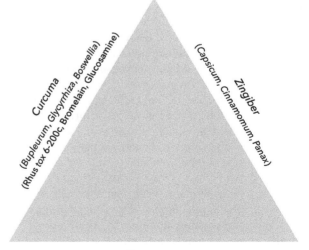

Figure 1.7. Formula Possibilities for Rheumatoid Arthritis with Cold Constitution

Curcuma (*Bupleurum*, *Glycyrrhiza*, *Boswellia*) (Rhus tox 6-200c, Bromelain, Glucosamine)

Zingiber (*Capsicum*, *Cinnamomum*, *Panax*)

Taraxacum (Gotu kola, Nettles, *Equisetum*)

She has some minor arthritic and degenerative changes in the low back and some bony deformity beginning in the finger joints. The sensation is heavy, stiff, and aching and is better when resting the affected limb. She tends to be chilly, has chronic postnasal drip, is on hormone replacement therapy, retains fluid, has mild constipation, is slightly overweight, and reports low energy. Figure 1.7 shows formula possibilities for this case.

4 Elements	Actions of Herbs	Energetics
Seek to dry out excess water	Seek anti-inflammatories, anodynes	Seek warming and drying
Seek to warm, fire	Seek to move fluid, diurese	Seek stimulating, moving therapies
Seek to lift, lighten with airy or volatile compounds	Seek connective tissue tonic	
	Seek alterative detox therapies	

The Use of "Toxic" Herbs in Formulas

In the sample case of insomnia with exhaustion where *Zingiber* is used as a synergist because of its heating and moving energetic qualities, *Zingiber* might be used in a lesser proportion than the other two herbs, in order to ensure the formula is overall nourishing, restorative, and tonifying. *Zingiber* is not particularly toxic, but due to its strong energy, there is a minor concern over its proportion in a formula. This is especially true for caustic, irritating, or outright toxic botanicals. Some such toxic or otherwise powerful herbs have strong moving or driving action: *Phytolacca* is a lymph mover, *Gentiana* is a bile mover, *Rhamnus* is an intestinal smooth-muscle mover, *Sanguinaria* is a tissue mover, and *Pilocarpus* is a secretory stimulant. All are used in formulas in a small or lesser quantity than the lead herbs, and yet their presence is equally contributory and as powerful as the other primary ingredients.

The case is the same when energy or motion is excessive, and a goal of the formula is to calm and soothe or to quiet down an excessively hot or hyperfunctioning action in a tissue or organ. *Veratrum, Aconitum, Atropa belladonna, Digitalis, Lobelia,* and *Conium* all have powerful relaxing, sedating, diminishing effects on nerve, cardiac, respiratory, and musculoskeletal tissues, appropriate only in small amounts in the overall formula.

In many cases, herbs with such extremely strong energies are added to formulas only when extremely strong imbalances are occurring. In general, the stronger and potentially toxic or harsh the herb, the smaller the dose in the formula. And yet even that milliliter, or as little as five drops, drives the formula and contributes equally to the other herbs occurring at a dose of 10 or even 100 times the amount. The smaller the degree of atrophy/hypertropy or the more minor the degree of imbalance of hot/cold or excess/deficient status of the body, the less the need for any strong energy herbs, and the more the formula can be based on purely nourishing and restorative ingredients for the base, synergist, and specific alike.

In the previous example, the insomnia patient is so exhausted and so cold that a little bit of *Zingiber* to warm up our formula is appropriate. *Zingiber* is quite hot but overall nontoxic and safe, even long term. You would not include *Podophyllum* or *Iris* right off the bat to warm the bowels, and you would not choose *Thymus* or *Allium* (she is sensitive to onions and might react) to warm the lungs and reduce infections. None of those herbs are as indicated or specific as *Zingiber* to drive the formula into the proper locations. Thus, for this insomnia patient, you might end up mixing a formula of 30 milliliters *Panax,* 20 milliliters *Astragalus,* and 10 milliliters *Zingiber* to fill a 60-milliliter (2-ounce) dropper bottle of tincture.

Further Guidance on Creating Warming and Cooling Formulas

The actions of herbs—both traditional folkloric and mechanistic—can be organized into simple categories of warming or cooling. Vasodilators, for example, and the hot spicy "blood movers" are generally warming, and demulcent herbs are generally cooling.

Warming stimulants. Here are some general guidelines for the use of warming stimulants in treating acute conditions.

- Use for colds, fever, and chills with acute onset
- Use for abundant mucus, phlegm in the throat and lungs
- Use as diaphoretic for those who fail to mount a useful fever in acute illness
- Use for many common infections of childhood in small doses
- Discontinue therapy when improvement is achieved
- Contraindicated in yang constitutions, for those who feel uncomfortably hot
- Contraindicated in hemorrhage, free sweating, or night sweats

Warming Stimulants

ACTION	HERBS
Peripheral vasodilators	*Achillea millefolium* *Allium sativum* *Cinnamomum* spp. *Ginkgo biloba* *Zingiber officinale*
Secretory stimulants	*Armoracia rusticana* *Capsicum* spp. *Iris versicolor* *Pilocarpus microphyllus*
Diaphoretics	*Achillea millefolium* *Capsicum annuum* *Zingiber officinale*
Warm antimicrobials	*Allium sativum* *Curcuma longa* *Origanum* spp. *Thymus vulgaris*
Chi tonics	*Eleutherococcus senticosus* *Panax* spp. *Withania somnifera*

Cooling Remedies

ACTION	HERBS
Demulcents	*Aloe vera, A. barbadensis* *Althaea officinalis* *Ulmus fulva* *Verbascum thapsus*
Bitters/alteratives	*Arctium lappa* *Berberis aquifolium* *Rumex crispus* *Taraxacum officinale*
Diuretics	*Apium graveolens* *Equisetum arvense* *Galium aparine* *Petroselinum* spp.
Cooling antimicrobials	*Echinacea purpurea* *Hydrastis canadensis* *Lomatium* spp. *Mentha piperita*
Astringents	*Geranium maculatum* *Hamamelis virginiana* *Rubus idaeus*
Energy dispersants	*Achillea millefolium* *Galium aparine* *Iris tenax, I. versicolor* *Mentha piperita*

In cases of chronic disease, observe these guidelines for warming stimulants.

- Use for circulatory enhancement, excessive clotting, blood stasis
- Use for fluid accumulation, poor circulation to organs
- Use for chronic inflammation with stiff and swollen character that feels better in hot weather or after a hot bath
- Use for cold hands and feet
- Use for atonic, feeble, sluggish constitutions
- Don't use for high fevers due to chronic debility, rather for weakness and sense of chill

Cooling remedies. These are guidelines for the use of cooling remedies in treating acute conditions.

- Use for acute mucosal inflammations, cool tissues with demulcent anti-inflammatories
- Use for burning sensations in the throat, intestines, or skin
- Use for tight, dry sensations in the mucous membranes or joints

- Use for acute infections with fever or sensation of heat; cool with antimicrobials and alteratives
- Use for acute toxicity states, joint pain, or headache; cool with bitters, alteratives, diuretics
- Contraindicated for acute disease with chills, abundant watery mucus, consolidation in the lungs

In cases of chronic disease, follow these guidelines when using cooling remedies.

- Use for dry, hot constitutions; use bitters, demulcents, alteratives, and diuretics
- Use for general yang states with warmth, redness, and heat in the body
- Use for tendency to acute fevers, infections, inflammation, and joint pain
- Contraindicated for chronic diseases associated with excessive dampness, coldness, deficiency, and fluid stasis

Types of Herbal Medicines

Herbs are available as teas, tinctures, powders, encapsulations, syrups, solid extracts, and other forms. Following are pros, cons, and indications for the most common types of these herbal preparations.

Herbal Teas

Teas are especially indicated for individuals with digestive and bowel weakness to avoid irritation by pills or sometimes tinctures and to improve absorption of desired substances. Teas are also indicated for urinary ailments where substances in water quickly reach the urinary passages. Demulcents to soothe inflamed mucosal surfaces such as the esophagus, stomach, intestines, and bladder are best delivered via teas. Teas are also relatively inexpensive and, when of good quality, can be a very effective method of getting nourishing herbs into the body, such as *Pueraria* as a daily circulatory and reproductive tonic or dried berries such as *Crataegus*, *Vaccinium*, *Sambucus*, or Goji to help protect the blood vessels from oxidative stress when diabetes or metabolic syndrome are present. When the desired herbs are particularly unpalatable, tinctures or pills would be friendlier. Or when a high dosage of an herb, action, or chemical is desired, it may be easier to accomplish that with pills or tinctures.

Tinctures

Tinctures are especially handy when fresh plant juices are desired for preservation. For the home herbalist, tinctures are a practical way to stock the medicine chest with inexpensive, valuable medicines with a long shelf life. Tinctures will store for many years, whereas dried herbs age as the months go by and should be replaced about every two years to be of medicinal quality.

To any clinical herbalist, another important virtue of tinctures is that they can be combined into formulas as precisely indicated for a given individual, where commercially available pills are a one-size-fits-all formula. Tincture prescriptions can be formulated in an exacting manner to reduce the need to take many different medications to address many different organs, imbalances, and pathologies. Tinctures, when well thought out, can address many different pathologies, levels, and energies all at once.

Pills

Tablets and capsules can be convenient and practical ways to combine herbs with vitamins, minerals, amino acids, protomorphogens, bile, and numerous other diverse substances. Formulas designed for treating everything from prenatal nutrition to congestive heart failure or hypothyroidism are available, and when well indicated they are very helpful. However, unless you are producing your own encapsulations, you are usually limited to prescribe the pill combinations as the proprietor has formulated them. Because they might not be as specific for particular presentations as your own formulas, you might complement proprietary encapsulations with more specific herbs in tea or tincture form as needed to round out the herbal prescription.

Pills are also useful when high dosages are being pushed for urgent health problems or when numerous different medications are being employed at once. Herbal pills may be simply dried and powdered plant material, or they may contain purified and concentrated plant constituents, such as curcumin, manipulated to boost its absorption and assimilation.

Knowing When to Use What Form of Medicine

When there are urgent circumstances, standardized or concentrated medicines might be desirable for their concentration and known potency, such as using standardized *Ginkgo* for serious ischemic disease or *Convallaria* for heart failure or silymarin concentrate for acute *Amanita* toxicity. On the other hand, there is no reason to take expensive standardized *Matricaria* when a dollar's worth of good quality chamomile tea would likely do the trick for a flare-up of indigestion or irritable bowel.

When treating complex and chronic disorders, it is often appropriate to use many different herbs or supplements at once. For example, a protocol might call for ginkgo, garlic, milk thistle, passionflower, and echinacea, along with a multivitamin, flaxseed oil, and some additional beta-carotene, zinc, magnesium, and calcium. A person on a budget (which is almost everyone) would not be able to keep up with such a program for very long if each item had to be purchased as a separate bottle of pills. Not to mention how difficult it is for many people to take handfuls of pills day in and day out. However, if the herbs could be combined in a tincture and a tasty tea and supported with the perfect vitamin and herbal/nutritional combination, the cost and convenience could both be improved.

Preparing Formulas for Use

Teas and tinctures are used throughout this formulary. With regard to herbal teas, as a general rule, delicate plant parts such as flowers are steeped rather than boiled, because the aromas and flavors can be lost or destroyed with vigorous boiling. Green leafy herbs such as nettles, alfalfa, and mint are also gently steeped, but harder, denser plant parts such as roots, barks, dried rose hips or hawthorn berries, and seeds are best simmered to extract the medicinal components. Simply steeping herbs is referred to as preparing an infusion, while gently simmering herbs is referred to as a decoction. Unless otherwise specified, it is ideal to infuse teas for 10 minutes. For herbs that are decocted, simmer them gently for 10 minutes and then remove from the heat to stand 10 to 15 minutes more. In a few cases in this text, I recommend steeping or simmering for a longer period, when the intention is to liberate minerals or other compounds that are not readily released with simple infusions or decoctions. In some cases, it is recommended that vinegar or lemon juice be added to the water to best extract minerals. In other cases, it is recommended that mucilaginous herbs be macerated in cold water for many hours or overnight before bringing to a brief simmer to best extract mucilage.

A general dosage for a tea is a minimum of 3 cups a day. For an acute or urgent situation, however, such as a chest cold threatening to turn into pneumonia, a more aggressive approach would be recommended, such as preparing an 8- or 10-cup pot of tea with a goal of drinking 1 cup every hour for several days.

Herbal tinctures are also used throughout this formulary with recipes given for the proportions and ratios. Because commercial tinctures are readily available, I do not go into detail discussing the making of tinctures. However, in some cases, herbal oils or vinegars that are not readily available are used in formulas, and in these cases, I offer brief details on how to prepare them. I also provide a few unique methods of preparing formulas, such as using solid extracts to thicken formulas to help them cling to the esophagus, or placing the formula in a spray bottle to use as a throat spray.

A general dosage strategy for tinctures is to use 1 dropperful (½ to 1 teaspoon) three times a day when treating chronic conditions. Most people have difficulty taking medicines any more than this, unless they are acutely ill or uncomfortable and motivation is high. In acute situations, tinctures may be taken as often as hourly for acute pain or infectious illnesses, or even as often as every 10 minutes. When tinctures or other herbal medicines are taken at this aggressive dosage, it is always for a very limited length of time, such as every 10 minutes for an hour, or every hour for a day, with instructions to reduce the frequency as symptoms improve.

Although the focus of the book is to create effective herbal formulas, in some cases, protocols are offered to create comprehensive therapies for chronic conditions such as thyroiditis or benign prostatic hyperplasia (BPH). In such cases, I may suggest the use of a tea, a tincture, medicinal foods, and herbal encapsulations to treat such challenging conditions; additional examples are circulatory insufficiency related to diabetes or prostate cancer, which, because it is slow to metastasize, may respond to herbal therapies requiring a long-term approach to see results. When herbal capsules are employed, they are often dosed as little as two per day or as much as two or three at a time, three times a day. While teas and tinctures can be taken with or without food, herbal capsules are most often taken with meals to enhance their digestion and to prevent them from being nauseating on an empty stomach.

Extending the Triangle Philosophy

As explained, herbal formulas exemplified throughout this book aim to use nourishing herbs as base ingredients in all formulas, complementing them with specifics and synergists. All the formulas include an introductory sentence or two explaining why the herbs are chosen, cite any research to support their use in specific conditions and situations, or comment on why they are appropriate for a specific presentation of the condition being discussed. I encourage readers to refer to "Specific Indications" at the end of each chapter to help master *materia medica* knowledge. Staying focused on the desired actions can help you avoid creating a formula that is too broad. With so many available herbs to choose from, knowledge of niche indications for various herbs helps prevent a "kitchen sink" approach. Instead of choosing every herb you know of that may be effective for hypertension or respiratory viruses, you can select the best choices with a solid knowledge of *materia medica*. It is a common mistake to place too many herbs with the exact same function in a formula and miss including herbs that address the underlying cause or other special considerations. For example, beginning herbalists may attempt to put half a dozen herbs known to help regulate blood glucose in a tincture, where some of those herbs would actually be more effectively given as medicinal foods (garlic, berries, fiber). The more medicine-like herbs could thereby be taken at a higher dose when just two or three are chosen for a tincture. In other cases, herbalists may place two or three adaptogens in a formula but overlook and omit hepatotonics, or use two blood movers that have similar action and forget the high-flavonoid vascular protectants. Another common mistake is to choose herbs based on a research study and be wooed by the newness of the information, thereby paying top dollar for a proprietary product when common alterative herbs and comparatively mundane herbs may be just as effective at a fraction of the cost. By choosing herbs with a desired action in mind or by listing all of the actions that one wishes an herbal formula to perform, such redundancies and omissions are less likely. And by considering which herbs with the desired action best fit the person and energetic state—warming or cooling, stimulating or sedating, moistening or cooling, and so on—one can craft a finely tuned formula that addresses the individual person rather than making a kitchen-sink attempt to treat the diagnosis. Once one has a solid knowledge of the *materia medica*, the specific indications of the herbs, prescribing and formulating become somewhat intuitive. This becomes easier, and even second nature, the longer one focuses on making formulas and the more experience one obtains.

Many herbalists believe that the plants are their teachers and that they learn from the plants themselves over time, as the plants teach us what they are capable of. Many herbalists are deeply aligned with nature and often have cultivated nature-based spirituality. Herbalists may embrace the notion that clinical intuition is evidence of a connection with the plants, and that this is the real root of herbal medicine as a healing discipline.

CHAPTER TWO

Creating Herbal Formulas for Endocrine Conditions

Thyroid and adrenal dysfunction are both very common issues in the general population, and alternative practitioners treat these conditions on a regular basis. Alternative medicine often gives more attention to the subtler presentations of adrenal and thyroid disorders, recognizing symptoms even when lab work is within normal limits. Thus, many patients who are told everything is normal yet continue to feel unwell often find their way to naturopathic physicians, herbalists, and other alternative practitioners, who employ herbal and nutritional approaches with excellent results. Herbalists often first aim to restore thyroid and adrenal function, if possible, rather than immediately recommending thyroid replacement therapy or corticoid prescriptions. Because all the endocrine organs (pituitary, adrenal, thyroid, parathyroid, and pineal glands) are synergistic, alternative practitioners approach adrenal and thyroid imbalances wholistically, supporting entire systems, such as the hypothalamic-pituitary-adrenal (HPA) axis, rather than focusing on single hormones. Naming a formal diagnosis is less important to alternative practitioners. They focus on recognizing organ system strengths and weaknesses and making an effort to restore balance, feedback loops, and synergism. The adrenal glands and the thyroid gland are a synergistic pair when functioning properly: They both promote metabolism and they both burn fuel to provide energy. If either the thyroid gland or the adrenal glands are weak, the other will also be affected. As the brain attempts to stimulate the malfunctioning organ, what endocrinologists refer to as "crosstalk" ensues, as feedback loops with the hypothalamus and pituitary become involved and hormonal signals are altered in numerous complex ways.

Altered thyroid and adrenal function also lead to metabolic changes, and both increased and decreased metabolic function will increase oxidative and inflammatory stress in the entire body. Hypothyroidism contributes to high cholesterol, diabetes, and polycystic ovarian syndrome (PCOS), highlighting the synergism between endocrine and metabolic systems. Adrenal dysfunction can also involve glucocorticoid and mineralocorticoid abnormalities and can contribute to blood sugar, blood pressure, and electrolyte imbalances.

In this chapter I explore the most common thyroid disorders including thyroiditis, goiter, hypothyroidism and hyperthyroidism, as well as the most common adrenal disorders, Addison's disease and Cushing's syndrome. This chapter will also share treatment strategies for subclinical adrenal and thyroid disorders that do not strictly fit the common disease categories.

An Overview of Thyroid Hormones

Thyroid hormones are important to embryologic and adolescent growth and development. In adulthood, these hormones help to regulate the function and metabolism of virtually every organ system. Tissue-specific modulation of thyroid hormone action is achieved by complex and redundant control systems that include thyroid hormone secretion, plasma transport, transmembrane transport, hormone receptor up- and downregulation, synergism with other hormones, and coregulation at the level of the nuclear membrane. The detailed mechanisms of these systems are beyond the scope of what is covered in this chapter.

The thyroid gland's primary secretion is thyroxine, which serves to regulate basic metabolic rate. Thyroxine causes stored energy to be metabolized for energy and to maintain body heat. Thyroxine, also called T4, is involved in basic metabolic balance and helps regulate the pulse and the respiratory rate. The thyroid also produces triiodothyronine, or T3. T3 is at least three times more powerful in its ability to stimulate metabolism in target cells than T4.

The thyroid synthesizes thyroid hormones in response to thyroid-stimulating hormone (TSH), which is produced in the pituitary gland and released into systemic circulation. TSH travels in the bloodstream to ultimately bind to TSH receptors on thyroid cells. The hormones are synthesized from tyrosine and iodine with the help of iodine-processing enzymes called iodinases. The thyroid secretes T4 at substantially greater levels than T3, relying on peripheral mechanisms to convert T4 to T3.

TSH receptors are noted to occur in many tissues besides the thyroid gland including on the thymus, adrenal glands, kidneys,[1] brain, adipose cells, and bone cells,[2] all of which also respond in organ-specific ways when bound by TSH. Estrogen and corticosteroids also play a role in the production of thyroid hormones.[3]

Factors influencing the peripheral conversion and cellular responses to thyroid hormones continue to be elucidated and help explain why many patients experience symptoms consistent with hypothyroidism, including low body temperature, even when their thyroid blood tests are normal.

Common Thyroid Disorders

Thyroid disorders are extremely common in the general population and are 8 times more common in women than in men. Onset of both hyper- and hypothyroid disorders may begin during the menopausal transition as the pituitary-thyroid-gonadal axis goes through changes. The hypothalamus-pituitary-adrenal (HPA) axis is more widely known, but similar hormonal cascades involve the brain's feedback loops with the

thyroid and reproductive organs, and this is referred to as the pituitary-thyroid-gonadal axis. There are many well-known conditions where thyroid disorders affect the reproductive system and lead to low sperm counts in men and menstrual irregularities in women. For example, women with hypothyroidism can be amenorrheic. Transient hypo- and hyperthyroidism are also common postpartum, all indicating synergism between the thyroid and reproductive hormones.

Excessive thyroxine (hyperthyroidism) can lead to hyperthermia, tachycardia, and excessive caloric use and anabolism that induces thin, restless individuals prone to hypoglycemia. Conversely, deficient thyroxine output (hypothyroidism) can lead to obesity and hyperlipidemia as excessive nutrients are stored instead of being metabolized.

The typical signs and symptoms of thyroid insufficiency include:

- Fatigue, lethargy, poor stamina
- Coldness, sensation of chill, cold hands and feet
- Overweight or difficulty maintaining weight
- Elevated lipids, possibly glucose
- Constipation, digestive dysfunction
- Dry skin, dry brittle hair and nails, poor healing wounds
- Slow pulse
- Allergies, frequent upper respiratory infections
- Menstrual irregularities or amenorrhea
- Low sperm counts

Patients with hypothyroidism display a higher prevalence of diabetes, hyperlipidemia, coronary artery disease, and possibly depression. Herbal formulas for hypothyroidism may include glucose-regulating, hypolipidemic, and antidepressant herbs, depending on the person and the presentation.

Hypothyroidism involves an inadequate production of thyroid hormones, due either to poor hormonal regulation or to the inability of the thyroid gland to respond to TSH. Reduced conversion of T4 to T3 may contribute to a variety of chronic diseases.[4] The deiodinase enzymes responsible for the conversion may be up- or downregulated in different disease states or may be affected by oxidative stress. Some drugs may alter peripheral conversion of T4 to T3, including corticosteroids, propylthiouracil, radiographic contrast dyes, amiodarone, and the beta blocker propranolol. Target tissue may also be resistant to thyroxine, even in the presence of seemingly normal lab tests. Resistance to thyroid hormones may present with elevated levels of thyroid hormone, with normal or slightly elevated levels of TSH, with goiter, and in the absence of symptoms of thyrotoxicosis.[5] Although resistance to thyroid hormones was first described in the 1960s, the disorder remains underrecognized, poorly studied, and without a standard treatment protocol. Inflammation in the thyroid and autoimmune reactivity can also contribute to both hypo- and hyperthyroidism. The body may produce antithyroid peroxidase antibodies (anti-TPO antibodies), thyrotropin receptor antibodies (TRAbs), and thyroglobulin antibodies that can contribute to both

Dietary Seaweed for Thyroid and Metabolic Function

Seaweeds are some of the richest sources of organic iodine compounds, and these compounds are shown to be readily absorbed by the human body. In Japan, where seaweed is a dietary staple, the daily average intake of iodine has been estimated at 1.2 milligrams per day,[6] which is well above the RDA. *Fucus vesiculosus*, commonly referred to as kelp or bladderwrack, is an edible brown alga and an excellent source of bioavailable iodine.[7]

In addition to iodine, *Fucus* also contains the flavonoid fucoxanthin,[8] which has antioxidant activity,[9] and fucoidans, which are reported to have anticancer activity.[10] Other brown seaweeds include *Undaria pinnatifida* (wakame), *Sargassum fusiforme* (also known as *Hizikia fusiformis*, common name hijiki), *Laminaria japonica* (ma-kombu), and *Sargassum fulvellum* (brown algae). Fucoxanthin is an orange-pigmented marine xanthophyll carotenoid found in brown seaweed that is credited with antiobesity effects as well as benefits to a wide variety of related pathologies, including metabolic syndrome, type 2 diabetes, and heart disease. Fucoxanthin supports metabolism in white adipose tissue cells, which helps to attenuate inflammation in metabolic syndrome[11] and optimize cytokine production.[12] Weight loss is supported via thermogenic effects,[13] improved insulin

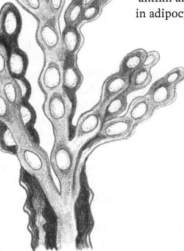

Fucus vesiculosus, bladderwrack

receptor response,[14] and increased uptake of glucose by skeletal muscle.[15] Fucoxanthin also optimizes lipid metabolism in adipocytes,[16] helping to lower serum and hepatic lipids when elevated.[17] In addition, fucoxanthin increases hepatic docosahexaenoic acid (DHA) levels;[18] DHA is an important fatty acid useful to both prevent

and treat metabolic syndrome.[19] Seaweed fucoxanthins are best taken with oil because lipids increase their absorption,[20] and fucoxanthin may in turn boost the activity of fish oils.[21] See also the *Fucus vesiculosus* entry on page 58.

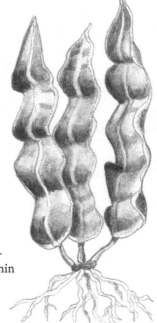

Laminaria japonica, Japanese kelp

oxidative stress and inflammation in the thyroid gland, as well as impair hormonal regulation. There is overlap in that both Hashimoto's and Graves' disease involve these antibodies, but anti-TPO antibodies are most commonly associated with Hashimoto's thyroiditis, a type of autoimmune-driven hypothyroid state.

Hyperthyroidism involves excess thyroid hormone being synthesized and secreted by the thyroid gland and can be overt or subclinical. Types of hyperthyroidism include thyrotoxicosis, which is associated with a high iodine uptake into an inflamed thyroid, and Graves' disease, a condition in which autoantibodies overstimulate the thyroid. The body may also produce thyroid hormones outside of the thyroid gland, and this is a less common cause of hyperthyroidism.

All types of hyperthyroidism may benefit from ample doses of antioxidants in the form of encapsulations, herbal teas and tinctures, and foods. The hyperthyroid state is accompanied with an increase in the ratio of prooxidant to antioxidant compounds, and this allows the accumulation of oxidatively damaged molecules, leading to oxidative stress in the thyroid gland and in the entire body. Hyperthyroidism can therefore lead to oxidative damage of liver, bones, and heart and contribute to liver disease, osteoporosis, and an increased risk of heart attack. The liver is particularly subject to oxidative stress, and thus liver support, alterative herbs, and lipotropic agents to improve metabolic clearance of oxidized fats and other inflammatory compounds are indicated. Many herbal tinctures or teas would be complemented by the use of lipotropic nutrients, berberine, silymarin, curcumin supplements, and other liver-supportive measures. Flavonoids may increase the uptake of iodine into the thyroid gland,[22] perhaps due to antioxidant effects that reduce inflammation. Notice that many formulas for thyroid imbalance include a liver herb such as *Curcuma longa* or *Silybum marianum*.

The typical signs and symptoms of thyroid excess include:

- Tremor, agitation, and anxiety
- Sweating, hot sensation, heat intolerance
- Hyperdefecation or diarrhea
- Heart palpitations, tachycardia, and arrhythmia
- Skin rash, dermatitis
- Insomnia, poor sleep, restlessness, anxiety

Rosmarinic acid (a phenolic compound derived from caffeic acid) from mint-family plants is noted to reduce stimulation of the thyroid by Graves' autoantibodies, explaining the traditional use of *Leonurus cardiaca*, *Lycopus virginicus*, *Melissa officinalis*, and other Labiatae family plants for the symptoms of hyperthyroidism. These herbs are prominently featured in formulas for hyperthyroidism and thyroiditis throughout this chapter. Rosmarinic acid has been shown to reduce thyroxine output via a variety of mechanisms including effects on iodinase and adenylyl cyclase enzyme systems inside thyroid cells. Furthermore, TSH and thyroid-stimulating immunoglobulins (TSI), which overstimulate the thyroid, are deterred by rosmarinic acid, helping to normalize thyroid function. Patients with severe hyperthyroid symptoms might be given pills, teas, tinctures, powders for drinks, and medicinal foods containing mint-family plants in tandem to create powerful protocols when trying to avoid thyroid-suppressing drugs.

Because viral infections can contribute to thyroiditis, patients might be given antiviral herbs such as *Andrographis paniculata*, *Hypericum perforatum*, *Phytolacca americana*, *Lomatium dissectum*, *Curcuma longa*, or other specifics to help reduce underlying immune triggers. Thionamide drugs can reduce thyrotoxicosis. The drugs accumulate in the thyroid follicular cells over time, eventually leading to a complete blockade of thyroid hormone production. As thyroid hormone levels decline, so do the autoantibodies produced against them. Forty to 50 percent of those using thionamide will experience lasting remission of Graves' disease over a 6- to 18-month course.

Formulas for Thyroiditis

Thyroiditis, an inflammation of the thyroid gland, may cause a goiter and may also cause the gland to atrophy. Thyroiditis may occur in euthyroid, hypothyroid, or (less commonly) hyperthyroid individuals and may be transient or chronic.[23] The cause of thyroiditis is not always apparent, but can be due to autoimmune disorders, infections, trauma, toxicity, and other underlying factors. Thyroiditis can be a mild inflammation that is not noticed by an individual, may cause diffuse enlargement of the gland that is visibly noticeable or that may cause symptoms such as a hoarse voice or dysphagia, or may be nodular, where discrete lumps develop (such lumps should be evaluated for possible cancer).

The autoimmune diseases of the thyroid—Graves' disease and Hashimoto's thyroiditis—both involve T cell-mediated inflammatory processes. These diseases affect approximately 5 percent of the population and may increase the risk of papillary cancer of the thyroid. Autoimmune thyroid disease is on the rise worldwide, believed due to nutritional and environmental factors including exposure to heavy metals, organochlorides, pesticides,[24] and tobacco smoke.[25] Hashimoto's thyroiditis, also called chronic lymphocytic thyroiditis, is the most common type of thyroiditis and is named after the Japanese physician, Hakaru Hashimoto, who first described this type of thyroid disease in 1912. The condition is most common in postmenopausal women.

Helicobacter pylori infection, a bacterium previously known to contribute to stomach ulcers and gastro-esophogeal reflux disease, has been shown to increase the incidence of autoimmune thyroid disease and thyroid nodules.[26] Viruses may also be involved with thyroiditis, triggering hypersensitivity and autoantibody production.

As noted above, thyroid hormones control basal metabolism, both catabolic and anabolic reactions, and can induce oxidative stress. Even normal synthesis of thyroid hormones liberates oxidizing substances in the healthiest of people, and in a vicious cycle, accelerated metabolism, especially in tandem with low antioxidant levels, exacerbates inflammation in the thyroid. Thus, alleviating oxidative stress with nutritional and herbal antioxidants should be the first order of business for treating thyroiditis.

Immune-modulating and anti-inflammatory herbs are appropriate in treatment protocols for thyroiditis, and selenium acts as a thyroid-specific antioxidant. Selenium supplementation has been shown to reduce antithyroid peroxidase antibodies and antithyroglobulin antibodies in patients with autoimmune thyroiditis. Autoimmune thyroiditis of Graves' disease can involve extensive lymphocyte infiltration, as well as follicular collapse and degeneration within the thyroid and is usually associated with enlargement of the thyroid gland and elevated serum thyroglobulin antibody (TgAb) and thyroid peroxidase antibody (TPOAb). Although these antibodies are also seen in Hashimoto's disease and may suppress thyroid function, in Graves' disease the inflammatory processes result in hyperfunction of the thyroid.

Thyrotoxic Substances

Many commonly occurring compounds are known to be thyrotoxic and implicated in creating oxidative stress in the thyroid, which may contribute to inflammation or trigger autoimmune reactivity in the thyroid gland.

Organophosphates are pesticides that decrease TSH, T4, and T3 production by thyroid cells.

Perchlorates are fertilizers that may interfere with the iodine symporter.

Nitrites are used in cured meats and may increase risk of thyroid and other cancer.

Heavy metals create oxidative stress and lipid peroxidation, which leads to inflammation of the thyroid.

Alcohol can suppress thyroid function directly due to toxic effects on cells and indirectly by blunting hypothalamic feedback loops.

Thiocyanates occur in *Brassica* family vegetables and can interfere with iodine uptake when excessive.

Soy foods may impair T3 to T4 conversion in susceptible people.

If left untreated, thyroiditis can permanently damage the thyroid gland, or it can progress to other serious diseases including hyperthyroidism, hypothyroidism, and thyroid nodules. There is also increased risk of thyroid cancer, especially in those with Hashimoto's thyroiditis. Therefore, treating even mild symptoms or mild elevations in thyroid antibodies is warranted. Even though Hashimoto's disease is associated with low thyroid hormones and Graves' disease is a condition of high thyroid hormone levels, both involve inflammation of the thyroid gland, and antithyroid antibodies may be present in both conditions. Thyroid-stimulating immunoglobulin (TSI) is also produced in Graves' disease. Therapies aimed at reducing inflammation, toxicity, or oxidative stress in the gland can be appropriate for both conditions and can be used independently of therapies aimed at normalizing thyroid hormones.

Artemisinin Simple for Autoimmune Thyroiditis

Autoimmune-driven inflammation occurs in Hashimoto's thyroiditis, Graves' disease, and atrophic thyroiditis. Artemisinin, a compound from *Artemisia annua*, has previously been studied as a therapy for malaria and some types of cancer, and products are available. Artemisinin may also help control acute and chronic thyroiditis.[27]

Artemisinin capsules

Take 2 capsules, 3 times per day.

Tincture for Graves' Disease with Thyroiditis

This tincture for Graves' disease aims to block thyroid-stimulating antibodies with *Leonurus* and *Melissa*, both of which contain rosmarinic acid. These herbs may also reduce excessive release of TSH from the pituitary due to dopaminergic effects. *Curcuma* offers systemic anti-inflammatory and hepatoprotective effects against the high level of oxidative stress.

Leonurus cardiaca
Melissa officinalis
Curcuma longa

Combine in equal parts. Take 1 to 2 dropperfuls, 3 or more times daily for many months.

Nutrients for Optimal Thyroid Function

Tyrosine is an amino acid required to synthesize thyroid hormones and may be supplemented using 500 milligrams, 2 or 3 times per day.

Iodine is an essential component of thyroxine, and 150 micrograms to 50 milligrams per day are needed for optimal thyroid function.

Selenium is a trace mineral and antioxidant. The deiodinase enzymes required to synthesize and metabolize thyroid hormones are selenium dependent; 100 to 500 micrograms per day of selenium are needed to support deiodinase enzymes.

Tincture for High Viral Loads and Thyroiditis

Viral infections may trigger autoimmune reactions when antibodies produced to fight the virus cross-react with TSH or thyroxine receptors and stimulate excessive thyroid activity, a phenomenon referred to as automimicry.[28] Increased interferon production in response to viral infections amplifies thyroid inflammation,[29] and interferon therapy for hepatitis patients can induce autoimmune thyroid disorders, which may persist for life.[30] Herpes, hepatitis,[31] echo, coxsackie,[32] and likely other viruses are noted to increase thyroid inflammation. Bacterial dysbiosis may also predispose, as in the case of *Helicobacter* found to be associated with an increased incidence of thyroiditis. Some patients with autoimmune thyroiditis can be stable for a decade or more upon experiencing a reactivation, perhaps due to an infection, increased viral load, or other inflammatory challenges, most often presenting as new onset hypothyroidism.[33] Consider including immune supportive and antiviral herbs in such patients.

Astragalus membranaceus
Andrographis paniculata
Melissa officinalis
Curcuma longa
Phytolacca americana (also known as *P. decandra*)
Glycyrrhiza glabra

Combine in equal parts. Take 1 to 2 dropperfuls of the combined tincture, 3 or more times daily for many months.

Tincture for Hashimoto's Thyroiditis

Hashimoto's thyroiditis causes primary hypothyroidism as thyroid tissue is destroyed. In Hashimoto's we are aiming to reduce inflammation in the thyroid while simultaneously improving the output of deficient thyroid hormones.

Commiphora mukul
Fucus vesiculosus
Glycyrrhiza glabra
Panax ginseng
Iris versicolor

Combine in equal parts. Take 1 to 2 dropperfuls, 3 or more times daily for many months.

Tincture for Postpartum Thyroiditis

Postpartum thyroiditis, also known as acute lymphocytic thyroiditis, occurs in 5 to 10 percent of all deliveries[34] and can be transient or long-lasting. Symptoms usually begin

Herbal Options for Thyroiditis

The following herbs are possible considerations in formulations for thyroiditis based on folkloric traditions and modern research.

Andrographis paniculata. Due to broad antiviral effects, andrographolide may be included in formulas for thyroiditis triggered by underlying viral infections.

Artemisia annua. *Artemisia* species have numerous antimicrobial activities, and artemisinin occurring in *A. annua* is shown to have multiple anticancer mechanisms of action. *Artemisia* occurs in TCM formulations for thyroid disease and may help deter the initiation of cancer in chronic thyroiditis.

Astragalus membranaceus. *Astragalus* is a widely used immune-modulating agent and may be considered in formulas when a high viral load is present and contributory to thyroid inflammation.

Fucus vesiculosus. The flavonoid fucoxanthin may decrease oxidative stress in the thyroid gland and reduce the risk of thyroid cancer in states of chronic thyroiditis.

Hypericum perforatum. Consider *Hypericum* in formulas for underlying viral infections that contribute to thyroiditis and when depression and mood disturbances accompany thyroid disorders.

Iris versicolor. *Iris* can be included in formulas for thyroiditis when there is tissue congestion, fluid retention, and soft diffuse enlargement of the thyroid gland.

Leonurus cardiaca. Rosmarinic acid in the plant has a powerful antioxidant effect that may reduce oxidative stress in an inflamed thyroid gland, and limited research suggests the plant may reduce heavy metal-induced inflammation and toxicity.

Lithospermum officinale. Rosmarinic acid in the plant acts as a strong antioxidant and may help reduce oxidative stress in the thyroid, and the plant is also noted to reduce hypothalamic thyrotropin-releasing hormone (TRH) and pituitary TSH stimulation of the thyroid.

Melissa officinalis. Lemon balm also contains rosmarinic acid, which can reduce oxidative stress, has antiviral activity, and may inhibit binding of TSH and autoantibodies to receptors on the thyroid gland.

---

very soon after the delivery. In many cases, hormones seem to be fluctuating based on associated symptoms of hot flashes, depression, anxiety, sleep disturbance, and blood pressure and pulse fluctuations. The greater the number and severity of the symptoms, the more likely thyroid disease is to be involved.[35] Multiparous women are at increased risk for postpartum thyroiditis compared with nulliparous women.[36] The clinical course of postpartum thyroiditis is characterized by three phases: thyrotoxic, hypothyroid, and euthyroid phase, but approximately half of affected women will have permanent hypothyroidism.[37] Selenium supplementation during pregnancy is reported to significantly decrease the occurrence of postpartum thyroiditis.[38] In this formula, *Glycyrrhiza glabra* has general anti-inflammatory, adaptogenic, and antiviral properties. *Curcuma* and *Silybum* are included as broad-acting anti-inflammatories that may also help the

liver to clear the high hormones of pregnancy and delivery. *Panax ginseng* may help support optimal hormone levels.

Curcuma longa
Silybum marianum
Glycyrrhiza glabra
Panax ginseng

Combine in equal parts. Take 2 dropperfuls, 3 or more times daily for many months.

Formula to Reduce Cancer Risk with Hashimoto's

To reduce the risk of cancer seen with Hashimoto's thyroiditis, aim to detoxify the body wherever and however possible and to reduce inflammation with the long-term use of antioxidant herbs and nutrients. These herbs all have various anticancer and immune-modulating effects and would complement the Tincture for Hashimoto's Thyroiditis.

Curcuma longa powder
Astragalus membranaceus powder
Ganoderma lucidum powder
Camellia sinensis powder

Combine some or all of the powders in equal parts. Add the mixture by the teaspoon to tablespoon to smoothies, warm milks, or medicinal truffles several times per week to daily. A single powder can also be used; making combinations is optional. This is a long-term therapy to provide immune-modulating, anti-inflammatory, and anticancer support.

Formulas for Goiter

Goiter is a pathologic enlargement of the thyroid characterized by proliferation of thyroid follicular cells, neoangiogenesis, and the hyperplasia of connective tissues and colloid. Goiter may result from both hyperthyroidism and inflammation in the gland as well as from hypothyroidism, where the functioning cells enlarge in an attempt to compensate for low output of thyroid hormones. Autoimmune thyroid disease, both Graves' disease and Hashimoto's thyroiditis, can result in goiter. However, iodine-deficient goiter is the most common type, occurring most often in iodine-deficient areas of the world. Thyroid cancer can also present as an enlarged gland.

Endemic goiter is defined as 10 percent or more of a local population having enlarged thyroid glands; this condition is associated with iodine deficiency in the local soil. In certain parts of the world, however, endemic goiter can also result from excessive iodine intake, usually where seaweeds are a staple part of the diet. The combination of high cruciferous vegetable intake and low iodine is particularly harmful to the thyroid and may result in goiter.

Whole Foods Therapy for Iodine-Deficient Goiter

Seaweeds are excellent sources of iodine and Brazil nuts are an excellent source of selenium, but so high in the trace mineral that only a few per day should be eaten. Include the following foods in the weekly diet to provide iodine and selenium.

Edible seaweeds such as kelp, nori, dulse,
  or sea lettuce
Fish, especially cod
Brazil nuts

Tincture for Goiter with Hypothyroidism

This formula is based on herbs emphasized in the folkloric literature for goiter. Iodine and selenium supplementation would be a good complement to this formula, as well as treating directly any possible contributors—such as environmental toxicity or nutritional deficiencies. Historic herbals claim that *Iris versicolor* and *Phytolacca* can help to decongest an enlarged thyroid, but that these herbs in and of themselves are insufficient to correct thyroid dysfunction. *Iris* is a folkloric remedy for goiter and tissue congestion due to its ability to move saliva, lymph, bile, and digestive secretions. There is very little modern research on *Iris*. Due to its heating and stimulating qualities, however, *Iris* would be best for hypofunction of the gland, rather than hyperthyroidism.

Commiphora mukul	20 ml
Fucus vesiculosus	20 ml
Iris versicolor	10 ml
Phytolacca americana (also known as *P. decandra*)	10 ml

Take 1 or 2 dropperfuls of the combined tincture, 3 times daily. Dilute with water before ingesting to reduce potential irritating effects on oral and digestive mucosa.

Herbal Options for Goiter

These herbs are emphasized in folkloric herbals for goiter and may be considered to include in oral formulations or possibly as topical applications for treating thyroid nodules, cysts, and diffuse goiter.

Commiphora mukul	*Lycopus virginicus*
Fucus vesiculosus	*Phytolacca americana*,
Iris versicolor	also known as
Lithospermum officinale	*P. decandra*

Topical Oil for Goiter

Older herbal texts report that the topical application of Iris medications will decongest the thyroid, but there is no modern research to support this. I have also searched for any research on topical applications for goiter—herbal, nutritional, or pharmaceutical—and have found no such claims. However, increasingly many agents are noted to be absorbed transdermally, and it may be possible to impact goiter with topical application. This would at least do no harm.

Iris versicolor powder	1 ounce (30 g)
Phytolacca americana (also known as *P. decandra*) oil	2 ounces (480 ml)
Ricinus communis (castor) oil	2 ounces (480 ml)

Place all ingredients in a pint canning jar and blend by shaking vigorously. Continue to shake daily for 4 to 6 weeks, and then strain. Use the strained oil topically by gently rubbing into the skin over a goitrous thyroid 3 times per day.

Tea for Iodine-Deficient Goiter

This formula is based on a TCM combination in use for more than 500 years to treat iodine-deficient goiter.[39] It contains seaweeds as a source of iodine mixed with anti-inflammatory herbs to help treat oxidative stress in the thyroid and reduce lipid or connective tissue accumulation. *Citrus reticulata* peels may protect against oxidative stress caused by hyperthyroidism, and animal studies have shown it to reduce elevated T4 and T3.[40] This formula has the property of moving stagnant fluids and resolving "qi" stagnation and phlegm coagulation, seen in TCM as a cause of goiter.

Sargassum spp. thallus
Laminaria japonica thallus
Pinellia ternata roots
Fritillaria thunbergii bulbs
Ligusticum striatum (also known as *L. chuanxiong*) roots
Angelica sinensis roots
Forsythia suspensa fruits
Glycyrrhiza glabra root
Citrus reticulata peels

Combine equal parts of the herbs; 2 ounces (60 g) of each ingredient will yield 1 pound (540 g) of dried herb blend. Simmer 1 tablespoon in 4 cups (960 ml) of water, gently reducing to approximately 3 cups (720 ml), and then strain. Drink 2 or 3 cups (480 or 720 ml) each day.

Iris versicolor for Goiter

Iris versicolor (wild iris) was recommended for goiter by the early American eclectic physicians, a group of MDs practicing in the late 1800s and early 1990s who used herbal medicines extensively. *Iris* was traditionally used both topically and internally for goiter. Little to no research, however, has been conducted on *Iris* in the modern era. Because wild iris roots stimulate saliva, bile, intestinal secretions, and perspiration, I suspect the plant to be a muscarinic agonist. *Iris* contains iridin, an isoflavone, volatile oils, resins, and alkaloids. *Iris* has historically been recommended for sluggish metabolic functions, including the thyroid. Iris was used to clear congested tissues, move sluggish bowels, decongest the lymph, and reduce enlargements of the spleen, liver, and thyroid gland. It was believed that *Iris* was capable of decongesting thyroid tissues, as well as stimulating lymphatic circulation and detoxification.

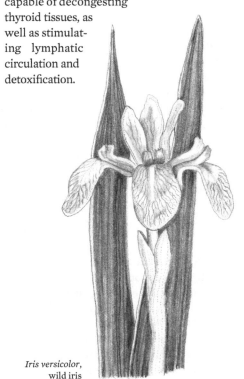

Iris versicolor,
wild iris

Iodine and the Thyroid

Iodine is essential to thyroid function and is a component of thyroid hormones. However, both too little and too much iodine can suppress the thyroid gland or, less often, overstimulate the thyroid. The World Health Organization uses the term "iodine deficiency disorder" and estimates that more than 800 million people are affected and at risk of health problems due to inadequate iodine intake. Ongoing iodine deficiency will typically result in goiter and low thyroid function. Systemic symptoms can also occur and be especially devastating to health when iodine is deficient during gestation and infancy, resulting in cretinism, mental retardation, and congenital heart, reproductive, and other defects. Iodine deficiency can also cause oxidative stress in the body, particularly when selenium and general antioxidant nutrients are low.[41] Selenium and iodine work synergistically[42] and all the iodinase, deiodinase, and peroxidase enzymes are selenium dependent.[43] In addition to seaweeds, some of the richest sources of iodine are animal products such as milk, fish, and eggs. Therefore, vegans may have a low intake of iodine and are most at risk for iodine deficiencies.

Tincture with Iodine for Goiter

Seaweed and licorice are particularly emphasized as a duo to treat iodine-deficient goiter in TCM. This formula calls for potassium iodide, also known as Lugol's solution, a strong iodine solution used topically as a disinfectant and internally to treat thyrotoxicosis and iodine deficiency. Named after a French physician, Jean Lugol, Lugol's solution is commercially available and is on the list of the World Health Organization's essential medicines. While allergies to iodine are occasionally seen, iodine solutions such as this are generally well tolerated.

Fucus vesiculosus	20 ml
Glycyrrhiza glabra	20 ml
Potassium iodide (Lugol's 2 percent solution)	20 ml

Take 1 dropperful of the combined tincture in a sip of water, twice daily.

Tincture for Goiter Associated with Thyroiditis

Rosmarinic acid in *Lycopus* and *Melissa* may form electromagnetic adducts with TSH and thereby reduce physiologic activity and thyroxine output.[44] *Astragalus* and *Curcuma* offer additional immunomodulating and anti-inflammatory activity, and *Phytolacca* may exert antiviral effects.

Lycopus virginicus	15 ml
Astragalus membranaceus	15 ml
Curcuma longa	15 ml
Melissa officinalis	8 ml
Iris versicolor	7 ml

Dilute the combined tincture with water before ingesting to reduce potential irritating effects on oral and digestive mucosa. Take 1 or 2 dropperfuls at a time, 3 or 4 times a day.

Formulas for Hypothyroidism

Hypothyroid patients are either deficient in endogenously produced thyroid hormone or are producing thyroid hormones but the tissues are not responding. Ninety percent of Hashimoto's thyroiditis patients have high antithyroid peroxidase (anti-TPO) and antithyroglobulin antibodies.[45] Levothyroxine to replace the inadequate thyroid hormone levels has been considered the standard of care for the treatment of hypothyroidism for many decades, and both natural and synthetic forms exist. Herbal approaches can be used in tandem with levothyroxine to help reduce inflammation or mitigate possible side effects. Both *Withania somnifera* and *Commiphora mukul* have been shown to improve hypothyroidism by elevating T3 levels and increasing T3:T4 ratios through peripheral metabolic pathways. In some cases, nutritional support of thyroid gland function may help restore thyroid function without the need for replacement hormone therapy. Several published studies report that mango, watermelon, and cucumber peels have stimulating effects on the thyroid,[46] and an ability to limit lipid peroxidation and

Soy and the Thyroid

There has been concern that soy, *Glycine max*, is toxic to the thyroid, but the isolated cases reported do not seem to be cause for alarm. Seaweed appears more likely to suppress the thyroid than soy. One of the commonly cited anecdotal reports of soy inducing thyroid dysfunction concerned a soy product that was mixed with kombu seaweed extract.[48] Because high doses of seaweed can have suppressive effects on the thyroid,[49] this begs the question whether the soy, the iodine, or the duo suppressed thyroid function. Dietary soy *may* interfere with the absorption of synthetic hormones and necessitate a higher dose of levothyroxine, but a review of 14 human clinical studies reported that soy showed no disruption or suppression of thyroid function.[50]

Genistein, a primary isoflavone in legumes, upregulates hepatic iodothyronine deiodinase, reducing fat deposition in the liver.[51] This is one reason why consuming a lot of beans supports weight loss. While some researchers have found that genistein and other isoflavones inhibit deiodinase and thyroperoxidase enzymes necessary

Glycine max, soybean

for thyroxine synthesis,[52] other researchers report that a diet high in soy leads to small but significant *increases* in thyroxine in primates.[53] These opposing findings may have to do with the amphoteric effect of some plant molecules in differing tissue states, as with the adaptogens (See "Licorice: A Model Adaptogen" on page 53).

Many studies show either no effect of dietary soy on thyroid function or a positive effect. A large meta-analysis on postmenopausal women confirmed a variety of beneficial estrogenic and reproductive hormonal effects from soy isoflavones, without significant effects on thyroid function.[54] Several studies investigating soy versus placebo showed no changes were observed in TSH, free T3, T4, autoantibodies, or thyroid hormone receptor assays in menopausal women.[55] One short-term study reported that daily consumption of boiled soybeans produced a transient increase in TSH in men but not women,[56] and another showed no differences in total T3, free T3, total T4, free T4, TSH, or thyroxine-binding globulin (TBG) in a soy-based diet, compared to a soy-free milk protein diet.[57]

---

help resist tissue changes seen with thyroid dysfunction.[47] However, I have not seen these foods emphasized folklorically or tried them for myself. Supplementing with all the nutrients needed to synthesize thyroid hormones—tyrosine, selenium, copper, and iodine—is often very helpful for those with hypothyroidism.

Tincture for Hypothyroidism

Fucus seaweed contains organic iodine while *Commiphora* is known to enhance the uptake of iodine into the thyroid gland. *Coleus forskohlii* stimulates cell function and may thereby support metabolic rate in a manner

different from the other ingredients in this formula, and *Panax* may enhance thyroid function as well as general energy. Other herbs can be added when high cholesterol, diabetes, or other metabolic and endocrine disorders accompany.

Commiphora mukul
Fucus vesiculosus
Coleus forskohlii
Panax ginseng

Combine in equal parts. Take 1 to 2 dropperfuls, 3 or 4 times a day long-term.

Tincture for Elevated Lipids with Hypothyroidism

Hypothyroidism is associated with an increased risk of coronary artery disease, beyond that which can be explained by its association with conventional cardiovascular risk factors. Coronary endothelial dysfunction precedes atherosclerosis, has been linked to adverse cardiovascular events, and may account for some of the increased risk in patients with hypothyroidism.[58] Thyroid hormones affect the cardiovascular system, and the hypothyroid state may decrease cardiac output due to impaired relaxation of vascular smooth muscle and decreased availability of endothelial nitric oxide. This produces a cascade effect of increased arterial stiffness that leads to increased systemic vascular resistance. This formula retains the *Commiphora* and *Coleus* used in the Tincture for Hypothyroidism and combines it with *Allium sativum* and *Vaccinium myrtillus* for their protective effects on the endothelium and blood vessels in general.

Allium sativum
Commiphora mukul
Coleus forskohlii
Vaccinium myrtillus

Combine in equal parts. Take 2 dropperfuls, twice a day. If there are no improvements after 3 months, the dose may be doubled or even tripled. Run labs again in another 3 months.

Guggul to Support Thyroid Function

Commiphora mukul, commonly called guggul, is a relative of the myrrh gum tree, traditionally used in Ayurvedic medicine for obesity, high cholesterol, diabetes, and hypothyroidism. These uses may stem from guggul's thyroid-supportive effects.[59] Guggul may also improve blood lipids—sterols in the plant, referred to as guggulsterones, act on bile acid receptors and enhance lipid processing[60]— and human clinical trials report improved cholesterol and glucose levels with guggul supplementation.[61] Guggulsterones may increase the thyroid's uptake of iodine and enhance the activity of thyroid peroxidase enzymes[62] and nuclear receptors contributing to the plant's hypolipidemic effect.[63] Guggulsterones may also have direct effect on adipocytes, preventing maturation of new fat cells from immature precursor cells[64] and promoting apoptosis of mature fat cells.[65] These effects all reduce the capacity to store excessive fats.

    Animal studies have shown *Commiphora* to reduce the effects of thyroid-suppressive drugs, indicating a thyroid hormone–enhancing effect and possible utility for hypothyroidism. A ketosteroid found in the oleoresin of *Commiphora* is reported to increase the uptake of iodine by the thyroid gland and enhance the activity of thyroid peroxidase

Commiphora mukul, guggul

enzymes. T3 has been shown to be promoted along with healthy alterations in the ratio of T3 to T4 indicating a thyroid-supportive effect.

Tincture for Hypothyroidism in Women at Perimenopause

The perimenopausal years are a common time for women to develop hyper- or hypothyroidism. In such cases, herbs that work on the pituitary-thyroid-reproductive axis may prove helpful.

Commiphora mukul
Fucus vesiculosus
Vitex agnus-castus
Panax ginseng

Combine in equal parts. Take ½ to 1 teaspoon, 3 or more times a day.

Tincture for Hashimoto's at Perimenopause

Hashimoto's may emerge or worsen in the perimenopausal years, sometimes presenting with alternating hyper- and hypothyroidism.[66] Progesterone levels have an effect on thyroid hormones and TSH release and regulation,[67] therefore, balancing female reproductive hormones may be beneficial.

Leonurus cardiaca
Vitex agnus-castus
Pueraria candollei var. *mirifica*
Withania somnifera

Combine in equal parts. Take 1 to 2 dropperfuls, 3 times per day.

Tincture for Sheehan's Syndrome

Glycyrrhiza and *Panax* are used as an emergency remedy in China for acute pituitary failure such as postpartum Sheehan's syndrome,[68] an acute deterioration of the pituitary gland leading to severe endocrine crises.

Glycyrrhiza glabra
Panax ginseng

Combine in equal parts. Take 1 dropperful every 20 to 60 minutes, all waking hours. Reduce gradually if hormone output is restored.

Fucus Snack for Hypothyroidism

Seaweed is an excellent source of iodine and other minerals. For those who enjoy, or at least tolerate, the flavor of seaweed, this snack can help replace chips, crackers, and other starches to help those with hypothyroidism, metabolic syndrome, frank diabetes, or PCOS lose weight. *Fucus* can be purchased dry from herb suppliers and eaten as is, as an easy source of iodine, or mixed with nuts and seeds as in this recipe.

2 cups (100 g) *Fucus vesiculosus*, large pieces if possible
1 cup (140 g) slivered raw almonds
1 cup (125 g) raw pistachios, crushed

Simply combine the seaweed and unsalted raw nuts in a bowl, blend, and transfer to small snack bags for storage. Eat 1 to 2 handfuls (½ cup or 45 g) each day.

Salad of the Sea

The salty, slightly fishy flavor of seaweed lends itself well to seafood recipes. This recipe also includes medicinal spices and plenty of vegetables. Quinoa, a gluten-free "pseudograin," takes the place of rice or noodles used in traditional Asian stir-fries.

1 cup (200 g) quinoa, cooked
3 cups (720 ml) water
½ cup (100 g) grated carrots
1 yellow onion, chopped
3 cloves garlic, minced
1 tablespoon minced fresh ginger
1 cup (300 g) cooked shrimp (may substitute scallops)
¼ cup (50 g) diced red bell pepper
¼ cup (50 g) diced yellow bell pepper
¼ cup (20 to 30 g) dried seaweed (such as dulse)
  or ½ cup (20 to 30 g) fresh
½ cup (50 g) fresh mung bean or other sprouts
¼ cup (60 ml) toasted sesame oil
¼ cup (60 ml) tamari
1 tablespoon honey or other sweetener
Juice and zest of 1 lime
Pinch cayenne
½ cup (50 g) toasted nuts

Simmer the quinoa in the water for roughly 10 minutes until soft. Remove from the heat, rinse, strain, and set aside. Place the carrots, onion, garlic, and ginger in a wok or heavy frying pan with several teaspoons water and gently simmer until the onions are translucent. Add the cooked quinoa, shrimp, bell peppers, seaweed, mung bean sprouts, along with the oil, tamari, honey, lime juice and zest, and cayenne, and continue to sauté for 10 minutes more, stirring frequently. Add the toasted nuts to each plate at the time of serving. Makes 4 servings.

Fucus Coconut Oil Dressing

Medium-chain triglycerides (MCTs) may boost the anti-obesity effect of fucoxanthin, a flavonoid in seaweed.[69]

Disrupting the Sodium-Iodide Symporter

The thyroid gland is particularly vulnerable to adverse effects from endocrine-disrupting chemicals. Many such chemicals interfere with the sodium-iodide symporter, which is an iodine-transporting system crucial to iodine uptake and the synthesis of thyroid hormones and is located on the basolateral membranes of the follicular cells of the thyroid.

Perchlorate, nitrate, and thiocyanates may inhibit the sodium-iodide symporter, and exposure to all three may cause significant suppression of thyroid function.[70] Perchlorate is found in combustible substances such as rocket fuel, fireworks, and explosives, as well as in some fertilizers. Nitrate can disrupt thyroid homeostasis by inhibiting iodide uptake. While vegetable-based nitrates such as those found in beets have not been implicated in thyroid disease, the consumption of chemical nitrates in cured meats is associated with an increased incidence of thyroid cancer.[71] Some municipal water sources are contaminated with perchlorates, which are associated with altered thyroid function, and high nitrate levels in public water supplies resulting from use of nitrogen-rich fertilizers also appears problematic.[72]

Thiocyanates are found in cruciferous vegetables, and although such foods are credited with numerous detoxifying and anticancer effects, consuming large amounts of raw crucifers may inhibit metabolic functioning in thyroid follicular cells. Due to suppressive effects on the thyroid, these vegetables have sometimes been referred to as *goitrogens*, meaning they are capable of inducing goiter. Many naturally occurring isothiocyanates can also arrest growth of cancer cells and promote cell death selectively in cancer cells. However, the research shows that subtle differences in the molecular structure of isothiocyanates gives them profoundly different physiologic actions, making blanket statements about closely related compounds impossible. Among the most common mechanisms of action is the ability of isothiocyanates to suppress diverse oncogenic signaling pathways often hyperactive in human cancers.

Cruciferous vegetables include collards, brussels sprouts, and kale, all of which contain glucosinolates, a parent storage molecule that can yield sulforaphane, progoitrin, and thiocyanates. Progoitrin inhibits iodine utilization by the thyroid when consumed at a dose of 25 millgrams or more per day.[73] All of these sulfur-containing compounds have a hot spicy flavor, and their release has been referred to as a "mustard oil bomb," a method by which such plants protect themselves against predation and infection.

Glucosinolates are stable even when cooked and are ingested intact in cooked vegetables such as broccoli and kale. Boiling cruciferous vegetables will leach some of the glucosinolates into the cooking water and reduce the content left in the vegetables by 20 percent to more than 50 percent. However, chewing and chopping the raw vegetables allows the enzyme myrosinase in the cells to come into contact with the glucosinolates and metabolize them into isothiocyanates. Thus, cooked crucifers provide glucosinolates, (and their anticancer and detoxifying effects), while raw crucifers provide isothiocyanates, which have thyroid-suppressive effects if consumed in large amounts.

While high intake of cruciferous vegetables is associated with cancer protection in general, especially hormone-related cancers and lung and colorectal cancer,[74] cruciferous vegetables may induce inflammation in the thyroid in a manner that increases the risk of thyroid cancer. This is especially true when the vegetables are consumed raw, because suppression of thyroid function can lead to inflammation in the thyroid gland, and chronic inflammation can lead to cancer initiation. Yet, when consumed at a low level, isothiocyanate is shown to help detoxify tobacco smoke,[75] a known lung carcinogen and contributor to Graves' disease. Unusually high rates of thyroid cancer are reported in New Caledonia, and investigations suggest that a high intake of cruciferous vegetables combined with low iodine intake may be responsible.[76] The combination of low iodine and high isothiocyanate intake appears particularly likely to induce iodine-deficient goiter.[77] Canola (*Brassica rapa*, *B.*

napus, B. juncea) is a cruciferous species high in glucosinolates and erucic acid, both thought to be responsible for detrimental health effects, sometimes referred to as antinutritional factors.[78] In one study, animals fed on canola seed meal suffered numerous toxic effects,[79] but agriculturists have bred the glucosinolates and erucic acid out of canola to make it better suited for animal consumption.

The crucifers highest in thiocyanates are brussels sprouts, mustard, radishes, kale, horseradish, and turnips, with other crucifers having roughly half the amount.[80] The seeds, sprouts, and microgreens have much higher concentrations per gram than do the mature plants. Those with hyperthyroidism might make an effort to eat such seeds and sprouts. Broccoli and radish seeds are readily available for sprouting or for grinding into mustard-like pastes. Those with hypothyroidism should *avoid* large daily doses of raw crucifers and their seeds and sprouts.

Coconut oil is a readily available natural source of MCT, but purified MCT oil products are commercially available and may be superior for weight-loss efforts. This recipe calls for fire cider, which is spicy herbs extracted in apple cider vinegar. (See Volume 1, page 74, for a digestive vinegar recipe example.) Regular vinegar may be substituted.

½ large avocado, chopped
½ cup (75 g) coarsely chopped onions
⅛ cup (5 g) *Fucus vesiculosus* or other edible seaweed

2 tablespoons fire cider
1 tablespoon MCT oil
1 tablespoon olive oil
2 raw Brazil nuts

Place all ingredients in a blender and liquefy. Taste and add any additional oil, vinegar, or other spices as desired or to suit the intended use of the dressing. Use on salads or drizzle over steamed vegetables, fish, or other appropriate dishes.

Formulas for Hyperthyroidism

Hyperfunctioning thyroid conditions include Graves' disease, toxic multinodular goiter, inappropriate and excessive release of TSH from the pituitary, and hyperfunctioning cysts or cells in the thyroid. Toxic multinodular hyperthyroid conditions occur most often in older people and commonly follow a history of the persistent presence of thyroid nodules. It is unknown what causes some of these cases to develop into hyperthyroid states.

Graves' disease is the most common cause of hyperthyroidism. Graves' disease is an autoimmune disease in which autoantibodies bind to thyroid-stimulating hormone (TSH) receptors on follicular cells of the thyroid gland, resulting in proliferation and enlargement of the thyroid gland, with excessive activation and production of thyroid hormones T4 and T3. Although autoantibodies can be observed in healthy individuals, in roughly 2.5 percent of the population, their presence triggers frank autoimmune disease such as Graves' disease.[81] Graves'

disease immunoglobulins are a type of immunoglobulin G (IgG) and are referred to as thyroid-stimulating immunoglobulins (TSIs), which may be detected in the blood and confirm the diagnosis of Graves' disease. However, other types of thyroid-suppressive antibodies may also be present and contribute to the oxidative stress and inflammation in the thyroid gland, explaining some of the overlap between Hashimoto's thyroiditis and Graves' disease.

The worldwide incidence of Graves' disease has seen annual increases, possibly due to the influence of environmental factors.[82] There are many triggers and contributors to autoimmunity; stress, environmental toxins, and viruses top the list. Several studies have provided evidence for a genetic predisposition to Graves' disease, and psychological stress, smoking, and being female also increase the risk. Other factors such as infection, especially with *Yersinia enterocolitica*, or adenoviruses may also promote some cases of Graves' disease via T helper cell ratios and their role in immune responses. Ionizing radiation or a deficiency of vitamin D or selenium may also predispose. It is proposed that these insults injure cells, leading to release of thyroid peroxidase enzymes and thyroglobulin protein into the bloodstream, which in turn elicits an immune response

Thyroid Storm

Thyroid storm is a fairly rare and severe form of thyrotoxicosis, necessitating emergency treatment because the condition is possibly fatal when a greatly accelerated heart rate and pressure place a burden on the heart. The condition is not entirely understood because the severity of the symptoms do not directly correlate with the degree of elevated thyroid hormones. Rather, the condition can be precipitated by cessation of thyroid-suppressing medication as well as by infection, trauma, stress, or pregnancy. Beta blockers are used to manage heart symptoms acutely, and radioactive iodine may be employed to destroy the thyroid. Thyroidectomy surgery may be utilized in urgent situations. Steroids and high doses of iodine may also suppress symptoms and serve as an alternative to surgical or radioactive procedures. Herbal therapies may not be strong enough to attenuate acute thyroid storm but are appropriate in emergent situations threatening to move toward thyroid storm.

Lugol's solution, a liquid form of potassium iodide, is sometimes used to suppress the thyroid gland. A 2 percent Lugol's solution supplies 8 milligrams of iodide per drop and is taken at a dose of 3 to 5 drops, 3 times a day, to inhibit the production of thyroid hormones. The effects last several weeks and can be employed while surgical or other interventions are being arranged. This therapy should be implemented in an in-patient setting, typically under the supervision of an endocrinologist-and-cardiologist team.

to the proteins. All the potential triggers can promote oxidative stress in the thyroid, and antioxidants may slow the progression of the disease.

Marked elevations in the output of thyroid hormones is also referred to as thyrotoxicosis. Graves' patients often have a diffuse goiter and symptoms typical of increased basal metabolic rate such as tachycardia, increased sweating and heat intolerance, fine tremor, insomnia, fatigue, and diarrhea. Graves' disease is also often associated with leukopenia, which can increase infection susceptibility as well as contribute to inflammation.

In tandem with increasing occurrence of autoimmune disorders of the thyroid, Graves' exophthalmia is now a significant cause of blindness worldwide. Graves' orbitopathy is a complication in which the upper eyelids retract in association with erythema and edema of periorbital tissue. In a small number of people, the inflammation may be so severe as to progress to optic neuropathy and compromise sight. Periorbital inflammation involves a proliferation of fatty and connective tissue behind the orbit, and a transformation of preadipocytes into adipocytes that push the eyeball outward. Graves' exophthalmia is a protrusion of the eyeballs that occurs in roughly half of Graves' subjects and is severe in 3 to 5 percent.

Conventional treatments for Graves' disease include antithyroid drugs such as methimazole, radioactive iodine to ablate the thyroid, and surgical approaches. One survey reported that 74 percent of those treated with antithyroid drugs were still in remission 5 years after the therapy, and 36 percent relapsed after the drug was withdrawn.[83] For those who do not respond to the drug, radioactive iodine may induce remission roughly two-thirds of the time, but radiation carries the risk of developing cancer at a later time. Herbal therapies for Graves' disease can be symptomatic, aiming to help reduce sweating, heart palpitations, and insomnia, as well as treat the disorder more deeply, aiming to reduce inflammation in the thyroid and calm autoimmune reactivity leading to antibody production.

Atrial fibrillation occurs in 15 percent of patients with hyperthyroidism and carries an increased risk of thromboembolism. Therefore, some experts recommend the use of anticoagulant medications when chest pain or signs of heart failure accompany.[84] Simple aspirin as a platelet antiaggregator may be a good choice in patients without other risk factors for thromboembolitic diseases. *Allium sativum*, *Zingiber officinale*, bromelain, nattokinase, and other natural platelet antiaggregators may also help protect against clotting risk. (See "Platelet Antiaggregators and PAF Inhibitors" on page 77 of Volume 2.) In the absence of cardiac symptoms, beta blockers and antithyroid agents are standard care. Because the hyperthyroid state increases metabolism, drugs and herbs may be rapidly cleared, and larger than normal dosages are often required. Beta blockers are ideal because of the anxiolytic effects but may excessively weaken heart action in those with preexisting heart failure. Calcium channel blockers and digoxin are alternatives in such cases. Amiodarone can control atrial arrhythmias with the added benefits of inhibiting the conversion of T4 to T3 and inhibiting T3 from binding to nuclear receptors. Plus, its high iodine content inhibits thyroid hormone synthesis. However, amiodarone has many toxic side effects. The formulation of amiodarone was inspired by the naturally occurring molecule khellin in *Ammi visnaga*, which was noted to alleviate anginal pain. While *Ammi* will not replace expert cardiac care in urgent situations, those with mild episodes of arrhythmia may respond to herbal therapies while Lugol's solution and/or methimazole is initiated to control thyrotoxicosis. *Leonurus* is a mild beta blocker and may complement *Ammi* in such situations.

Tincture for Hyperthyroidism

This is a formula to address the main symptoms of hyperthyroidism: *Leonurus* for heart palpitations, *Withania* for insomnia, and *Melissa* for anxiety and tremulousness. *Melissa* can reduce excessive thyroid activity and possibly autoantibody-driven inflammation, as do other herbs containing rosmarinic acid, such as *Lycopus*, *Rosmarinus officinalis*, and *Perilla frutescens*.

Leonurus cardiaca	20 ml
Withania somnifera	20 ml
Melissa officinalis	20 ml

Take 1 to 2 dropperfuls of the combined tincture, 3 or more times daily for many months.

Tincture for Hyperthyroidism Due to Excess TSH

Lithospermum officinale and *L. ruderale* have historically been recommended for the symptoms of hyperthyroidism and modern research finds the plant to be high in rosmarinic and lithospermic acids,[85] which may interfere with the binding of TSH receptors on the thyroid. Furthermore, *Lithospermum* may reduce the release of all pituitrophins, being helpful for thyrotoxicosis, but possibly too suppressive to other hormones. Choose *Lithospermum* for overall excess situations, such as

> ## Include Adrenal Support for Thyroid Disorders
>
> Since the thyroid works synergistically with the adrenal glands and reproductive organs, include adaptogens in formulas for thyroid imbalance or inflammation. Note how *Withania*, *Panax*, and *Glycyrrhiza* are included in hypothyroid, hyperthyroid, and thyroiditis formulas.

hyperthyroidism associated with hyperestrogenism. *Raphanus* is high in isothiocyanates that may help reduce excessive production of thyroxine.

Lithospermum ruderale	25 ml
Melissa officinalis	25 ml
Raphanus sativus var. *niger*	10 ml

Take 2 dropperfuls of the combined tincture, 2 to 4 times daily. Check lab results in 3 to 6 months and continue if effective. This tincture may complement other therapies to create an aggressive protocol.

Tea for Hyperthyroidism

Choosing from the tastiest herbal options, this tea would complement the Tincture for Hyperthyroidism. Consider taking a daily selenium supplement as well.

Melissa officinalis	2 ounces (60 g)
Rosmarinus officinalis	2 ounces (60 g)
Lithospermum ruderale	2 ounces (60 g)
Glycyrrhiza glabra	2 ounces (60 g)

Mix the dry herbs together. Steep 1 tablespoon of the mixture per cup of hot water and then strain. Drink freely, 2 or more cups (480 ml) per day.

Bone Protective Tea for Hyperthyroidism

Bone loss may occur with hyperthyroidism, a condition sometimes called thyroid bone disease. Even though thyroxine has bone-building effects, complex endocrine imbalances leave Graves' disease patients at an increased risk of bone loss, in part due to suppression of hypothalamic and pituitary hormones that occur when T3 and T4 levels are elevated. Elevations in serum T3, in particular, are associated with loss of bone density in the spine, hip, and femur.[86] Herbs and nutrients typically used to treat osteoporosis may also be appropriate for Graves' patients. This formula is a bone-building mineral tonic and may be complemented by other therapies that more directly treat thyrotoxicosis.

Epimedium spp.
Centella asiatica
Equisetum arvense
Pueraria montana var. *lobata*
Medicago sativa
Urtica urens or *U. dioica*
Glycyrrhiza glabra

Combine 2 or 3 ounces (60 or 90 g) of each herb, blend them together, and store in glass jars in a dark cupboard.

Hashimoto's and Graves' Interrelationship

Even though at first glance Graves' disease and Hashimoto's thyroiditis appear to be opposite disorders—hyper- and hypofunction of the thyroid respectively—they might also be seen as the same condition manifesting in different ways. These autoimmune conditions may both involve loss of tolerance to thyroid antigens and toxins leading to white blood cell activation and infiltration in the thyroid glands. A type of white blood cell known as T helper 1 cell is involved with thyroid cell apoptosis and hypothyroidism of Hashimoto's, while T helper 2 cells become hyperreactive and are involved with antibody production in Graves' thyrotoxicosis.[87] Hashimoto's and Graves' disease share a common genetic predisposition and can occur in the same family.[88] Both Graves' disease and Hashimoto's patients display a functional defect in some types of natural killer (NK) cells, contributing to imbalances in cytokine profiles and autoantibody generation. Supplementation with dehydroepiandrosterone (DHEA) may improve NK cell function in this population.[89]

Prepare a tea by gently simmering a heaping tablespoon of the mixture in 4 cups (960 ml) of water for 10 minutes and then strain. Drink the brew daily, or at least 3 times a week, long term, to support the bones.

Hepatoprotective Tincture for Hyperthyroidism

Bupleurum is used in numerous traditional formulas for inflammation and pain, especially for the liver and abdominal organs. Recent animal research shows *Bupleurum* to possibly reduce hyperthyroidism, improving oxidative stress and tissue changes in a manner equal to thyroid-suppressing drugs.[90] *Curcuma* is an all-purpose antioxidant and anti-inflammatory and can also help protect the liver. It is combined here with *Silybum marianum* and *Berberis aquifolium*, all among the most studied hepatoprotective agents. Anyone with significantly excessive metabolic rate could benefit from a tincture such as this, but it would be especially important when elevated liver enzymes, or hyperlipidemia are present.

Bupleurum falcatum
Curcuma longa
Silybum marianum
Berberis aquifolium

Combine in equal parts. Take 1 to 2 dropperfuls 2 to 4 times daily.

Pingmu Decoction for Graves' Disease

This formula is based on a TCM formula called Pingmu, traditionally used for the symptoms of Graves' disease. Some of the herbs in classic Pingmu are not commonly available in the Western herbal marketplace, so I have omitted them and replaced them in this formula with herbs more readily available but with the same action and intent in mind. Cell culture studies have shown the original formula to induce apoptosis of adipocytes and prevent differentiation and maturation of preadipocytes.[91] *Plantago* seeds are more well-known as a digestive demulcent in Western herbalism but are also emphasized to "brighten the eyes" in TCM and included for pain and redness of the eyes in this formula. It is uncommon to use *Sinapis alba* (also known as *Brassica alba*) seeds in decoctions, but the hot spicy herb is high in selenium, which may help explain their inclusion in this formula for the thyroid, plus it is in the cabbage family and contains isothiocyanates known to sometimes suppress thyroid function if eaten raw in significant quantity. Another classic formula for Graves' symptoms, Yinglu, also includes *Astragalus* and *Sinapis alba* seeds.[92]

Plantago indica seeds	1 ounce (30 g)
Astragalus membranaceus	1 ounce (30 g)
Epimedium spp.	½ ounce (15 g)
Ligusticum striatum (also known as *L. wallichii*)	½ ounce (15 g)
Sinapis alba	½ ounce (15 g)
Salvia miltiorrhiza	½ ounce (15 g)

Combine the dry herbs and seeds and mix well. Use 1 teaspoon of the mixture per cup of water and simmer gently for 15 minutes. These herbs were traditionally decocted twice by straining the first decoction, adding fresh herbs and decocting a second time, and then preserving with alcohol to create a shelf-stable medicine to use over several weeks. This could also be prepared as a simple decoction to consume over the course of 1 to 2 days and prepare anew.

Rosmarinic Acid Tincture for Hyperthyroidism

Rosmarinic acid is most common in Lamiaceae family plants including *Melissa*[93] and *Lycopus* species including *L. europaeus* (gypsywort) and *L. virginicus* (bugleweed). *Melissa* and *Lycopus* are placed in a soothing base of licorice in this formula.

Melissa officinalis
Lycopus virginicus
Glycyrrhiza glabra

Combine in equal parts. Take 3 dropperfuls, 3 times each day.

Rosmarinic Acid for Hyperthyroidism

Rosmarinic acid is found in plants of the Nepetoideae subfamily of the Lamiaceae (mint) family including *Rosmarinus* (rosemary)[94], *Prunella vulgaris*,[95] *Melissa officinalis*,[96] *Coleus* species,[97] *Salvia officinalis*,[98] and *Lycopus*. Rosmarinic acid is known to inhibit the thyroid's response to TSH.[99] Rosmarinic acid may also form adducts with TSH, meaning that due to particular electromagnetic affinities, rosmarinic acid can form loose bonds with endogenous TSH, thereby affecting its physiologic activity.[100] Researchers have reported that when rosmarinic acid forms adducts with TSH, the hormone is prevented from stimulating the TSH receptor, adenylate cyclase is not activated, and thyroxine output is greatly diminished. *Melissa*, *Lycopus*, and *Lithospermum* have also been shown to inhibit the ability of Graves' autoantibodies to bind TSH receptors and promote intracellular cyclic AMP responses.[101] These same plants and rosmarinic acid also inhibit deiodinase enzymes and thus interfere with activation of thyroid hormones.[102]

Although there is a long-standing historical tradition of using these plants in hyperthyroidism, there are very few human clinical experiments. One clinical study reported *Lycopus europaeus* to reduce cardiac symptoms in hyperthyroid patients, without noticeable effects on serum thyroid hormones or TSH.[103] Another study showed *L. europaeus* to greatly increase urinary excretion of T4 while improving tachycardia in hyperthyroid subjects.[104]

Rosmarinus officinalis, rosemary

Medicinal Mustard for Hyperthyroidism

Mustard seeds are high in isothiocyanate, as are broccoli and radish seeds. Choose one or a combination of all three to create a mustard for use in the diet.

2 tablespoons brown mustard seeds
2 tablespoons broccoli seeds
2 tablespoons radish seeds
1 tablespoon coarsely chopped fresh turmeric root
1 tablespoon coarsely chopped fresh ginger root
¼ cup (50 g) fresh or frozen cranberries
¼ cup (60 ml) apple cider vinegar
2 to 3 tablespoons (30 to 45 ml) water
1 teaspoon fine sea salt

Combine all in a blender and puree, using a spatula as necessary to move the material toward the blades. Transfer to a glass jar and refrigerate. Use in salad dressings and in thick pastes for marinating fish or poultry. For the highest dose, spread this mustard straight on a cucumber slice or two to consume twice a day.

Tincture for Hyperthyroidism with Hypertension and Palpitations

Leonurus is a weak beta blocker that also can block excessive TSH stimulation of the thyroid. *Viburnum opulus* or *V. prunifolium* and *Lobelia* are also natural beta blockers and can promote vasodilation, with *Lobelia* chosen for this formula. *Rauvolfia* is both a nervine and a hypotensive agent, and one of the more powerful herbal tools to bring down high blood pressure quickly. *Valeriana* also has hypotensive effects and can also offer a calming action when blood pressure is elevated due to stress.

Leonurus cardiaca	15 ml
Ammi visnaga	15 ml
Valeriana officinalis	10 ml
Lobelia inflata	10 ml
Rauvolfia serpentina	10 ml

Take ½ to 1 teaspoon of the combined tincture, 3 or 4 times a day. Take a full teaspoon as often as hourly for severe symptoms, reducing as symptoms improve.

Coumarin Tea for Hyperthyroidism

Animal studies have shown isolated coumarin to limit excessive thyroid function as well as the resulting hepatotoxicity. Coumarin also reduces peripheral conversion of T4 to active T3.[105] Apiaceae family plants are high in coumarins, and *Foeniculum, Bupleurum, Angelica,* and others have historically been used in a wide variety of inflammatory diseases. There is limited research on the effectiveness of whole plants that contain coumarin compounds for hyperthyroidism, but the herbs are safe, gentle, and even palatable, so can do no harm as a complementary tea. Preparations that concentrate coumarin constituents from Apiaceae plants warrant further investigation given the promising outcome for treating a variety of ailments in animal studies.

Foeniculum vulgare
Bupleurum falcatum
Angelica sinensis
Glycyrrhiza glabra

Combine 4 ounces (120 g) of each herb, blend them well, and store in an airtight container. Gently simmer 1 tablespoon of the mixture in 4 cups (960 ml) of water for 10 minutes and then strain. Drink each day, or at least 4 or 5 days a week.

Tincture for Hyperthyroidism with Insomnia and Jitters

Melissa and *Leonurus* both contain rosmarinic acid, shown to reduce excessive thyroid activity by several mechanisms and are both calming to the nervous system via dopaminergic activity. *Withania* and *Valeriana* might be complementary choices to improve sleep and reduce tremors.

Melissa officinalis	15 ml
Leonurus cardiaca	15 ml
Withania somnifera	15 ml
Valeriana officinalis	15 ml

Take ½ to 1 teaspoon of the combined tincture, 3 or 4 times a day for tremors and anxiety. For insomnia, take additional dosages hourly starting at 8 p.m. until bedtime.

Tincture for Exophthalmia

Due to the presence of thyroglobulin receptors on the back of the eye, 25 to 50 percent of patients with Graves' disease and 2 percent of patients with chronic thyroiditis[106] will develop exophthalmia; advanced age, smoking, and radioiodine treatment increase the risk. There is no research on herbs to treat exophthalmia, so this formula is my own best effort. *Salvia* and *Ginkgo* are chosen to reduce vascular reactivity and oxidative stress, *Vaccinium* is to protect the vasculature from oxidative stress, and *Centella* and *Curcuma* are aimed at limiting sclerosis and tissue proliferation. Selenium supplementation may improve

Herbal Medicines for Exophthalmia

The following herbs are mentioned in folklore, TCM formulas, or modern research for reducing eye inflammation and exophthalmia in thyroiditis.

Artemisia annua. Artemisinin in the plant may reduce T cell-driven autoimmune inflammation.

Centella asiatica. The plant may protect against excessive proliferation of connective tissues that may contribute to protrusion of the eyeballs.

Epimedium species. The flavonoid icariin is shown to promote apoptosis in adipocytes that may accumulate in the orbit and contribute to exophthalmia.

Fucus vesiculosus. Kelp is mentioned in folkloric herbals for thyroid disorders in general as well as is a standard ingredient in formulas for goiter and exophthalmia.

Ginkgo biloba. Gingko leaf medications may improve retinal circulation, reduce oxidative stress in the eye, and help reduce chronic conjunctival and orbital inflammation.

Rosmarinus officinalis. Rosemary eyewashes may reduce redness and inflammation of the corneal epithelia and inhibit fibrosis in the orbit.

Salvia miltiorrhiza. Red sage may improve microcirculation in the eye and reduce fibrotic processes contributing to exophthalmia.

---

exophthalmia, may reduce antithyroid antibody levels,[107] and can be supplemented along with *Epimedium grandiflorum*, as detailed in the *Epimedium* Simple for Graves' Exophthalmia.

Salvia miltiorrhiza
Ginkgo biloba
Vaccinium myrtillus
Centella asiatica
Curcuma longa

Combine in equal parts. Take 1 to 2 dropperfuls, 3 to 6 times daily, reducing over several months' time.

Tincture for Hyperthyroidism with Heat Symptoms

Increased perspiration, a sense of heat, or extreme sensitivity to heat may occur in hyperthyroidism. Since salicylates have been shown to break the binding of thyroxine from its transport enzymes,[108] herbs high in salicylates such as *Actaea*, *Filipendula*, and *Salix* species might be employed to reduce bodily heat. Because night sweats occur in adrenal insufficiency, an adrenal-supportive herb such as *Glycyrrhiza* might be complementary in the formula. *Salvia officinalis* is often highly effective in reducing hot flashes and excessive perspiration.

Salix alba
Glycyrrhiza glabra
Salvia officinalis

Combine in equal parts. Take 1 to 2 dropperfuls, 3 or 4 times daily.

Epimedium Simple for Graves' Exophthalmia

Graves' exophthalmia involves a proliferation of preadipocytes into mature cells in the retro-orbital region, due to the stimulating effects of elevated thyroid hormones. Icariin, a flavonoid that occurs in many *Epimedium* species, has a wide range of biological and pharmacological effects and has been shown to induce apoptosis or autophagy in fat cells behind the eye and thus may be therapeutic to Graves' orbitopathy.[109]

Epimedium grandiflorum

Take 2 dropperfuls, 3 or 4 times daily. If capsules are readily available, 2 capsules may be taken 3 times daily.

Formulas for Adrenal Disorders

The hypothalamic-pituitary-adrenal (HPA) axis has been discussed for more than 50 years, beginning with Hans Selye's initial research showing that the adrenal glands could be affected by chronic stress and that the feedback loops between stress hormones and the neuroendocrine system of the brain are intimately connected. Because the research is vast (even the botanical research), I do not delve into the broad complexities of the biochemical and physiologic underpinning of adrenal disorders in this chapter, but instead focus on the most clinically relevant syndromes and their botanical therapies.

Cortisol is the principal glucocorticoid produced by the human adrenal cortex. Excess cortisol can result in Cushing's syndrome. Cortisol deficiency can result in Addison's disease, which is potentially fatal. Although other disorders and syndromes exist, herbalists and naturopathic physicians primarily treat these most common disorders, and especially the subtler presentations of these disorders that defy typical diagnostic categories. Cortisol often begins to rise around 4 a.m., to help boost blood sugar that begins to wane by that hour of the night. Many people awaken at this hour and have difficulty sleeping soundly thereafter, likely due to immediate activation of the sympathetic nervous system. Adaptogens, such as *Withania somnifera* and *Panax ginseng*, may improve cortisol rhythms and reduce nighttime waking as well as many other symptoms of adrenal disorders.

The use of steroid medications such as hydrocortisone will suppress adrenal function, and upon cessation, the adrenal gland often needs a considerable amount of time to recover. *Glycyrrhiza glabra* (licorice) is especially researched to improve adrenal downregulation, and other adaptogens may speed the restoration of adrenal function and the ability of the gland to produce its own cortisol.

The adrenal medulla's primary secretion is adrenaline, which like thyroxine, serves to provide energy for immediate use via catabolic metabolism of energy stores. Adrenaline promotes the catabolism of starch and fat into sugar, increasing the blood pressure and pulse and shunting blood from the digestive organs to the muscles and lungs. The pupils and bronchi dilate under the direction of adrenaline and other catecholamines. Situations inducing fear create an adrenaline response, and abnormal adrenal functions (such as panic attacks) may induce anxiety or sensations of fear, accompanied by the physical sensations of tachycardia, heat, and asphyxiation as adrenaline surges. Panic attacks and anxiety occur when adrenaline and sympathetic system neurotransmitters are easily triggered, which, in turn, activates cortisol release. Nerve-calming herbs such as *Matricaria chamomilla* or *Scutellaria lateriflora* may complement adaptogenic herbs in such cases, and cultivating meditation, relaxation, and mindfulness practices is also recommended.

Adrenal excess is not common as a chronic disorder, but rather tends to occur as bursts of hyper-responsiveness in an otherwise deficient organ. Adrenal insufficiency is extremely common in the modern population, perhaps due to effects of chronic stress and fear, as has been proposed by Hans Selye and many others over the last century. Adrenal insufficiency is associated with lower metabolism, decreased body heat, difficulty with blood sugar regulation, and weakness and atrophy of the skeletal muscles. Low blood pressure, slow pulse, anemia, leukopenia, and low immunity may also occur. Brain fog or fuzzy thinking is another common symptom of stressed adrenals.

Individuals with adrenal dominance tend to be of solid muscular build, with a great deal of stamina and endurance. Since adrenal hormones are made from cholesterol, adrenal types may crave animal foods rich in cholesterol. When the adrenal glands are insufficient, individuals find it hard to maintain muscle mass, difficult to exert themselves, and prone to blood sugar instability. Those accustomed to adrenal dominance often seek out coffee, sugary snacks in the afternoon, heavy exercise, and other stimulants to maintain a state of adrenal excess. Avoidance of stimulation is difficult for adrenal types who may tend to be "adrenaline junkies" but find their stamina and stress tolerance waning as they age. Even the most resilient personalities may experience some emotional instability as the adrenals weaken, and anxiety may manifest. Immune deficiency states such as chronic fatigue syndrome (CFS) and fibromyalgia may follow. Clinicians must keep in mind that supermoms and weekend warriors may also present with their own brand of adrenal weakness, often presenting adrenal challenge in an earlier stage than those with fully developed exhaustion or CFS.

Other than frank Addison's disease, conventional medicine does not recognize adrenal insufficiency, much

The Ginsengs: *Panax*, *Eleutherococcus*, and *Withania*

Ginseng is probably the most well-known vitality tonic in the world, due to the roots' use in TCM to tonify chi, improve resistance to stressors and infectious agents, and support longevity. *Panax ginseng* is so revered in China that single prized roots may sell for thousands of dollars, and others are sold in specially made velvet lined boxes to give the gifts of chi—the vital energy of life. *Panax* and other adaptogenic herbs are noted to improve both the release of adrenocorticotropic hormone (ACTH) from the pituitary and corticotropic-releasing hormone (CRH) from the hypothalamus, as well as to optimize the response from the adrenal gland.[110] The research on *Panax* is vast, showing broad tonifying effects on the immune system, metabolic function,[111] hormone regulation, and numerous other antioxidant, neuroprotective, hepatoprotective, and renoprotective actions. The voluminous research details molecular and physiologic details whereby *Panax* protects, restores, and vitalizes all organ systems. Animal studies have noted *Panax* administration to enhance energy metabolism during exercise.

A relative of *Panax*, *Eleutherococcus* is also of the Araliaceae family and is used in manners similar to *Panax*. *Eleutherococcus* is native to far northern regions of Eurasia and is commonly called Siberian ginseng. *Eleutherococcus* may also improve adrenal function and stress tolerance[112] and enhance immune function and resistance to infection. *Eleutherococcus* may also reduce the side effects of radiation or chemotherapy therapy.[113] The plant contains medicinally active coumarins, glycosides, and a group of immune stimulating polysaccharides known as eleutherosides A, B, C, D, and E.[114]

Withania somnifera goes by the common name ashwagandha in India. Because it is a key adaptogenic herb in Ayurvedic medicine, it has also been referred to as Indian ginseng, even though it belongs to the Solanaceae family rather than Araliaceae. *Withania* has also been researched for its ability to resist mental-emotional, chemical, and other stressors.[115] The withanolides, a group of steroidal lactones, are credited with adrenal supportive effects.[116] *Withania* is also shown to have neuroprotective effects, supporting memory retention in stressful situations[117] and protecting neuronal cells from oxidative and inflammatory damage.[118] *Withania* is also shown to support optimal immune regulation[119] and improve physical endurance and muscle mass.[120]

Eleutherococcus senticosus,
Siberian genseng

less propose a therapy. Patients presenting with weakness, fatigue, insomnia, anxiety, stress symptoms, and intolerance are typically prescribed antidepressants and anxiolytics. If hypertension and heart palpitations are most prominent, beta blockers or other cardiovascular drugs are prescribed. If low adrenal function has led to low progesterone and menstrual irregularities, birth control pills may be offered or, if perimenopasal, hormone replacement therapy may be offered. If dysglycemia and episodic blood sugar difficulties are the presenting symptoms, the case is often misunderstood because the blood sugar may run normal most of the time but be prone to sudden dips and imbalances. Vital signs and routine blood work may often be normal, and the individual is typically told all is fine, or perhaps offered psychiatric medications.

In Traditional Chinese Medicine and modern herbal medicine, tonics are agents capable of restoring normal tone and function to an organ or the entire body. Tonics are normalizing and health restoring to both hyperfunction and hypofunction, because they do not have single molecular mechanisms; rather, they affect the function of organs through supporting homeostatic and regulatory mechanisms. Adrenal tonics optimize tone and function of the adrenal glands and can normalize both excess or deficiency of adrenaline. The term *adaptogen* is used to refer to those herbs capable of improving adrenal regulation, hypothalamic and pituitary feedback loops, and normalizing up- or downregulated adrenal function. Researchers in the Soviet Union coined the term after studying tonic botanicals, particularly *Panax ginseng* and *Eleutherococcus senticosus* (Siberian ginseng), capable of improving general immunity and stress tolerance. Adaptogens are defined as safe innocuous agents that optimize and restore adrenal, immune, and metabolic function via numerous broad and nonspecific actions. *Panax* and other adaptogenic herbs improve both the release of adrenocorticotropic hormone (ACTH) from the pituitary and corticotropic-releasing hormone (CRH) from the hypothalamus, as well as optimize responsiveness of the adrenal gland itself. Overall, adrenal tonics appear to improve adrenal response to stress and physical challenges, as well as have specific immune-enhancing effects.

Panax, Eleutherococcus, Withania somnifera, and *Glycyrrhiza* are among many efficacious adrenal tonic herbs used by modern herbalists and naturopathic physicians. These starchy roots are taken for at least several months, possibly a year, for depleted individuals. *Withania* is used by herbalists as an anxiolytic and hormonal balancing agent, especially when such symptoms occur as a result of adrenal insufficiency. *Withania* has long been used for stress and anxiety and much contemporary research on this herb has focused on its optimizing effects on adrenal cortisol. *Withania* is also known to promote the neurotransmitter γ-aminobutyric acid (GABA), a neurotransmitter noted to have significant calming, relaxing effects.

Panax and *Eleutherococcus* are both believed to improve stress response by reducing excessive sympathetic response that promotes a fight-or-flight cascade. *Panax, Withania,* and other herbs may take the place of benzodiazepines and have a calming effect on the nerves without the side effects of the pharmaceuticals. Several human trials have shown clinically useful antianxiety effects with the use of *Panax* preparations, without any adverse side effects reported.[121]

The typical signs and symptoms of adrenal insufficiency include:

- Fatigue
- Stress intolerance, nervousness, anxiety
- Muscle weakness, exercise intolerance
- Light headedness, dizziness, orthostatic hypotension
- Poor concentration, confusion, brain fog
- Dysglycemia, hypoglycemia, sugar and alcohol cravings
- Allergies, dermatitis, atopic tendencies
- Thin, dry skin, with sparse perspiration
- Hypothenar erythema
- Thin, fine-textured, shiny hair
- Raynaud's-like symptoms: numbness and blanching of extremities when exposed to cold temperatures
- Postural hypotension
- Thin, asthenic build; history of overwork or high stress

These symptoms may occur in a vague or minor way, or they may be substantial and the presenting chief complaint of a medical visit. For more information about adrenal insufficiency, see "Formulas for Addison's Disease" on page 50.

Tincture for Adrenal Dysfunction with Anxiety

This formula is based on three widely available adaptogenic herbs, *Withania* (ashwagandha), *Eleutherococcus,* and *Glycyrrhiza,* all shown to support adrenal dysfunction, helping to allay stress, promote restful sleep, and normalize cortisol rhythm. (For more about *Withania* and *Eleutherococcus,* refer to "The Ginsengs: *Panax, Eleutherococcus,* and *Withania*" on page 47.) These herbs are complemented by the nervous system trophorestorative *Avena* that can help relieve anxiety and nervous symptoms.

Withania somnifera	15 ml
Eleutherococcus senticosus	15 ml
Glycyrrhiza glabra	15 ml
Avena sativa	15 ml

Take ½ to 1 teaspoon of the combined tincture, 3 or more times daily.

Tincture for Exhaustion and Debility

Prolonged stress and acute illnesses are very taxing to the adrenal system. When fatigue, poor concentration, nervous debility, or mood disorders begin under such circumstances, adrenal support for several months will often improve the situation. This formula uses *Panax* and *Glycyrrhiza* as the main adaptogens and complements them with *Rhodiola* to provide a mild stimulating effect to address apathy and exhaustion. *Avena*, especially medicine prepared from "milky" oats, is emphasized in folkloric herbalism for nervous debility, shattered nerves, and exhaustion—a description sounding much like the modern diagnosis of posttraumatic stress disorder.

Panax ginseng	15 ml
Glycyrrhiza glabra	15 ml
Rhodiola rosea	15 ml
Avena sativa	15 ml

Take ½ to 1 teaspoon of the combined tincture, 3 or more times daily.

Tincture for Debility with Frequent Infections

Many cases of poor immune resistance will improve under the combination of immune-modulating and adaptogenic herbs. Because stress and high cortisol can suppress normal white blood cell activity and immune function, adrenal support can improve immune function in those with frequent infections.

Panax ginseng	15 ml
Glycyrrhiza glabra	15 ml
Astragalus membranaceus	15 ml
Echinacea purpurea	10 ml
Allium sativum	3 ml
Zingiber officinale	2 ml

Take ½ to 2 teaspoons of the combined tincture, 3 or more times daily.

Rhodiola Simple to Improve Stamina

Adrenal downregulation is associated with fatigue, poor stamina, and exhaustion or poor recovery following rigorous exercise. All of the adaptogens can help with these and other symptoms of adrenal downregulation, but *Rhodiola* is particularly well studied for improving endurance and exercise tolerance.

Rhodiola rosea capsules

Take 2 capsules, 3 times daily, reducing as symptoms improve.

Herbal Mechanisms of Action for Adrenal Insufficiency

Adaptogenic and other herbs are useful in formulas for adrenal insufficiency. The following herbs have been the subject of research elucidating molecular mechanisms of action involving adrenal activity and explaining folkloric uses for stress, fatigue, and weakness.

***Bupleurum* species.** Saikosaponins in *Bupleurum chinense*, *B. flavum*, and *B. falcatum* increase cortisol synthesis and improve tissue response to corticoids.

Coleus forskohlii. Forskohlin increases cAMP-regulated signaling in adrenal cortex cells.

Cordyceps militaris. This mushroom, as well as *C. sinensis*, stimulates adrenal gland cells and helps to increase corticosterone production when deficient.

Glycyrrhiza glabra. *Glycyrrhiza* inhibits 11-beta-hydroxysteroid dehydrogenase, reducing the metabolic breakdown and urinary clearance of cortisol, thereby prolonging its half-life and boosting tissue levels.

Panax ginseng. Ginsenosides bind glycocorticoid receptors and may optimize steroidogenesis, having a tonifying affect on the HPA axis.

Adaptogenic Herbs

These herbs can be considered to have adaptogenic properties based on traditional writings and modern research.

Eleutherococcus senticosus	*Rhodiola rosea*
Glycyrrhiza glabra	*Smilax* species
Panax ginseng	*Withania somnifera*

Tincture for Fibromyalgia

There may be some overlap between endocrine dys-regulation and chronic myalgia. Many with cortisol imbalances present with muscle stiffness that may be diagnosed as fibromyalgia or myositis due to systemic inflammation.

Eleutherococcus senticosus
Actaea racemosa
Curcuma longa
Piper methysticum

Combine in equal parts. Take 1 to 2 dropperfuls at a time, 3 or more times a day.

Tincture for Adrenal Hyperfunction with Poor Sleep

Cortisol level normally rises in the morning and falls at dusk. In people with acute stress or adrenal hyperfunc-tion, cortisol may begin rising at midnight, leading to restless sleep and early waking. This formula is based on *Withania somnifera*, which is often effective for cor-tisol-related insomnia. These herbs are not sedatives and are taken during the day rather than right before bed and tend to help increasingly over time, rather than immediately.

Withania somnifera	60 ml
Glycyrrhiza glabra	30 ml
Passiflora incarnata	30 ml

Take ½ to 1 teaspoon of the combined tincture, 3 or more times daily.

Tincture for Adrenal Hyperfunction and Elevated Cortisol

Acute stress can elevate cortisol, suppressing immune func-tion, disrupting glucose regulation, and interfering with sleep. General adrenal support can improve many of these symptoms. This tincture includes *Withania* for its calming effect and combines it with the nervine agent *Melissa*.

Withania somnifera	15 ml
Glycyrrhiza glabra	15 ml
Panax ginseng	15 ml
Melissa officinalis	15 ml

Take ½ to 1 teaspoon of the combined tincture, 3 or more times daily.

Tincture for Adrenal Virilism in Women

Excessive secretion of cortisol can commonly be caused by stress alone. However, excessive secretion of androgens from the adrenal gland can be due to adrenal tumors. If tumors are suspected, a thorough workup with diagnostic imaging is called for. Adrenal virilism can cause hirsutism and infertility, and a variation of this condition may occur in women with polycystic ovarian syndrome. Spearmint tea can reduce elevated androgens in women and would complement this tincture.

Glycyrrhiza glabra	60 ml
Serenoa repens	60 ml

Take 1 teaspoon of the combined tincture, 3 times a day for several months and then begin reducing the dose as symptoms improve.

Formulas for Addison's Disease

I discussed adrenal insufficiency in the section on Formulas for Adrenal Disorders earlier in this chapter. Addison's disease is a severe form of adrenal insuffi-ciency (inadequate secretion of cortisol from the adrenal gland). The condition can be classified as *primary* adre-nal insufficiency, in which a problem exists within the adrenal cortex, or *secondary* adrenal insufficiency, in which the essential problem involves decreased adre-nocorticotropic hormone (ACTH) stimulation of the adrenal cortex. With prolonged deprivation of ACTH, the adrenal cortex will atrophy and thereby appear to be a problem with the adrenal tissue itself.

Addison's disease is primary adrenal insufficiency, and it may be due to various causes including autoimmune destruction of the gland or infectious destruction of func-tioning adrenal cells. In all cases, the failure to produce adequate amounts of cortisol causes severe symptoms, and Addison's disease can be fatal if left untreated. Because adrenal androgens (e.g., dehydroepiandrosterone or DHEA) and mineralocorticoids (aldosterone) will decline along with cortisol, reproductive and blood pressure issues are common symptoms of adrenal insufficiency.

Adrenal insufficiency is sometimes triggered by abrupt cessation of steroid medications, and a tapered withdrawal schedule along with adaptogenic herbs can often avoid this scenario.

Early and minor symptoms of adrenal insufficiency are often better addressed by herbalists, because until the diagnosis is obvious and laboratory tests show abnormal results, some physicians may fail to recognize subtle changes. The symptoms can indeed be vague and include general malaise, weakness, fatigue, anorexia, weight loss, nausea, vomiting, abdominal pain, or diarrhea that alternates with constipation. Many sufferers may be incorrectly diagnosed as having irritable bowel syndrome or depression. Hyponatremia is observed in 85 percent to 90 percent of patients due to mineralocorticoid deficiency, and dizziness and orthostatic hypertension may occur due to vasopressin insufficiency. Hypoglycemia occurs more commonly in children with primary adrenal insufficiency but can be promoted in all affected people at the onset of an infection or following the ingestion of alcohol. Many with Addison's disease may crave salt or develop odd cravings for salty foods from chips, to soy sauce, to pickle juice, a symptom related to aldosterone deficiency. Hyperpigmentation occurs in nearly all patients due to the increased production of pro-opiomelanocortin, a prohormone, which is then cleaved into ACTH releasing melanocyte-stimulating hormone (MSH) along with it. The hyperpigmentation is most noticeable on high-friction, light-exposed areas including the face and neck, the dorsal surface of the hands (especially the knuckles), and the palmar creases. Other high-friction areas include the elbows, knees, and waist. In some patients the oral and vaginal mucosal and the nipples become hyperpigmented.

Tincture for Addison's Disease

Addison's disease is a severe form of adrenal insufficiency, and some cases may require cortisol injections or medications. Large repetitive doses of adaptogenic herbs are necessary, as well as nervine herbs to support the nervous system, and avoidance of all stress possible to help the adrenal glands to recover. Saikosaponins in *Bupleurum* have been found to enhance the action of endogenous corticoids as well as promote cortisol synthesis in the adrenal glands. *Bupleurum* saikosaponins induce the pituitary to release ACTH. Saikosaponins also inhibit the inflammatory prostaglandins and promote protein synthesis in the liver. *Coleus* may help upregulate poorly responding adrenal cortical cells, and *Panax* and *Glycyrrhiza* provide a restorative adaptogenic base.

Bupleurum falcatum
Coleus forskohlii
Panax ginseng
Glycyrrhiza glabra

Combine in equal parts. Take 1 to 2 dropperfuls, 3 or 4 times daily.

Salty Solution for Addison's-Related Orthostatic Hypotension

The orthostatic hypotension and dizziness of Addison's disease can often be effectively remedied by salting food to taste and using various forms of licorice. Unlike health recommendations given for most healthy people, it's important to encourage patients with Addison's disease to take in adequate sodium, especially when exercising or sweating. The salt and the licorice can be consumed in many ways, but here is one simple idea.

Glycyrrhiza glabra solid extract	½ teaspoon
Soy sauce	1 teaspoon
Fruit juice	1 teaspoon

Place the *Glycyrrhiza* in the bottom of a small shot glass. Add the soy and fruit juice and stir vigorously to thin down the solid extract. Drink all at once. Consume 2 to 4 times a day.

Tea for Addison's and Adrenal Insufficiency

This tea combines the adaptogen *Glycyrrhiza* with many nervine herbs to help reduce stress and calm HPA activation from nervous tension. This tea would complement the Salty Solution for Addison's-Related Orthostatic Hypotension.

Glycyrrhiza glabra	2 ounces (60 g)
Avena sativa	1 ounce (30 g)
Schisandra chinensis	1 ounce (30 g)
Hypericum perforatum	1 ounce (30 g)
Melissa officinalis	1 ounce (30 g)

Combine the dry herbs and mix well. Steep 1 tablespoon of the mixture per cup of hot water and then strain. Drink freely, 3 or more cups (720 ml) a day.

Formulas for Cushing's Syndrome

Cushing's syndrome is generally defined as a state of glucocorticoid excess, regardless of the reason—be it the use of exogenous synthetic glucocorticoids or the inappropriate production of endogenous glucocorticoids due to an adrenocorticoid tumor or other reason. Ectopic production of ACTH, such as from a neuroendocrine tumor, may also induce Cushing's syndrome, as can faulty HPA-axis regulation and pituitary lesions. Glucocorticoid drugs are potent anti-inflammatory and immunosuppressive agents used for many inflammatory diseases but in cases when these medications induce Cushing's syndrome, there may be herbal and other alternatives.

Elevated cortisol stimulates the appetite, and weight gain is common. Additional symptoms of Cushing's syndrome include proximal muscle weakness, easy bruising, and violaceous striae on the skin of the belly and elsewhere. Excess glucocorticoids have a catabolic effect on skeletal muscle with increased activity of myofibrillar proteinases and reduced uptake and conversion of amino acids into proteins. This negative nitrogen balance is associated with increased protein wasting, type II muscle fiber atrophy, and significant muscle weakness predominantly in the large core muscles of the pelvic girdle. Decreased bone mineral density, osteoporosis, and related fragility fractures have been reported in 60 to 80 percent of patients with glucocorticoid excess. The pathogenesis of this bone loss is multifactorial.

Psychiatric disturbances are present in up to 85 percent of patients with Cushing's syndrome. Depression and irritability are the most common manifestations (51 to 86 percent); others include emotional lability, mania, paranoia, acute psychosis, anxiety, and panic attacks.

Hypercortisolemia is associated with a decrease in apparent brain volume, particularly the hippocampus, and related impairment in learning, cognition, and short-term memory. The mechanism underlying structural atrophy is not fully understood but is thought to include reversible atrophy of the dendritic processes, inhibition of neurogenesis, and increased susceptibility to cell injury and death due to increased levels of excitatory amino acid neurotransmitters such as glutamate.

Chronic glucocorticoid excess is associated with dyslipidemia and glucose intolerance. Increased hepatic gluconeogenesis, peripheral insulin resistance, and direct suppression of insulin release together contribute to the development of impaired glucose tolerance in 30 to 60 percent of patients and overt diabetes in 20 to 50 percent of patients.

In addition to dyslipidemia and glucose intolerance, Cushing's sufferers can develop hypertension (related to upregulation of the renin-angiotensin system and the mineralocorticoid effects of cortisol described earlier) and a hypercoagulable state (related to increased synthesis of clotting factors such as fibrinogen). These conditions lead to an additional increase in cardiovascular risk that may not return to baseline even after remission of hypercortisolemia.

Tincture for Cushing's Syndrome

Cushing's syndrome is the most common type of adrenal hyperfunction and can be due to overfunctioning adrenal glands, excessive ACTH stimulation by the pituitary such as with a pituitary adenoma, oat cell types of lung cancer, and direct cortisol excess with adrenal lesions. Each cause will need to be addressed directly, and in some cases, surgically. This is an all-purpose formula to help reduce excess cortisol and protect the tissues from the damaging effects. Because the liver is responsible for metabolizing hormones, including cortisol, *Curcuma* is included as a liver-supportive herb.

Glycyrrhiza glabra	15 ml
Panax ginseng	15 ml
Withania somnifera	15 ml
Curcuma longa	15 ml

Take ½ to 1 teaspoon of the combined tincture, 3 times or more each day.

Tincture for Cushing's Syndrome with Pituitary Adenoma

When no underlying cause is found for Cushing's syndrome and/or the condition is unresponsive to pharmaceutical agents, botanical agents are certainly worth a try. This formula is purely theoretical based on a known ability of *Vitex* to reduce excessive prolactin in cases of pituitary adenomas. *Hypericum* and *Melissa* also affect central neurotransmittors and may complement adaptogenic herbs.

Vitex agnus-castus	15 ml
Hypericum perforatum	15 ml
Melissa officinalis	15 ml
Curcuma longa	15 ml

Take ½ to 1 teaspoon of the combined tincture, 3 times or more each day.

Licorice: A Model Adaptogen

Glycyrrhiza (licorice) is one of the most widely used herbs in Traditional Chinese Medicine and has many hormonal and metabolic regulatory properties. Licorice affects the hypothalamic-pituitary-adrenal axis and impacts adrenal hormones, in part, by affecting cortisol metabolizing dehydrogenase enzymes,[122] increasing serum cortisol when low,[123] and decreasing urinary excretion of cortisol, which boosts cortisol activity in the tissue. Due to these effects, licorice may support weak or aging adrenal glands, reducing gradual loss of strength and vitality by helping to maintain responsiveness of the adrenal gland and to maintain feedback loops involving neuro-endocrine regulation in the brain.[124] A decline in adrenal output of DHEA and a reduction in DHEA relative to cortisol is associated with general aging[125] as well as atherosclerosis, immune dysregulation,[126] and cognitive and mood disorders.[127] Because of this, licorice may be included in herbal formulas for the elderly or for adrenal insufficiency states.

At the same time, licorice might act in an opposite manner and decrease elevated serum cortisol.[128] Thus, licorice fulfills the definition of an adaptogen, helping adrenal balance in states of both excess and deficiency. For example, licorice may increase tissue sensitivity to glucocorticoids in women with polycystic ovarian syndrome (PCOS),[129] while decreasing sensitivity in women with PMS.[130] Licorice decreases the excretion of cortisol when the body needs to conserve it, yet licorice will also increase the urinary excretion in Cushing's patients, as much as double.[131] Licorice may reduce elevated androgens,[132] yet not suppress androgens when they are within normal levels.[133] (For more information on licorice, see the *Glycyrrhiza glabra* entry on page 58.)

Adaptogens may normalize adrenal responsiveness, and single research studies must be interpreted with wholistic understanding that such herbs might be shown to have differing, and even opposite, activity, given the disease or tissue state being investigated.

Glycyrrhiza and many other adaptogenic plants contain steroidal saponins shown to exert

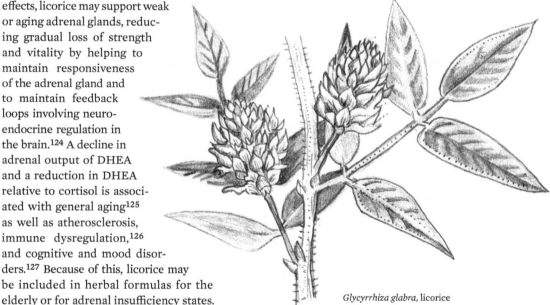

Glycyrrhiza glabra, licorice

immune-modulating and anabolic activities.[134] Steroidal saponins are credited with many supportive effects on adrenal glands and HPA axis that are dysregulated due to acute and long-term stress[135] or due to faulty corticotrophin production. Though it may seem paradoxical, licorice is useful in formulas for both Addison's[136] and Cushing's patients.[137] Licorice is featured in many formulas for adrenal imbalance, and other adaptogens such as *Withania*, *Panax*, or *Eleutherococcus* may also have tonifying and optimizing effects on the adrenal glands and HPA axis and be useful in formulas for both hyper- and hypofunction.

Tea for Cushing's Disease with Thinning Skin

Elevated cortisol is known to thin the skin and subdermal connective tissue. Many patients with Cushing's syndrome will develop purplish discoloration on the abdomen and suffer from poor healing wounds. This formula employs herbs that support connective tissue, *Centella* and *Equisetum*, along with anti-inflammatory *Calendula* and *Curcuma*, which also enhance healing. The use of a lemon water soak in preparing this tea can help extract more minerals from the herbs, which can help build tissue strength.

Centella asiatica	3 ounces (90 g)
Equisetum arvense	2 ounces (60 g)
Calendula officinalis	2 ounces (60 g)
Glycyrrhiza glabra	2 ounces (60 g)
Hypericum perforatum	2 ounces (60 g)
Curcuma longa	1 ounce (30 g)
Lemon	

Combine the dry herbs and mix well. Place 3 heaping tablespoons of the mixture and 4 cups (960 ml) of water in a saucepan. Add the juice of one lemon and allow the herbs to soak in the lemon water overnight. In the morning, bring to a gentle simmer for several minutes only in a covered pan; let stand for 15 minutes more, and then strain. Drink the entire amount daily. The *Curcuma* may be omitted for those who dislike the flavor.

Specific Indications: Herbs for Endocrine Conditions

Most of the folkloric and ancient herbal texts do not use the terms *thyroid* or *adrenal* when discussing symptoms and diseases. However, many such texts will describe the symptom picture such as cold hands and feet, with constipation and obesity, or heart palpitations with sweating and insomnia, suggesting recognition of hypo- and hyperthyroid states, well before an understanding of hormones and the roles that hormone-producing organs played in the body. The following herbs have been mentioned in herbal traditions throughout the world for symptom constellations matching endocrine disorders formally recognized today. Many of them have been researched and molecular mechanisms affecting the adrenal and thyroid systems have been identified.

Andrographis paniculata • King of Bitters

The entire plant—leaves, roots, and flowering tops—has been used in Ayurvedic medicine where *Andrographis* goes by the common name kalmegh or kalamegha. Include *Andrographis* in formulas for infections, whether acute, chronic, or lingering in hypothyroid and diabetic patients. *Andrographis* is thought to have a "cold property" in various energetic systems of herbal medicine, useful to rid the body of heat and fever and to dispel toxins. Consider *Andrographis* in formulas for thyroiditis suspected to be associated with underlying herpes or hepatitis infections and when acute viral infections cause a worsening of thyroid or metabolic symptoms. *Andrographis* is shown to improve the fatigue that often accompanies autoimmune diseases.[138] *Andrographis* has been the subject of significant research, and many studies credit the labdane diterpene andrographolide with immune-modulating activity. *Andrographis* is credited with anti-inflammatory, antioxidant, antidiabetic, anticancer, immunemodulatory, sex hormone modulatory, antiangiogenic, and hepatorenal protective activities, all of which may benefit patients with thyroid and adrenal imbalance and metabolic dysfunction.

Arctium lappa • Burdock

Include *Arctium* root to support hepatic clearance of hormones, lipids, and carbohydrates, to help optimize the intestinal microbial ecosystem, and to prevent fatty degeneration in the liver. Arctigenin in the roots is shown to enhance energy uptake and utilization in skeletal muscle that may improve stamina and endurance in those with fatigue and poor exercise tolerance.[139]

Artemisia annua • Sweet Annie

The aerial parts of *Artemisia annua* contain artemisinin, a sesquiterpene lactone, widely studied for malaria and other purposes. Artemisinin may also improve inflammatory and autoimmune diseases and is shown to suppress some hypersensitivity reactions. Artemisinin has shown immune regulatory actions via effects on T cell imbalances.[140] It has been shown to reduce Graves' exophthalmia, and animal studies have shown immunosuppressive activity.[141] *Artemisia annua* is known as qinghao in TCM and also goes by the common name sweet wormwood. Isolated artemisinin is referred to as qinghaosu in TCM and used to treat malaria, among other ailments.

Astragalus membranaceus • Milk Vetch

Astragalus roots have a starchy palatable flavor, making the powder useful to include in medicinal foods. Root decoctions and other preparations can be included in formulas for frequent common infections that linger and are associated with fatigue and exhaustion. *Astragalus* is in the legume family and the phytosterol isoflavones that legumes contain may improve hormonal balance. *Astragalus* has immune-modulating properties, making it useful for thyroiditis or instances when high viral loads trigger flare-ups of thyroid or metabolic symptoms. *Astragalus* may also reduce the risk of cancer in chronic thyroiditis. *Astragalus* may improve insulin response and is also useful for renal insufficiency. The roots are available as large slices that can be decocted in soups and removed at the time of serving. *Astragalus* can improve metabolic function.

Avena sativa • Milky Oats

The flowering tops of oats are gathered at the milky stage to make the best nerve-restoring medications. *Avena* is useful in formulas for nervous exhaustion and debility and to support recovery following exhausting illnesses, malnutrition, and addictions. *Avena* is specific for fatigue following overwork, loss of vital fluids such as acute hemorrhage, and as a supportive ingredient in formulas for adrenal dysfunction. Forty different phenolic compounds, the avenanthramides, have been identified in *Avena*, showing anti-inflammatory and antiatherogenic effects, decreasing plaque formation on endothelial cells via effects on adhesion molecules and proinflammatory cytokines. *Avena* can also, therefore, help protect against metabolic stress in hypothyroid disorders and other situations of metabolic stress.[142]

Berberis aquifolium • Oregon Grape

Berberis is the source of berberine, found in the yellow inner root bark of this Pacific Northwest shrub, which is widely studied for numerous benefits to metabolism. Berberine improves hyperlipidemia and hyperglycemia, enhances insulin sensitivity, and is credited with thermogenic action due to effects on brown adipose cells. Include *Berberis* in formulas for fungal infection in diabetes and as a general antimicrobial and hepatotonic in cases of intestinal dysbiosis related to hyperglycemia or sluggish elimination in hypothyroidism. *Berberis* may also be included in formulas for thyroiditis, Hashimoto's disease, and hyperthyroidism. It can help control lipids as one aspect of metabolic syndrome therapy. This plant is also known as *Mahonia aquifolium*.

Bupleurum falcatum • Chinese Thoroughwax, Saiko

Bupleurum is used in numerous traditional formulas for inflammation and pain, and modern research has revealed numerous anti-inflammatory and antioxidant actions. The roots are available to decoct or use in tincture. Recent animal research shows *Bupleurum* to possibly reduce hyperthyroidism, improving oxidative stress and tissue changes in a manner equal to thyroid-suppressing drugs.[143] *Bupleurum* has been used in Traditional Chinese Medicine to help the liver "smooth the chi" and is appropriate for acute inflammatory reactions in the body. *Bupleurum* may also be included in formulas to help protect the liver and kidneys in cases of Graves' disease when oxidative stress is high. *Bupleurum* saikosaponins have immunomodulatory activity and may inhibit platelet activation while enhancing cortisol synthesis and corticosterone secretion, supporting the pituitary's production of ACTH and offering hepatoprotective and nephroprotective effects in situations of oxidative stress in hyperthyroidism.[144] Saikosaponins also inhibit the inflammatory prostaglandins and promote protein synthesis in the liver, contributing to anti-inflammatory and hepatoprotective effects. *Bupleurum* is specific for organomegaly, especially hepatomegaly, and abdominal pain associated with liver and digestive disease. *Bupleurum* is often combined with *Paeonia* to treat liver congestion and disease in Traditional Chinese Medicine. *Bupleurum chinense* is also used medicinally.

Calendula officinalis • Pot Marigold

Calendula flowers are a traditional medicine for skin wounds and lesions, both topically and internally. *Calendula* can be included in formulas for thinning and pigmentary changes in the skin associated with Cushing's syndrome and to help treat itching and skin lesions that may occur with thyroid disease.

Camellia sinensis • Green Tea

Green tea leaves have broad antioxidant mechanisms and can be a gentle metabolic stimulant for hypothyroidism and deficiency states. Green tea can be part of a daily protocol to reduce oxidative stress in the liver and to reduce the risk of developing thyroid cancer in those with thyroiditis.

Centella asiatica • Gotu Kola

Centella leaves may help protect tissues including the liver, nerves, heart, and blood vessels from fibrotic changes due to inflammation. *Centella* may also prevent tissue proliferation behind the eyes in Graves' exoph-thalmia and protect the connective tissue from excessive thinning in cases of Cushing's disease.

Citrus sinensis • Orange

The peels of oranges and sometimes other citrus fruits are frequently used in TCM formulas for a broad variety of actions, including suppressive effects on the thyroid. *Citrus*, including orange juice, is shown to reduce T4 and T3 levels in animal studies, resulting in an increase in thyroid-stimulating hormones (TSH).[145] *Citrus* may ameliorate hyperthyroid symptoms, improving both elevated thyroid hormones and protecting the tissue against oxidative stress, however, one should not expect too much effect from consumption of *Citrus* on its own.

Cnicus benedictus • Blessed Thistle

Cnicus leaves are specific for sluggish liver function. Like its Thistle-family relatives *Silybum* and *Cynara*, *Cnicus* may improve the liver's processing of hormones and help protect the liver from fatty degeneration.

Coleus forskohlii • Coleus

The roots of *Coleus* are specific for atonic organ function, sluggishness of general metabolic functions, and excessive inflammatory responses. *Coleus forskohlii* extract has been used for centuries in Ayurvedic medicine to treat various diseases including hypothyroidism, and its use may support weight loss in obese patients with low metabolic rate or metabolic syndrome. *Coleus* has direct effects on cells by enhancing cyclic adenosine monophosphate (cAMP), often referred to as the "second messenger" due to its role in supporting cells' signal transduction, helping to pass along cell membrane signals to the interior of the cell. Research on *Coleus* has mainly focused on forskolin, a uniquely structured labdane-type diterpenoid in roots of the plant. Because cAMP promotes the breakdown of stored fats in animal and human fat cells, regulates the body's thermogenic response to food, increases the body's basal metabolic rate, and increases utilization of body fat, *Coleus* may be helpful in weight loss formulas.[146] Animal research suggests that *Coleus* has antiobesogenic and metabolic benefits to rats fed a poor diet, reducing food intake and blood lipids, and supporting weight loss.[147] Limited human studies have shown *Coleus* to favorably impact changes to body composition, reducing body fat percentage in overweight and obese subjects.[148] Another diterpene, coleonol, may stimulate the release of insulin and glucagon from pancreatic islets.[149] When combined with a low-calorie diet and exercise, *Coleus* may improve insulin resistance and metabolic risk factors. *Coleus forskohlii* is also known as *Plectranthus forskohlii*, and some sources also list *Plectranthus barbatus* as an alternate name for *Coleus forskohlii*.

Commiphora mukul • Guggul

Guggul is a traditional medicine for obesity, diabetes, and hyperlipidemia used in Ayurvedic medicine for more than 2,000 years, and guggulsterones in the plant's resin have been shown to have numerous direct effects on fat cells. The resin is collected from incisions in the trunk of guggul trees (which are related to myrrh trees) and ground down into a powder to use in medicine and in dressing wounds. Guggul is specifically indicated for elevated lipids due to hypothyroidism or slow metabolism, a sense of weight in the pelvis, tissue congestion, fat accumulation, and constipation. Guggul may support weight loss and lipid metabolism via enhancing thyroid uptake of iodine, supporting thyroxine output, and metabolic rate.[150] Animal studies have shown guggul to increase T3 concentration and T3/T4 ratio,[151] but this area of research has not matured in the 30 years since first reported. Guggul's hypolipidemic effects have received more research attention in recent decades, and guggulsterones are shown to help improve fat metabolism by the liver and to help resist or reverse insulin resistance due to dietary stressors.[152] *C. myrrha* (myrrh) and *C. wightii* are also used medicinally. The plant is referred to as guggulu in Ayurvedic medicine.

Coptis trifolia • Goldthread

Coptis roots are high in berberine and related isoquinoline alkaloids, useful for chronic infections in mucous membranes. *Coptis* may be included in formulas for diabetic and hypothyroid patients who have frequent sore throats or other common infections, with slow recovery time. Due to its many metabolic effects, *Coptis* may help deter endothelial atherosclerotic plaque deposition and may help protect the brain against neurodegenerative effects due to metabolic syndrome. *Coptis* also supports liver metabolism, improving carbohydrate and lipid processing,[153] and offers cardioprotection.[154] *Coptis chinensis* is one of the 50 fundamental herbs used in TCM, where it is called duan e huang lian.

Crataegus species • Hawthorn

Various species of hawthorn trees are used medicinally, including *Crataegus monogyna, C. oxyacantha*, and *C.*

laevigata. Both the young flower buds with their leaves and the ripe berries are used medicinally. *Crataegus* berries are high in flavonoids that may help protect the heart and blood vessels from inflammatory damage in situations of high oxidative stress with hypermetabolic syndromes such as Graves' disease.

Curcuma longa • Turmeric

The bright yellow tuberous roots of *Curcuma* are widely used for many medicinal as well as culinary purposes. Much of the research on *Curcuma* is focused on a group of golden-colored flavonoids known as curcuminoids, especially curcumin. *Curcuma* has antioxidant and anti-inflammatory properties that may help protect the liver from fatty degeneration in cases of hyperlipidemia and metabolic syndrome. It offers general hepatic support and may assist the liver in clearing hormones. *Curcuma* can improve lipid and carbohydrate metabolism in cases of diabetes and hyperlipidemia and may improve intestinal dysbiosis and propensity to fungal and other infections. *Curcuma* may also help protect connective tissue and organs from oxidative stress and fibrotic demise. In addition, it may help protect the liver from fibrosis in chronic thyroiditis, reduce the risk of emergent cancers, and protect the connective tissue from thinning in Cushing's syndrome.

Eleutherococcus senticosus • Siberian Ginseng

Eleutherococcus roots are a traditional medicine for relieving stress, treating mental and emotional disorders, and improving stamina and general energy in cases of chronic fatigue. It is also used for those recovering from debilitating illnesses. *Eleutherococcus* acts as an adrenal tonic in both hypo- and hyperfunction of the adrenal gland, and it is specifically indicated for long-term stress resulting in nervous symptoms, fatigue, poor sleep, and poor stamina, with easy exhaustion following minor exertion. *Eleutherococcus* may be a supportive ingredient in formulas for endocrine imbalances. Siberian ginseng, also called eleuthero, may help modulate the hypothalamic-pituitary-adrenal (HPA) axis response to stress, as indicated by an attenuated corticosterone response, and reduce mental-emotional consequences such as anxiety or depression.[155] Animal research has shown *Eleutherococcus* to improve immune response when the immune system has been suppressed by chronic stress and to prevent metabolic syndrome in situations of metabolic stress, improving skeletal muscle energy metabolism.[156] Numerous anti-inflammatory and tissue-protective

mechanisms have been demonstrated, and *Eleutherococcus* may be helpful in protocols for hypothyroidism and hyperthyroidism where high levels of oxidative metabolites may damage tissues.[157] The plant is also known as *Acanthopanax senticosus* and referred to by the common name ciwujia in China.

Epimedium grandiflorum • Horny Goatweed

Epimedium is a traditional Chinese herb whose leaves and roots are said to invigorate yang and often used in skin, bone, and reproductive hormonal-balancing formulas. Other *Epimedium* species are also used medicinally and go by the same common name in the West. *Epimedium* is said to be a kidney yang tonic in TCM, and kidney yang deficiency syndrome shares clinical similarities with glucocorticoid withdrawal. *Epimedium* may also protect against developing Cushing's syndrome in those who require long-term glucocorticoid therapy.[158] Some of *Epimedium*'s immunomodulating and HPA axis–stabilizing effects may be via estrogen agonism.[159] *Epimedium* has bone-building effects and may protect against thyroid-related bone disease, possibly by estrogenic effects on bone cells.[160] *Epimedium* may reduce atopic dermatitis that frequently occurs with HPA axis dysfunction, particularly seen in those with low serum cortisol.[161]

Equisetum arvense • Horsetail

Equisetum's aerial parts are an ancient medicine for wound healing, trauma recovery, and to support the excretory function of the urinary system. *Equisetum* is high in silica and other minerals, contributing to its reputation for healing fractures and alleviating musculoskeletal pain. Include *Equisetum* in formulas to support connective tissue; it may help wound healing in chronic diabetic ulcers or to strengthen the skin in patients with Cushing's syndrome. Many related species are used in the same manner, including *E. hyemale*. See also the *Equisetum arvense* entry on page 93.

Eschscholzia californica • California Poppy

Eschscholzia roots and young flower tops are indicated for symptoms of anxiety, muscular tension, and poor sleep, and the herb is appropriate to include in formulas for adrenal exhaustion due to long-term stress. Nervines such as *Eschscholzia* are often useful in protocols for adrenal dysfunction, helping to soothe mental anxiety and treat stress symptoms that lead to cortisol activation and contribute to adrenal dysfunction.

Fucus vesiculosus • Bladderwrack

Fucus is a genus of brown algae found in intertidal rocky seashores in temperate regions, especially the Pacific shores of North America. *Fucus* has been used as food and medicine for centuries, particularly in Asian cultures. *Fucus* also goes by the common name bladderwrack because the plant possesses small air-filled bladders or flotation devices that help keep it close to the water's surface. *Fucus* is also referred to as kelp, which was also the name given to the alkaline ashes that were produced by burning various seaweeds and used as an alkali agent in soap making. *Fucus vesiculosus* contains the flavonoid fucoxanthin and is reported to have the greatest anti-oxidant activity of all the edible seaweeds tested. *Fucus* and other brown seaweeds are specifically indicated for symptoms of hypothyroidism, including obesity, goiter, exophthalmia, constipation, and flatulence. *Fucus* and other seaweeds are a natural source of iodine, important to thyroid function, and the fucoxanthins may reduce the risk of thyroid cancer and hormonal cancer in general. Folklore often reports that *Fucus* protects a poorly functioning thyroid from further inflammatory stress.[162] See also "Dietary Seaweed for Thyroid and Metabolic Function" on page 27.

Ganoderma lucidum • Reishi

The entire woody reishi mushroom is used medicinally, processed into tinctures, powdered and encapsulated, and decocted in teas. Reishi and other medicinal and edible mushrooms contain β-glucans and immunemodulators credited with antiviral, tissue protective, and metabolic actions. These actions explain *Ganoderma*'s reputation as the "mushroom of immortality." *Ganoderma* may help protect the nerves from oxidative stress and support the immune system in situations of poor immunity, including in adrenal and thyroid disease. *Ganoderma* may also protect the HPA axis from downregulation in Cushing's syndrome, protecting against weight gain, insulin resistance, and altered lipid metabolism. *Ganoderma* may also contain enzymes that metabolize cortisol into the less active corticoids. This may also help to attenuate aldosterone activation that drives fluid retention and hypertension in Cushing's syndrome, reducing cortisol from binding mineralocorticoid receptors.[163] Bisphenol A and triclosan are known endocrine-disrupting chemicals, and *Ganoderma* species are shown to help metabolize these toxins in municipal water supplies employing water-filtering technologies.[164] It is unknown how this ability of *Ganoderma* may benefit the human body, but it is plausible that the mushroom may similarly support the enzymatic breakdown of endocrine disrupters, supporting endocrine health.

Ginkgo biloba • Ginkgo, Maidenhair Tree

Ginkgo leaves are specifically indicated for oxidative stress in the blood, clotting, and circulatory insufficiency. *Ginkgo* may support circulation in heart disease and protect the heart in situations of metabolic syndrome and hyperlipidemia. Include *Ginkgo* in formulas for diabetic retinopathy, nephropathy, and neuropathy, as well to treat impotence due to diabetes and poor circulation. The research on *Ginkgo* is extensive and shows numerous antioxidant, antiallergy, vasodilating, vascular-protective and other tissue-protective anti-inflammatory effects. *Ginkgo* may also enhance circulation to the eye in a manner that reduces oxidative stress to treat Graves' exophthalmia. Include *Gingko* in formulas for hypertension and inflammation in Graves' disease and metabolic distress. *Ginkgo* is reviewed in detail in Volume 2, chapter 2, of this set.

Glycine max • Soy

Soybeans have numerous tonifying effects on reproductive hormones and may improve many problems with metabolic distress ranging from diabetes to elevated lipids. Isoflavone phytosterols in soy are credited with numerous hormone-regulating effects on hormone sensitive tissues. Soy *may* interfere with the uptake of levothyroxine.[165]

Glycyrrhiza glabra • Licorice

Licorice is among the most versatile, popular, and widely used of all plant medicines in all cultures. Licorice roots are specifically indicated for exhaustion, loss of strength, weight loss, and debility. Licorice may be included in formulas for PCOS, Addison's disease, Cushing's syndrome, insulin resistance, adrenal virilism, orthostatic hypotension, and postpartum declines in hormone output and balance. Licorice can enhance urinary excretion of cortisol in situations of elevated cortisol, as well as enhance adrenal production of cortisol when low. Licorice also has antiviral properties and may help limit inflammation in the thyroid when thyroiditis is associated with underlying viral auto-immune reactivity. Licorice may also replace the pharmaceutical spironolactone for those who tolerate it poorly. Licorice is among the plants known to impact hormonal balance and the HPA axis. There has been extensive research on the ability of licorice to improve adrenal fatigue and stress intolerance, via enhanced cortisol response. Licorice contains a cortisone-like constituent and also inhibits the breakdown of cortisone in the liver.[166] A concomitant pseudoaldosterone action may cause retention of fluid and elevation of blood pressure,[167] yet improve symptoms of mineralocorticoid insufficiency such as Addison's disease. Glycyrrhizinic acid is a component of the roots shown to have hepatoprotective effects, preventing changes in cell membrane permeability by inhibiting phospholipase.[168] Licorice extracts may also improve dyslipidemia. Licorice may improve insulin resistance, and licorice flavonoids are shown to suppress abdominal fat accumulation in obese mice and rats.[169] The triterpenes in licorice roots are particularly credited with metabolic effects.[170] For more information on *Glycyrrhiza*, see "Licorice: A Model Adaptogen" on page 53.

Hypericum perforatum • St. Johnswort

St. Johnswort flower buds may be used as an ingredient in formulas to strengthen and reduce inflammation in all types of vascular tissue, from the heart muscle to the microvasculature, and as a circulatory and neural protectant in formulas for metabolic distress seen with hypo- and hyperthyroidism. *Hypericum* is specifically indicated for neuralgic pain and may be included in formulas for polyneuropathies in diabetic patients. *Hypericum* is traditionally emphasized for shooting, lancinating pain, or buzzing and tingling. It may help reduce neuronal inflammation, irritable bladder, and bladder weakness in diabetics, especially when used as a tea. *Hypericum* also has antiviral activity and may be included in formulas for thyroiditis. It may also be a supportive ingredient in formulas for adrenal dysfunction with peri- or postmenopausal onset, especially when associated with mood swings, anxiety, and depression.

Iris versicolor • Wild Iris

Herbal tinctures are prepared from the roots of wild iris, also called blue flag, for use as an overall secretory stimulant, specific for congested lymphatic tissues and glands—lymph nodes, spleen, liver, and thyroid. In thyroid disorders, *Iris* may be included in formulas for goiter or simply mild soft enlargements of the thyroid gland. *Iris* may also reduce congestion in the tissues that may lead to exophthalmia in hyperthyroidism. *Iris* increases the secretions of salivary and digestive glands and is useful as a complementary herb in formulas for hypothyroid–digestive insufficiency, poor digestive and

liver function, and fat intolerance with steatorrhea. Wild iris is not commonly used as a tea but is sometimes included as an ingredient in commercial capsules aimed at thyroid support. *Iris* tincture is potentially caustic and irritating to the oral and digestive tissues and should also be used in small amounts and diluted with water or other herbs.

Leonurus cardiaca • Motherwort

The flowering tops of *Leonurus* are specific for thyroid disorders that begin at the climacteric and for other endocrine imbalances associated with a restless anxiety. *Leonurus* may offer cardio- and vascular-protective effects in metabolic syndrome or in situations of hypertension and hyperlipidemia. *Leonurus* is specific for nervous debility and restlessness, for tics, twitches, and tremors, as well as for cardiac excitability and irregularity. *Leonurus* is specific for anxiety, restlessness, and heart palpitations associated with hyperthyroidism. *Leonurus* is a gentle beta blocker, useful to help treat tachycardia and hypertension, and contains rosmarinic acid, which may help reduce excessive stimulation of the thyroid by TSH or autoantibodies. *Leonurus* may also have weak heavy metal–chelating effects, contributing to possibly antitoxin activity in the thyroid.[171] *Leonurus cardiaca* may be included in formulas for those prone to atrial fibrillation such as Graves' disease patients. *Leonurus* calms turbulent and overstimulated heart action by inhibiting the inward flow of calcium and potassium ions at the sinoatrial node and individual myocytes, prolonging muscle contraction and repolarization interval. *Leonurus* also improves blood viscosity and fibrinogen volume via antioxidant and platelet antiaggregating effects.[172] Other species of *Leonurus* are also used medicinally in China.

Lithospermum species • Gromwell

The young leaves and flowering tops of various *Lithospermum* species are used to treat conditions associated with excessive hormone levels. Also known as stoneseed, *Lithospermum* may reduce elevated follicle-stimulating hormone (FSH), luteinizing hormone (LH), prolactin, or TSH from the pituitary, which may be helpful in formulas for treating thyroid storm in acute thyroiditis or severe hyperestrogenism. Rat studies on *Lithospermum officinale* have shown the plant to inactivate circulating TSH and prolactin, as well as reduce the output of these hormones from the pituitary.[173] *Lithospermum* extracts have been shown to reduce goiter mass in hyperthyroid animals. Shikonin, a naphthoquinone in the plant, is credited with hormonal antagonism and may protect against papillary carcinoma of the thyroid, for which Hashimoto's and all thyroiditis patients are at increased risk.[174] *Lithospermum officinale* goes by the common name European stoneseed, and *Lithospermum ruderale* goes by the name Western stoneseed; the common names are used interchangeably. *Lithospermum erythrorhizon* is an Asian species, sometimes referred to as purple gromwell or red-root gromwell, and has been used for centuries in anticancer formulas. This species also contains shikonin.

Lobelia inflata • Indian Tobacco

The flowering tops of *Lobelia* are an important bronchodilator, with expectorating action that may be an emetic action in the higher dosage ranges. *Lobelia* is both a partial agonist and partial antagonist of nicotinic acetylcholine receptors on the vasculature, but overall it will relax angina and vasospasms and calm tachycardia.[175] *Lobelia* may be used as a tool to treat severe hypertension or heart palpitations associated with hyperthyroidism. Lobeline, the primary alkaloid in *Lobelia*, may also calm hyperthyroid states via effects on dopaminergic pathways, inhibiting monoamine transport in vesicles in a manner that limits sympathetic symptoms.[176] Lobeline is a piperidine alkaloid, and along with a group of related alkaloids, is credited with general muscle-relaxing effects. *Lobelia* inhibits the reuptake of dopamine, prolonging its life in neural synapse.[177]

Lycopus virginicus • Bugleweed

The young leaves and flowering tops of various *Lycopus* species have been used medicinally in ancient European herbalism as a gentle heart sedative and to manage the symptoms suggestive of Graves' disease, such as tachycardia, exophthalmia, tremor, anxiety, and insomnia. *Lycopus virginicus* is specifically indicated for excessive thirst and hunger and is also specific for vascular excitation due to excessive sympathetic stimulation with rapid pulse or tumultuous heart action. *Lycopus* is occasionally helpful in cases of exophthalmia and goiter due to hyperthyroidism and can be one ingredient in broader protocols. *Lycopus* can block excessive stimulation of thyroid cells.[178] Animal studies have shown that *Lycopus europaeus* (gypsywort) may help control hyperthyroid-induced heart palpitations and hypertension, as well as reduce hyperthermia, yet without having a significant impact on blood levels of TSH or thyroid hormones.[179] Other animal studies, however, have reported that an ethanol extract (tincture) of *Lycopus europaeus* decreased levels of T3, T4, and TSH in the bloodstream for 24 hours or more after a single dose.[180]

Mahonia • See Berberis

Melissa officinalis • Lemon Balm

The lemony-smelling aerial parts of lemon balm have long been used as an antidepressant and antianxiety agent, especially in those with endocrine imbalances, including thyroid and female hormonal disorders. Old herbal texts report *Melissa* to be specific for mental apathy and lack of libido and for mental and physical weakness. *Melissa* can be included in formulas for Hashimoto's disease; it may have a suppressive effect on hyperthyroidism due to rosmarinic acid. One study investigating whole *Melissa* showed the extract to inhibit the binding of TSH to the TSH receptor, as well as to have direct effects on the TSH receptor itself in a manner that interrupts signals when ligands bind.[181]

Melissa officinalis may improve tachyarrhythmia, and even a single dose is shown to help protect the electrical rhythm of the heart in animal studies,[182] calming ventricular conductivity via sodium or potassium channel blockade. *Melissa* may be included in formulas for Graves' patients with tachycardia. One clinical trial showed that 500 milligrams of *Melissa* given twice a day reduced the frequency of heart palpitations in anxious subjects.[183]

Melissa also has antiviral activity and may improve thyroiditis, specifically when respiratory viruses induce a flare-up or aggravation of endocrine symptoms. *Melissa* also has significant antioxidant effects that may help protect tissues and organs in situations of metabolic distress.[184] *Melissa* may enhance cognition via cholinesterase inhibition and have anxiolytic effects via GABA agonism.[185]

Opuntia ficus-indica, O. streptacantha • Prickly Pear

The prickly pear cactus bears orange- and magenta-colored fruits high in betalains. See "*Opuntia* for Insulin Resistance" on page 83 for details on use of *Opuntia* to reduce insulin resistance in diabetics. The succulent pads of the prickly pear, referred to as nopales, are also used medicinally. Nopales are eaten as a food and also processed to make herbal medicines for treating high cholesterol, among other ailments. *Opuntia* can help regulate lipids, which are often elevated in hypothyroidism.

Panax ginseng • Ginseng

Ginseng root medications are traditional to support recovery from exhausting illnesses and to treat endocrine imbalances following long-term overwork and stress. *Panax* can be included in adrenal support formulas and may improve energy in hypothyroid, diabetic, and adrenal-insufficient patients and normalize both elevated and deficient cortisol. *Panax* is a warming herb best in cold or neutral constitutions and is specifically indicated for exhaustion, weight loss, and debility. Clinical trials have shown *Panax* to improve fatigue in cases of chronic fatigue syndrome or other unexplained cases of chronic fatigue.[186] *Panax quinquefolius*, North American ginseng, appears to act in similar ways.

Phytolacca americana, P. decandra • Pokeweed

Phytolacca roots are prepared into tinctures, and occasionally an oil, which is used topically to improve lymphatic circulation. *Phytolacca* is specific for fluid and lymphatic congestion and may be included in formulas for goiter. It has significant antiviral effects credited to lectins and other proteins[187] in the plant. *Phytolacca* contains ribosomal inactivating proteins that inhibit viral replication[188] and has activity against a wide range of viruses, including HIV-1, herpes simplex virus, cytomegalovirus, influenza virus, and polio virus,[189] several of which are known to trigger autoimmune thyroiditis. Pokeweed antiviral proteins have been produced from cell cultures, altered into immunotoxins, and investigated to treat virally induced cancers and chronic viral infections.[190]

Plantago species • Plantain

Plantago seeds have been traditionally used to treat blurred vision in Asia, and animal research shows oral consumption to reduce congestion and hyperpermeability of the retinal vessels in diabetic rats and significantly reduce oxidative stress. Some TCM formulas have featured the seeds in teas for eye disease including Graves' exophthalmia. Psyllium is the common name for *Plantago psyllium*, and the ground seed husks are readily available as a fiber supplement that can be supportive to weight loss, hyperlipidemia, and hyperglycemia. *Plantago ovata* is referred to as ispaghula in Ayurvedic medicine, and the seeds are sometimes referred to as blond psyllium.

Rauvolfia serpentina • Indian Snakeroot

Rauvolfia roots are one of the more powerful choices in the herbal armamentarium to bring down blood pressure quickly in cases of thyroid storm or in hyperthyroid states and vascular disease when there is a full bounding pulse. *Rauvolfia* is a powerful hypotensive, contraindicated in lethargy, circulatory weakness, and depressive states. Ajmaline is credited with antiarrhythmic activity that may calm atrial fibrillation. It is extremely helpful for vigorous and excessive tachycardic symptoms or for hyperthyroid and hypermetabolic states causing hypertension and arrhythmia.

Rhodiola rosea • Arctic Rose

Rhodiola is specifically indicated for fatigue, muscle weakness, inability to exert oneself, mental fatigue, poor stress tolerance, and nervous exhaustion. As such, *Rhodiola* may be considered an adaptogen. The astringent roots of *Rhodiola* are considered to have adaptogenic activity and are useful to support recovery from prolonged infection and for various states of debility. *Rhodiola* may prevent fatigue and other symptoms in hypothyroid patients.[191] *Rhodiola* is shown to improve oxygen utilization in the heart muscle, preventing oxidative stress and improving athletic performance and exercise tolerance. It may help prevent dysregulation of the HPA axis in situations of physical stress.[192] *Rhodiola* also protects against changes in the HPA axis in situations of chronic stress, attenuating excessive corticotropin-releasing hormone (CRH) expression in hypothalamus, as well as reducing significantly the levels of serum corticosterone. *Rhodiola rosea* can prevent the development of depression by preventing inflammatory and glucocorticoid-driven changes in the hippocampus.[193] *Rhodiola rosea* contains several compounds credited with anti-inflammatory and antioxidative actions and the ability to enhance endurance. There are at least 200 species of *Rhodiola* and more than 20 species have been used in a similar manner, including *R. crenulata*, *R. quadrifida*, and *R. kirilowii*.

Rosmarinus officinalis • Rosemary

The young leaves of rosemary are used as a culinary spice and to prepare a variety of herbal medicines. *Rosmarinus* contains rosmarinic acid, which may help reduce excessive TSH-driven stimulation of the thyroid gland, and rosemary may be included in formulas for hyperthyroidism. Rosmarinic acid has shown potent cardioprotective and antiarrhythmic effects due to both antioxidant and possibly antiadrenergic effects, making it useful in formulas for thyrotoxicosis.[194] Yet, *Rosmarinus* is also suitable in formulas for hypothyroidism and has shown positive weight loss results in animal models of metabolic insufficiency.[195] Rosemary has powerful antioxidant effects that may help both excess and deficient metabolic states. Oxidative stress occurs both in hyperthyroidism due to accelerated metabolic rate, and in hypothyroidism due to slow metabolism that allows accumulation of lipids and other inflammatory compounds. *Rosmarinus* is very high in carnosic acid, a phenolic diterpene specific to the Lamiaceae family, and is noted to protect lipids from oxidation including cell membrane linolenic acid.[196] Carnosic acid has been shown to protect the photoreceptors of the eye from neurodegenerative inflammation, slowing the visual loss in retinitis pigmentosa.[197] While no studies on rosemary for Graves' exophthalmia have been conducted, *Rosmarinus* is noted to deter fibrotic process in many organs, including epithelial cells of the eye.[198] It would do no harm to include rosemary in teas and tinctures or even prepare into soothing eyewashes for people with this condition.

Salvia miltiorrhiza • Red Sage

Also known as dan shen, red sage roots can be included in formulas to protect the blood vessels in patients with elevated lipids and glucose. Include *Salvia miltiorrhiza* in formulas for diabetes, metabolic syndrome, vascular reactivity, and Graves' exophthalmia. *S. miltiorrhiza* can improve circulation to the eyes in treating retinopathy, to the kidneys to protect against and treat nephropathy, and to the heart in cases of high lipids and oxidative stress.

Sargassum muticum • Japanese Wireweed

Sargassum is a brown seaweed with strong antioxidant properties, widely used in TCM for the treatment of numerous diseases due to its ability to protect cells from oxidative stress conditions.

Silybum marianum • Milk Thistle

Milk thistle seeds are specific for liver inflammation and can protect the liver from fatty degeneration. *Silybum* can improve hepatic metabolism of fats and sugars and can help protect the kidneys and liver in states of metabolic stress and improve cellular function of the organs. *Silybum* also protects the liver, pancreas, kidney, and other tissues from exogenous toxins and virally induced tissue changes in the case of hepatitis, plus it protects against immune-driven reactivity, so could do no harm in patients with thyroiditis.

Smilax ornata • Sarsaparilla

The roots of *Smilax* species can be used as aromatic alterative agents with hormonal- and adrenal-supportive properties. *Smilax* is specifically indicated for fatigue, muscle weakness, and poor digestion. *Smilax* has slightly warming properties and is especially useful in situations of metabolic insufficiency with coldness and weakness. The plant has formerly been classified as *Smilax regelii*, and other common names include Honduran sarsaparilla and Jamaican sarsaparilla. *Smilax* species are high in steroidal saponins credited with much of its hormonal and metabolic activities. The plants also contain flavonoids, phenylpropanoids, and stilbenoids. The antiobesity effect may be due to stimulation of lipolysis induced via activation of the β-adrenergic receptor.[199] The size and number of lipid droplets in fat stores is decreased by *Smilax*, and triglyceride accumulation is broken down and metabolized. Adrenergic signaling may increase the expression of adenylate cyclase and boost metabolic activity in adipose stores.[200]

Taraxacum officinale • Dandelion

Alterative herbs such as dandelion are included in some formulas in this chapter as complementary ingredients in endocrine formulas, supporting hepatic metabolism when hormones levels are excessively high and helping to protect the liver in cases of metabolic, oxidative, and toxic stress. Dandelion root may be used in formulas to support the liver and protect against fatty and cirrhotic degeneration. It is specifically indicated for gastric headaches, biliary insufficiency, jaundice, coated tongue, pain in the right upper quadrant of the abdomen, hepatic torpor and constipation, and night sweats due to liver disease. Include *Taraxacum* in formulas for hepatoprotective effects.

Valeriana officinalis • Valerian

Valeriana may be included in formulas for anxiety, insomnia, and heart palpitations due to hyperthyroidism. Iridoid valepotriates in valerian root are known to bind to GABA receptors, promote the release of GABA (γ-aminobutyric acid), decrease GABA's reuptake, and decrease GABA degradation, all contributing to the anxiolytic effects. Valerian may attenuate stress-induced increases in serum corticosterone,[201] and thereby help prevent adrenal downregulation in those struggling with chronic stress and anxiety. Valerian also has hypotensive and antiarrhythmic effects, reducing myocardial excitability due to effects on ion channels, and may help to treat elevated blood pressure and help prevent atrial fibrillation.[202] Due to valerian's sedating effects, it is best for hypervigilant states, insomnia, anxiety disorders, musculoskeletal tension, hypertension, and heart palpitations.

Vitex agnus-castus • Vitex, Chaste Tree Berry

The small round peppercorn-like fruits of *Vitex* are used medicinally, explaining the alternate common name monk's pepper, as the fruits were sometimes ground and sprinkled on food to help reduce the libido in religious orders. *Vitex* is specifically indicated for mood disorders associated with hormonal fluctuations, including premenstrual dysphoria, or other cyclical occurrence of depressive moods, irritability, anxiety, and confusion. *Vitex* is often highly effective for mastalgia, fatigue, and headache associated with hormonal fluctuation, all of which are symptoms that involve neuroendocrine activity of the central nervous system. *Vitex* has dopaminergic actions that may affect the regulation of pituitary hormones and in turn affect prolactin, LH, and FSH. *Vitex* reduces the release of prolactin in hyperestrogenic states, but less so in normal estrogen states.[203] *Vitex* also binds and activates μ- and δ-, but not κ-, opioid receptor subtypes, which contributes to mood-calming effects; it has shown no addictive or sedating tendencies. Include *Vitex* in thyroid and adrenal support formulas when hormone imbalance begins in the perimenopausal transition.

Withania somnifera • Ashwagandha

Withania roots are credited with adaptogenic and anxiolytic properties. *Withania* is specific for stress and adrenal disorders associated with high cortisol levels and poor sleep. *Withania* is specifically indicated for people who fall asleep easily but wake around midnight unable to sleep well thereafter. *Withania* is also specific for weakness, fatigue, and poor stamina and for anxiety, jitters, and insomnia in hyperthyroid and perimenopausal patients. *Withania* can help treat patients with and recovering from acute or prolonged stress. It also enhances memory and mental function.

Zingiber officinale • Ginger

Ginger can help improve digestion, reduce elevated lipids, and reduce inflammation in the blood and tissues. Ginger roots can be included in teas and tinctures for elevated lipids that may accompany hypothyroidism and be used as a medicinal food. Ginger is specifically indicated for cold and deficient constitutions, to stimulate circulation, and to bring more heat to the digestive system.

CHAPTER THREE

Creating Herbal Formulas for Metabolic Conditions

Diabetes and metabolic syndrome have been increasing worldwide in incidence for decades and are predicted to continue escalating with the coming generation. This epidemic is associated with a decreased life-span due to pervasive tissue inflammation created by elevated blood sugar and lipids. The epidemic prevalence of diabetes mellitus is thought to be largely due to poor quality diets, lack of exercise, and high levels of oxidative stress due to obesity, environmental toxins, and lack of antioxidant nutrients. Although hereditary predispositions, viral and bacterial afflictions of the pancreas, and pancreatic islet autoantibodies do contribute to the development of this disease, diet, lifestyle, and obesity are by far the most significant risk factors. High sugar and simple carbohydrate fare, especially when associated with "bad" fats and low amounts of nutrients and fiber, contributes to the development of diabetes. Soda pop and other foods containing high-fructose corn syrup are particularly harmful, and I believe that allowing children—or anyone—to drink these beverages is tantamount to condoning a slow form of poisoning. Numerous studies have shown that rats fed a high-fructose diet will develop insulin resistance and high triglycerides.[1] A high intake of concentrated fructose and sweetened fruit juices impair glucose tolerance in genetically susceptible individuals, whereas eating fresh fruit appears free of such consequences.[2] Heart and blood vessel damage occurs when a 10 percent fructose solution replaces plain water.

Type 1 diabetes is a condition in which the pancreas makes too little insulin, limiting the uptake of glucose by the body's cells. This type of diabetes may result from destruction of pancreatic β cells due to drug toxicity, an autoimmune disease, or an acute viral infection. Also referred to as *juvenile onset diabetes*, type 1 diabetes is less common than *type 2 diabetes*, accounting for only about 10 percent of all diabetes mellitus cases. Because type 1 individuals are unable to receive basic metabolic fuel, they may be underweight and chronically thirsty. Type 2 diabetes, in contrast, is not due to a lack of insulin, but rather occurs due to the cell's lack of response to insulin. Cells become resistant to insulin and hence, this type of diabetes is referred to as *insulin resistance* or *noninsulin-dependent diabetes*.

Metabolic syndrome is a particularly pronounced type of insulin resistance, where blood glucose, insulin,

The "Deadly Quartet"

The combination of diabetes, obesity, hypertension, and hyperlipidemia are referred to as the "deadly quartet" because this constellation leads to heart disease and impairs longevity. Elevated lipids and sugars induce changes in the vascular wall, damaging endothelial and vascular smooth muscle cell, contributing to hypertension, atherosclerosis, and an increased risk of heart attack, stroke, and clot formation.

and lipids are all elevated, leading to hypertension and an increased risk of heart disease and stroke. Metabolic syndrome may emerge as early as prepuberty and can overlap with hypothyroidism or polycystic ovarian syndrome, and the faulty metabolism also leaves those affected predisposed to gout, fatty liver, blood clots, cancer, Alzheimer's disease, and other forms of dementia.[3] There is a great deal of overlap between metabolic syndrome and cardiovascular disease, and additional herbal therapies for treating elevated lipids, hypertension, and heart disease are covered in Volume 2, chapter 2, of this set of formularies.

The central goals in treating diabetes and metabolic syndrome are to improve metabolism of sugar and fat while protecting the tissues from the ravages of elevated levels of these substances. Another way to state the goal is this: Protect the vasculature, neurons, liver, pancreas, kidneys, and, if possible, the entire rest of the body as well while aiming to reduce the oxidative stress itself by controlling elevated glucose, insulin, and lipids as best as possible. Innumerable studies describe elevated cytokines, inflammatory mediators, and enzyme systems involved with diabetic and the dysmetabolic damage they cause to organs, nerves, blood vessels, and general tissue. There are also countless studies on foods, herbs, fatty acids, antioxidants, and a long list of nutrients shown to help restore balance to all such inflammatory molecules. The data is so vast that it would be impossible for a diabetes patient to consume all of the substances that are shown to have a potentially beneficial effect. Fortunately, attempting to do so is not necessary. Instead, it is much simpler, not to mention more cost effective, to work on optimizing one's diet, aiming to consume as many of the "foodlike" medicines as feasible. Berries, green leafy vegetables, foods containing essential fatty acids such as fish and seeds, and culinary spices such as onions, garlic, turmeric, and ginger can all be powerful medicines when eaten in plentiful quantities, while eschewing problem foods such as sugar and starch-based food from breads, pastas, and rice to desserts, candy, and all sweetened products. Add in some of the best-indicated herbal teas and nutraceuticals as well as fiber and exercise. Making such changes can often greatly improve lab and physical indices of metabolic syndrome for those with prediabetes or other worrisome health trends. Those with advanced disease and those who are insulin dependent can still benefit from the same formulas and practices, perhaps being more aggressive with dosage and the number of therapies implemented.

And those with advanced metabolic impairment can still aim to slow the progression of the disease and improve longevity through lifestyle changes.

Dietary Considerations

Dietary approaches are always an essential part of therapy for diabetes and metabolic syndrome. Eliminating grains is ideal for diabetics, but not always easy for people to accomplish. Although foods are not strictly herbal medicines, I've included numerous medicinal food recipes in this chapter in addition to herbal teas and tinctures. For example, Flax Crax (page 79), Yacon Cookies (page 80), Chia Chocolate Pudding (page 80), and grain-free Peanut Butter Blondies (page 82) can serve as inspirations for creating grain-free and sugar-free bread, cracker, and dessert alternatives. Ironically, it is much better to eat vegetables, salads, and steamed greens than even these grain- and sugar-free but carbohydrate-rich foods. However, I offer the recipes due to my decades of working with patients and observing how difficult it is for people to stick to what one patient called a "strictly twigs and berries" diet. People often become diabetic because they would much rather eat toast, chips, and cookies than salads and stir-fries. It becomes a quality of life issue for them to abstain from their favorite foods month after month. Therefore, I have offered these recipes as examples of more innocent, if not downright healthy, alternatives to the problematic foods that so many of us crave. If we can help keep our patients sated with creative recipes such as these, adhering to the larger dietary scheme of a vegetable-based diet may be easier to comply with. Give patients the tools they need to succeed by including recipes such as these in patient handouts or even by offering classes to show and tell how to make these more innocent sweets. The ingredients they contain *do* have some redeeming medicinal qualities, which I detail in this chapter, especially in the "Specific Indications" section.

Intermittent fasting is also highly recommended. For example, a person can eat three healthy meals per day at roughly 10 a.m., 2 p.m., and 6 to 7 p.m., and then abstain from food for the rest of the evening. Holding off on breakfast until midmorning is important because insulin resistance results from constantly elevated glucose and insulin levels. Abstaining from all food for a 12-hour period gives the body a break from the continued presence of glucose and insulin in the blood and may help to improve insulin response. Intermittent fasting is also associated with lower levels of oxidative stress in the

Plantain for a High-Fiber Diet

A high-fiber diet is highly recommended for all diabetics. Consuming more than 25 grams of dietary fiber per day may reduce the dose of insulin required by type 1 diabetics,[4] however, 90 percent of people consume only 15 grams per day.[5] Numerous studies have shown high-fiber diets to reduce the risk of developing diabetes.[6] Guar gum and pectin, two common types of fiber, have been shown to have a positive effect on blood sugar control. Consumption of *Plantago* (plantain or psyllium) and *Linum usitatissimum* (flax) seeds slow gastric emptying, blunt the glycemic load and blood glucose curve, and thereby help to improve blood glucose and blood lipids. Fiber consumption also modifies gastrointestinal peptides involved in appetite and metabolic regulation and exerts prebiotic effects.[7] Regular consumption of legumes is recommended (see "Food as Medicine for Diabetics" sidebar on page 69 for more about legumes). High-fiber diets also support weight loss efforts.[8] Eating a diet high in fresh fruits and vegetables will naturally supply a large amount of fiber, and making a daily effort to eat oatmeal or take fiber supplements is helpful in treating diabetes and high cholesterol and supporting weight loss.[9]

Plantago ovata, plantain

body and longevity in general. Stretching the fasting period to 14 or even 16 hours may be even more effective, and yet it still allows people to eat three meals per day. Drinking herbal tea upon waking is permitted and encouraged, and going for a rigorous walk, practicing yoga, or engaging in some other physical activity in the morning before eating can help distract a person from hunger for an hour or two. Exercise is always essential for diabetics and for all of us working to stave off diabetes and heart disease and to support longevity.

The Role of Environmental Toxins

Many physicians will also address exposure to environmental toxins known to impair the liver's highly important metabolic role. Physicians must rule out obvious exposures to toxic substances through smoking, workplace exposure, household and lawn chemicals, chemical-laden drinking water, eating food stored in plastic packaging, polluted air, heavy metals, or any other stressor. When extensive and long-term exposures are suspected, or when there are toxicity symptoms in the body or compromised liver function, 3 to 6 months of a detoxification protocol may be warranted as the initial therapy. Precise and extensive protocols are outside the scope of this text but some general approaches involve fasting, sweating, chelating heavy metals where necessary, and using agents such as glutathione, *Silybum marianum*, and N-acetylcysteine (NAC) to support liver detoxification of fat-soluble toxins. NAC, a sulfur-containing compound, is supportive to liver-detoxification pathways, and it may also improve insulin resistance according to human trials using 1.8 to 3 grams per day depending on body weight.[10] Alkali minerals such as silica, zinc, calcium, and magnesium may help naturally chelate some toxic compounds, plus assist the kidney in excreting water-soluble toxins through the urinary system. *Equisetum arvense* and *Centella asiatica* are two nourishing herbs high in minerals and also known to bind heavy metals in soil. Research is lacking, but *Equisetum* and *Centella* are folkloric detoxification and tissue-restoration tonics.

There are some clinicians who specialize in environmental toxicity and a number of in-patient facilities that assist people with fasts and modified fasts. Several nutraceuticals are particularly well studied and shown to have broad anti-inflammatory and antioxidant effects on a wide array of inflammatory compounds in a wide assortment of tissue types and situations. Among these are resveratrol and alpha-lipoic acid as well as well-known antioxidant nutrients such as vitamins C and E. These are so widely applicable that one can't go wrong including these supplements in a protocol for diabetes, especially when attempting more aggressive protocols to treat uncomfortable paresthesias due to advancing neuropathy or to treat rising blood urea nitrogen (BUN) and creatinine levels to attempt to protect remaining glomeruli.

Formulas for Diabetes and Related Conditions

In the early stages of diabetes and metabolic imbalance, patients may not experience particularly uncomfortable or disturbing symptoms. Because of this, many diabetics are reluctant to take on challenging dietary and weight-reduction programs. Elevated blood glucose or glucosuria may be the only finding, and like hypertension, the condition may exist silently until some serious pathological changes have occurred. Such changes can result when excess glucose is converted to fat, sorbitol, and other substances, which accumulate in the tissues and insides of blood vessels. These serious changes include diabetic neuropathy, nephropathy leading to kidney failure and need for dialysis, retinopathy leading to blindness, and poor circulation that may cause chronic, poor healing ulcerations of the feet and may lead to gangrene and amputations.

Metabolic syndrome may be induced by long-term ingestion of a diet high in fat and refined carbohydrates,[11] and research is underway to determine how much impact adopting a healthier diet and lifestyle can have on reversing pathology. Herbs that are high in bioflavonoids such as the procyanidins and anthocyanins can protect the endothelium from sclerosis and should be used liberally in the diet and in herbal medicines. Berries are an excellent source of procyanidins and other polyphenolic compounds and can be used in the diet. Medicinal products such as blueberry solid extracts, beet powder, and pomegranate juice concentrates can be used in a variety of medicines, exemplified throughout this chapter. Numerous botanicals agents are reported to have lower blood glucose levels, including hundreds reported in ancient and traditional literature. Modern testing confirms that most of these botanicals are indeed hypoglycemic agents.[12]

Some types of diabetes involve a lack of pancreatic insulin production, but other types involve an excessive amount of insulin, and thus botanicals must be chosen appropriately for the specific condition. Herbs that protect the tissue from oxidative stress and inflammation are always appropriate and suitable for all types of diabetic cases.

Legume family herbs are featured prominently in this chapter due to their ability to improve insulin response and other mechanisms. Herbs used traditionally for diabetes include the legumes *Galega officinalis* (goat's rue) and *Trigonella foenum-graecum* (fenugreek). *Galega* is not readily available, however, and although *Trigonella* is used as a culinary herb, it is a challenge for most to enjoy as a tea. *Trigonella* capsules are available, and regular use of small amounts of the powder on a daily basis may be helpful. There is little modern human research, but studies show that fenugreek powder can reduce hyperglycemia in type 1 diabetic test animals almost as effectively as insulin[13] and equally to metformin (one of the leading drugs for insulin-resistant diabetes) in animal models of type 2 diabetes,[14] improving glucose tolerance and reducing hyperlipidemia.[15] One human clinical study reported *Trigonella* to improve fasting blood sugar levels and postprandial glucose control and to improve glycosylated hemoglobin[16] in diabetic patients who had not responded well to sulfonylurea drugs. One innovative study used fenugreek powder liberally in a bread recipe and reported the bread reduced insulin resistance for a 4-hour period immediately following consumption compared to a wheat bread control.[17]

Galega contains alkaloidal guanidine derivatives such as galegine. These compounds are shown to have hypoglycemic properties,[18] blocking gluconeogenesis and promoting glycolysis; all of these effects result in enhanced glucose utilization.[19] *Galega* may potentiate the action of insulin and inhibit absorption of glucose from the intestines. Early animal research reports the plant to regenerate pancreatic cells.[20] Some of the early research on *Galega* in the 1920s and '30s led to the development of the antidiabetic biguanide drugs, based on the galegine's mechanisms of action. These drugs however had numerous side effects, where *Galega* itself did not. Metformin (trade name Glucophage) is a biguanide agent.

Although the folkloric literature emphasized goat's rue and fenugreek for diabetes, other leguminous herbs have emerged in the modern era, have received more research, and are more readily available as quality products. Legumes in the form of beans, as well as legume family medicinal herbs, are rich in inositol compounds, carbohydrate-like molecules that occur naturally in the body and serve many important metabolic and reproductive roles. Legume family herbs can be used in teas, tinctures, and medicinal foods and are featured throughout this chapter.

The formulas in this chapter focus on those herbs able to improve insulin sensitivity as well as foodlike agents to be used on a daily basis to protect the tissues from the oxidative stress that metabolic disorders cause, and protect against organ and vascular damage caused by elevated blood glucose and lipids. The great diversity of anti-inflammatory herbs and medicines for neurons,

Food as Medicine for Diabetics

The following foods are emphasized in the recipes and herbal formulas throughout this chapter due to robust research suggesting beneficial effects on insulin response and overall blood sugar control.

Berries. Anthocyanosides in *Vaccinium* species such as blueberries inhibit free-radical damage via potent antioxidant activity and enhance perfusion via vasodilatory effect. *Vaccinium* helps protect the vascular endothelium, increases prostacyclin activity, and prevents platelet aggregation,[21] all serving to improve microcirculation.[22]

***Allium cepa* and *Allium sativum*.** Onions and garlic help lower blood glucose levels, as the sulfur-containing constituents allyl propyl disulfide and diallyl disulphide-oxide or allicin compete with insulin for binding sites within the liver,[23] resulting in increased insulin in free circulation. A minimum dosage of 25 grams of onions is needed to achieve any effect.[24]

Inulin. *Taraxacum, Arctium, Echinacea, Helianthus tuberosus* (Jerusalem artichokes), and *Inula* may help lower hyperglycemia due to the inulin they contain.[25] Inulin does not require insulin to be absorbed into cells and is credited with many beneficial effects on gut flora, which in turn has beneficial effects on metabolism.

Legumes. Beans are high in complex carbohydrates, fiber, and saponins, all of which are of value for diabetics.[26] The *Phaseolus* genus, which includes green, string, wax, mung, and other beans have been shown to have hypoglycemic activity.[27] Beans also contain inositol compounds that may reduce insulin resistance.

Mushrooms. Mushrooms are an excellent source of fiber, minerals, vitamin D precursors, and immune polysaccharides, noted to help modulate immune response and ameliorate inflammation to protect tissues from fibrotic and other pathologic changes. The edible mushroom *Grifola frondosa* has hypolipidemic and antiatherosclerotic effects and may help prevent and treat hyperlipidemia induced by atherosclerosis.[28] *Lentinus edodes*, the shiitake mushroom, improves lipid metabolism in a manner that helps ameliorate hypercholesterolemia.[29] *Ganoderma* and *Inonotus* are medicinal mushrooms, not edible mushrooms, but they are noted to have tissue-protective effects for diabetics, and other edible mushrooms may offer similar effects. *Ganoderma lucidum*, the reishi mushroom, contains β glucans credited with numerous anti-inflammatory and tissue-restorative effects, such as protecting the tissues from changes induced by hyperlipidemia.[30] *Inonotus obliquus* may ameliorate damage to the kidneys and protect against renal fibrosis and nephropathy induced by high glucose and lipid levels in the blood.[31]

Spices. Cinnamon, cloves, ginger, turmeric, black pepper, fenugreek, and many other common spices are antioxidant, anti-inflammatory, and circulatory-enhancing, and all have direct antidiabetic effects as well.

Legume Family Herbs for Insulin Resistance

Legume family herbs such as *Astragalus membranaceus*, *Medicago sativa*, and *Glycyrrhiza glabra* can be valuable ingredients in formulas for treating blood sugar, obesity, and metabolic syndrome.[32] These benefits occur when eating beans and all of the fiber, magnesium choline, lecithin, and phytosterols they contain, but benefits may also be seen when including legume family herbs in teas, or finding ways to include *Astragalus* or *Medicago* powders in the diet, such as the *Astragalus* Oatmeal for Diabetes on page 83.

Beans and legume family herbs contain isoflavone phytosterols, with numerous hormonal, metabolic, and medicinal benefits. (For more details, see "Phytosterols as Selective Estrogen Response Modifiers" on page 159.) Legumes also contain pinitol, an inositol-based compound that is converted into d-chiro-inositol in the human body. Pinitol enhances insulin response and may improve glycemic control, increases HDL levels, and promotes antioxidant enzymes in the liver.[33] Inositol compounds include d-chiro-inositol, myoinositol, and pinitol and are all well studied to improve insulin response. They act as second messengers—agents capable of enhancing signal transduction when insulin binds to the cell membrane. Purified inositol compounds are commercially available, can be supplemented in cases of metabolic syndrome, and are shown to help to lower triglycerides, cholesterol, and blood pressure. Soy lecithin may contain as much as 1,200 milligrams of d-chiro-inositol per 100 grams, but a reasonable quantity to consume each day is only 5 or 10 grams, 30 at the most, due to its thick sticky character. See the Better Butter recipe on page 84 for inspiration in using larger amounts of soy lecithin. Carob contains roughly 1,000 milligrams of inositol per 100 grams and garbanzo beans roughly 760 milligrams per 100 grams, and with some effort and creative use of recipes, reasonable dietary intake could approach these amounts. A low-effort method of increasing inositol in the diet is to aim to consume legumes as the carbohydrate portion of the diet instead of grain-based products. For cases in which even beans are too great a carbohydrate load, it is possible to use legume teas and inositol supplements to provide these compounds.

Including legume herbs in tinctures and teas for patients with diabetes, metabolic syndrome, obesity, and PCOS can be one part of a broader protocol aimed at treating insulin resistance. *Pueraria* species, *Astragalus*, *Medicago*, and *Glycyrrhiza* are among the legume family herbs tasty enough to use as teas, and even to use as powders in drinks, oatmeal, and medicinal truffle recipes. Puree cooked dried beans or canned beans of all sorts in a blender or grind them to pulp in a food mill for use in casseroles, dips, and baked goods. Many health food stores carry soy and garbanzo bean flours to use in baking; try substituting for one quarter of the regular wheat flour called for in a recipe.

nephrons, and glomeruli are covered in deeper detail in the chapters dealing specifically with renal disease, liver disease, cognitive impairment, and heart and blood vessel disease in other volumes of this set.

Tea for Diabetes and Obesity

Obesity has negative health consequences related to fat distribution, particularly the central or visceral accumulation of fat. Obesity is found in more than 90 percent of noninsulin-dependent diabetics, and weight loss achieves excellent results in improving carbohydrate tolerance in many cases. Dietary improvements and weight loss are the prime therapies in correcting this type of diabetes. Botanical agents may be of value in conjunction with dietary and exercise/weight-loss programs. This formula is not a weight loss tea, per se, but it improves the liver's ability to process fats and carbohydrates in a manner that may support exercise and calorie restriction. Licorice can improve insulin sensitivity, reduce fat deposition in tissues,[34] and suppress the accumulation of abdominal fat.[35] Licorice helps blend the flavors in this tea, and glycyrrhizic acid, a saponin in *Glycyrrhiza*, is credited with an ability to improve insulin resistance and fat metabolism in animal models of metabolic syndrome.[36] *Glycyrrhiza* has also been found to reduce inflammation and support function in animal models of diabetic kidney damage.[37] *Pueraria tuberosa* improves insulin sensitivity.[38] *Cinnamomum* (cinnamon) potentiates the action of insulin, and proanthocyanidins from cinnamon prevent the formation of advanced glycation end products. Some human clinical studies have shown cinnamon to reduce fasting insulin, glucose, total cholesterol, and LDL cholesterol and enhance insulin sensitivity in subjects with elevated blood glucose.[39] The Alterative Tea for Diabetes on page 74 has similar supportive effects as this tea.

Pueraria species
Cinnamomum verum or other species
Astragalus membranaceus
Berberis aquifolium
Glycyrrhiza glabra
Citrus paradisi, C. aurantium peel
Zingiber officinale

Combine the herbs in equal parts, or to taste. Decoct 1 teaspoon of the mixture per cup of water, simmer gently for 10 minutes, let stand in a covered pan for 15 minutes more, and strain. Drink 3 or more cups (720 ml) per day, long term.

Tincture for Diabetic Retinopathy

Damage to the tiny capillaries of the retina is common with hyperglycemia and hypertension and, when uncontrolled, can cause vision loss. Vascular tonics are appropriate for all diabetics, to help protect the blood vessels from damage. *Allium, Vaccinium* (blueberry, bilberry), and *Ginkgo* limit sorbitol formation, and each support circulation via different mechanisms. Blueberry and bilberry leaves appear useful in controlling hyperglycemia and have long been a folk remedy for diabetes. Anthocyanosides in *Vaccinium* berries stabilize capillaries, reduce oxidative damage to vascular walls, and may support tissue integrity and support circulation in the retina, helping to prevent and treat diabetic retinopathy.[40] *Ginkgo* enhances peripheral circulation and promotes blood flow in compromised blood vessels.[41]

Allium sativum
Vaccinium myrtillus
Ginkgo biloba

Combine in equal parts. Take 1 to 2 dropperfuls, 2 to 4 times per day. Another alternative is to purchase a commercial herbal encapsulated formula that contains *Vaccinium* and

Herbs That Reduce Insulin Resistance

Inositol compounds in *Pueraria*, chromium in *Stevia*, berberine in *Berberis*, and likely other compounds in these plants are shown to improve the cellular response to insulin, improving signal transduction when insulin binds.

Berberis aquifolium	*P. montana* var. *lobata*
Glycyrrhiza glabra	*P. tuberosa*
Opuntia ficus-indica	*Stevia rebaudiana*
Pueraria candollei var. *mirifica*	

Ginkgo concentrates and take it daily while also setting a goal to eat more garlic and berries. This formula would be complemented by a variety of the teas and dietary advice detailed throughout this chapter, as well as by the use of berberine, silymarin, and fiber products. Alpha-lipoic acid, resveratrol, and other all-purpose antioxidants are also appropriate and complementary.

Tincture for Chronic Infections in Diabetics

Because elevated blood sugar creates a hospitable environment for bacteria, viruses, and fungi, many diabetics are plagued by frequent infections. Skin infections and fungal diseases may occur. Frequent bladder infections may occur related to high amounts of glucose in the urine. And frequent colds and flus, sore throats, and vague malaise may occur. Controlling the elevated blood sugar as well as providing immune and nutritional support may help individuals with this complaint. Topical antifungal agents, such as tea tree oil may be useful for chronic skin fungus, and boric acid or herbal douches may help women suffering from chronic yeast vaginitis. *Andrographis paniculata* (king of bitters) is a traditional Ayurvedic herb for treating diabetes, hepatitis, and chronic infections, and modern research has demonstrated immunostimulant, hepatoprotective, antioxidant, and anti-inflammatory effects.

Andrographis paniculata
Berberis aquifolium
Echinacea purpurea or *E. angustifolia*
Astragalus membranaceus

Combine in equal parts. Take 1 to 2 dropperfuls several times a day for 2 to 6 months for those experiencing frequent infections with a slow recovery. Switch to a more specific UTI or URI formula at the onset of any new infections.

Diabetes and Related Conditions

Blueberries for Diabetic Retinopathy

Blueberries (*Vaccinium myrtillus*) are among the richest sources of anthocyanins. Many of the antioxidant and anti-inflammatory effects of berries are credited to anthocyanins, blue-purple pigmented flavonoids also found in mulberries, raspberries, concord grapes, acai berries, cherries, currants, black rice, black soybean, purple corn, and eggplant. Anthocyanins are often found bound to sugar groups in plants, referred to as glycosides. Anthocyanin glycosides are also referred to as anthocyanidins and are considered to be the parent compound and metabolized into free sugar and anthocyanins during the digestive process where the anthocyanins can travel in the blood stream and benefit many tissues. Anthocyanins enhance protein kinase activity and support blood vessel endothelial cells[42] in a manner that helps protect retinal blood vessels[43] and enhances ocular circulation.[44] Anthocyanins may also offer hypotensive effects and reduce platelet activation as additional circulatory-enhancing mechanisms.[45] In those with metabolic disorders, elevated blood lipids can become deposited in the blood vessels, impairing the elasticity of the vessel and eventually narrowing the vessels and impairing perfusion.

Vaccinium is shown to tighten the connective tissue structure of blood vessels and

limit the ability of fats and cholesterol in the blood to damage endothelium and underlying tissues.[46] Collagen fibers provide strength and structural integrity to tissues, particularly as the long, thin fibers cross-link into strong sheets, lending strength and flexibility to connective tissue. Anthocyanins and other flavonoids support cross-linking of collagen fibers and render them less easily damaged by inflammatory processes. Anthocyanins appear to have an affinity for the retinal blood vessels and the photoreceptors, supporting the integrity of tight junctions between pigmented epithelial cells of the retina, as well as the endoplasmic reticula of these cells, helping to protect against oxidative stress[47] and resisting apoptosis in situations of active inflammation.[48] Animal models of diabetes have shown *Vaccinium* to protect the vasculature of the retina via effects on vascular endothelial growth factor (VEGF) expression, an important cellular signaling protein involved in blood vessel growth and maintenance. VEGF is sometimes referred to as vascular permeability factor because it enhances vascular permeability, yet supports blood vessel integrity, preventing the degradation of occludin and claudin proteins and supporting the adherence of endothelial cells to one another. Via these mechanisms, *Vaccinium* helps resist the development of retinopathy in states of metabolic and oxidative stress.[49]

Including blueberries in the diet and the use of *Vaccinium* preparations such as leaf and dried berry teas, berry powders in smoothies and dressings, solid extracts, encapsulations, and tinctures can help protect the vasculature from oxidative stress, including the capillary bed of the fingers and toes, to help prevent peripheral nerve damage. *Vaccinium* has been particularly well studied to protect the capillary bed of the eye to help protect against diabetic retinopathy.

Vaccinium myrtillus, blueberry

Vaccinium Solid Extract for Diabetic Retinopathy

Blueberries are a folkloric remedy for failing vision, and modern research suggests the medicine is effective for diabetic retinopathy. *Vaccinium* may not be able to reverse retinopathy, but may be able to prevent and slow the progression of inflammatory damage to the blood vessels or the eyes, and the vasculature in general.

Vaccinium myrtillus solid extract

Take ¼ teaspoon, 3 times a day. The extract may also be put in smoothies, in the day's drinking water, or in salad dressings and sauces.

Formula for Nephropathy

Damage to the vascular tufts in the kidney can lead to destruction of individual glomeruli, which slowly impairs the glomerular filtration rate. Ultimately, destruction of the renal tissue can progress to kidney failure and necessitate periodic dialysis; the majority of patients in dialysis clinics are diabetics. Glycyrrhizic acid, a saponin in licorice, is noted to reduce insulin resistance and improve fat metabolism in animal models of metabolic syndrome.[50] It also reduces renal inflammation and supports renal function in animal models of diabetic glomerulonephropathy.[51] *Salvia miltiorrhiza* improves circulation to the kidneys and, along with *Silybum marianum*, may improve cellular functioning of the kidneys. *Bupleurum* may offer protective effects on the kidneys, and the saikosaponins found in *Bupleurum* may have immunomodulating properties. This formula can be prepared as a tincture or a tea.

Glycyrrhiza glabra
Salvia miltiorrhiza
Silybum marianum
Bupleurum falcatum
Hypericum perforatum
Ginkgo biloba
Pueraria montana var. *lobata*

To prepare as a tincture, combine the ingredients in equal parts, aiming to take a dropper or more, 3 times a day, long term. To prepare as a tea, start with equal parts of each dry herb and adjust to taste. Drink 3 cups of tea per day, at least 5 days per week long term. Complement this formula with encapsulated antioxidants such as alpha-lipoic acid, resveratrol, or benfotiamine, all shown to have neuroprotective effects in animal studies and possibly to be capable of improving neuropathic pain in

diabetics. *Curcuma* in capsules, foods, and a preparation of turmeric blended with nut or other milk may also offer neuroprotection and complement this formula. Seed oils such as black seed oil (*Nigella sativa*) and black sesame seed oil (*Sesamum indicum*) may also reduce inflammation in the neurons and can be consumed as a medicinal food. (There are additional formulas for treating nephropathy in Volume 1, chapter 4.)

Tincture for Diabetic Neuropathy

Elevated blood sugar and fats lead to deposits in the vasculature and damage to the endothelia, often first noticeable in our tiniest blood vessels, the terminal capillary beds. As capillaries are destroyed, the delivery of oxygen to the fingers, toes, lips, and nose is impaired, and nerve endings become inflamed and die painfully due to the lack of perfusion. *Salvia miltiorrhiza* and *Ginkgo* in this formula are aimed at enhancing circulation and protecting blood vessels, while the *Hypericum* and *Ganoderma* may reduce nerve inflammation. *Hypericum* is emphasized in folkloric traditions for nerve pain and tingling.

Salvia miltiorrhiza
Ginkgo biloba
Hypericum perforatum
Ganoderma lucidum

Combine in equal parts. Take 1 to 2 dropperful(s), twice a day to as often as hourly, reducing as symptoms improve. Complement with essential fatty acid supplements, additional teas, medicinal foods as exemplified in this chapter, and antioxidant nutrients.

Herbs with Hypoglycemic Effects

These hypoglycemic herbs work via a variety of mechanisms to improve elevated blood sugar, with clinical trials or molecular investigations showing beneficial effects for diabetics.

Allium species	*Momordica charantia*
Astragalus membranaceus	*Opuntia ficus-indica*
	Panax ginseng
Berberis aquifolium	*Pterocarpus marsupium*
Galega officinalis	*Stevia rebaudiana*
Gymnema sylvestre	*Trigonella*
Lepidium meyenii	*foenum-graecum*
Medicago sativa	

Formula for Diabetic Cutaneous Ulceration

Some diabetics develop chronic nonhealing ulcers in the lower limbs as excessive glucose inflames the tissues and the lack of oxygen and nervous regulation leads to ulceration of the dermis and subcutaneous tissues. This formula can be prepared as a tea or tincture and taken frequently throughout the day.

Calendula officinalis
Ginkgo biloba
Centella asiatica

To prepare as a tincture, combine equal parts and take 1 to 2 dropperfuls at a time, as often as hourly while helping a stasis ulcer to heal, or just 2 or 3 times a day as maintenance therapy. To prepare as a tea, combine 3 ounces of each herb and blend. Steep 1 heaping tablespoon per cup of hot water for 10 to15 minutes, and then strain. Drink 3 or more cups (720 ml) daily. Complement with topical compresses, antimicrobials, if needed, and vitamin A and zinc supplements.

Tincture for Vascular Damage

Elevated sugars, sorbitols, and lipids in the blood can be deposited on the endothelial lining of the blood vessels in diabetics, and the resulting inflammatory and oxidative state accelerates the process of atherosclerosis. As the blood vessels lose their elasticity, blood pressure may increase, and the risk of myocardial infarction, stroke, and peripheral vascular diseases increase greatly. Individuals with metabolic syndrome are particularly noted to have an increased risk of heart and vascular diseases. This tincture may help to reduce the risk of cardiovascular and peripheral vascular damage. (See also Volume 2, chapter 2, for further formulations for treating atherosclerosis.)

Ginkgo biloba
Angelica sinensis
Salvia miltiorrhiza

Combine in equal parts. Take 1 to 2 dropperful(s) at a time, twice a day for prevention, and as often as 5 or 6 times daily for more severe issues.

Alterative Tea for Diabetes

Digestive bitters are useful in diabetes for supporting digestion and helping to process fats. Berberine is increasingly being shown to reduce blood glucose and lipids, as well as blood pressure, and may promote regeneration of pancreatic β cells in those who have type 1 diabetes. *Berberis* is a source of berberine and can be included in formulas. *Pueraria* enhances insulin signal transduction,[52] thereby reducing insulin resistance in type 2 diabetes and metabolic syndrome.[53] Cinnamon improves insulin sensitivity, and *Arctium* and *Taraxacum* root contain inulin, useful to support intestinal flora, as well as to enhance glucose regulation. This tea is especially indicated when there are digestive difficulties such as constipation, gas, bloating, a coated tongue, or toxicity symptoms.

Berberis aquifolium root bark
Pueraria species
Cinnamomum verum bark
Arctium lappa
Taraxacum officinale root
Glycyrrhiza glabra

Combine the dry herbs in equal parts. Decoct 1 teaspoon of the mixture per cup of hot water. Drink 3 or more cups (720 ml) per day.

Milk Thistle Capsules for Metabolic Syndrome and Hyperlipidemia

Silymarin is a mixture of flavonolignans in milk thistle (*Silybum marianum*) seeds, long studied for hepatoprotective effects. Via numerous beneficial effects on the liver, *Silybum* can help lower cholesterol.[62] Unlike statin pharmaceuticals associated with muscle and liver toxicity, silymarin exerts significant antioxidant and membrane-stabilizing effects on hepatocytes,[63] reducing cholesterol deposition in the liver[64] and supporting

---

β (Beta) Cell Regenerators?

It is not widely thought that the pancreas can recover when damaged by drug reactions, infection, or other assault, or that humans can regenerate new pancreatic β cells; however, β cell regeneration has been reported a number of times in animal research. It may be possible to use herbs such as these in new onset cases of diabetes to rescue the remaining pancreatic β cells for type 1 diabetics. Epigallocatechin (sometimes referred to as EGC) supplements may also be helpful. EGC is found most notably in *Camellia sinensis*.

Camellia sinensis　　　*Gymnema sylvestre*
Galega officinalis　　　*Pterocarpus marsupium*

Gymnema sylvestre: The Sugar Destroyer

Animal studies and a few human studies have shown *Gymnema* to have antiobesity and antidiabetic properties and to significantly reduce body weight and improve laboratory markers of glycemic control. Dosages of 120 milligrams/kilogram to 200 milligrams/kilogram of various *Gymnema* extracts improve metabolic markers of diabetes in approximately 1 month's time, including positive effects on body mass index, blood pressure, heart rate, hemodynamic parameters, serum leptin, insulin, glucose,[54] blood lipids, and antioxidant enzyme status.[55] Human clinical studies have reported *Gymnema* to improve blood parameters in type 1 diabetes at a dose of 500 milligrams/day.[56] Several other studies have shown marked improvements in blood glucose and glycosylated hemoglobin in medicated diabetics, such that participants were able to reduce their medication dosage, and roughly 25 percent of subjects were able to discontinue pharmaceuticals altogether.[57] Beta cell regenerative effects are reported from *Gymnema*'s triterpene saponins,[58] known as gymnemic acids;[59] *Gymnema* extracts may promote the production of insulin in weak or damaged β cells[60] and possibly increase the number of pancreatic β cells and islets of Langerhans,[61] according to histological assessments in diabetic animals. *Gymnema* is not used as a tea due to the solid, rocklike nature of the plant resin, but is available as a tincture to use in obesity and diabetes formulas and as capsules to use in various protocols for treating diabetes.

Gymnema sylvestre, gurmar

---

healthy blood lipids.[65] Parameters indicative of the early stage of atherosclerosis may be lowered by taking milk thistle.[66] *Silybum* can help reduce hyperlipidemia and metabolic syndrome and is readily available as a single herb or in formulas for comprehensive liver support, such those combining *Berberis*, *Taraxacum*, *Arctium*, *Curcuma*, and other herbs. *Cynara scolymus* (artichoke) is also in the thistle family and has had similar research, showing similar benefits. Choline and ox bile is combined with *Silybum* in "lipotropic" formulas when the intended use is for gallstones or biliary insufficiency but may also be an excellent choice for patients with metabolic syndrome who are developing fatty liver or atherosclerosis.

Silybum marianum capsules

Take 1 or 2 pills at a time, 3 times a day. A combination lipotropic formula in capsule form can be taken at the same dosage.

Berberine Capsules for Metabolic Syndrome

Berberine is an isoquinoline alkaloid in *Berberis* and other plants that is being robustly researched for supportive effects on lipid and carbohydrate metabolism. Berberine may promote thermogenesis via direct metabolic effects on brown adipose cells[67] and by promoting glucose metabolism[68] and preventing the accumulation and storage of fats.[69] Berberine may improve glucose metabolism via increased insulin sensitivity in metabolic syndrome patients[70] and has been shown to improve body mass index and blood lipid levels in women with PCOS,[71] as well as in obese patients.[72]

Berberine capsules, 500 milligrams each

Take 1 to 3 capsules, 2 or 3 times daily. The dose can often be decreased as the broader protocol of diet, exercise, weight loss, and other supplements improve metabolic functioning.

Berberine and Silymarin Duo for Metabolic Syndrome

Berberine and silymarin in combination appear to impact metabolic syndrome more powerfully than either substance used in isolation, showing greater impact on triglycerides, low-density lipoprotein, cholesterol, fasting blood glucose, and glycosylated hemoglobin[73]—and without the muscle or liver toxicity seen with statin drugs.[74] Berberine has poor oral bioavailability[75] due to P-glycoprotein efflux in the enterocytes, while silymarin effects P-glycoproteins in a manner that improves the absorption of berberine, as well as offers its own metabolic support.[76] This combination may be a fundamental and all-purpose therapy for metabolic disorders to complement with other teas, tinctures, and dietary guidelines offered throughout this chapter.

Berberine and *Silybum marianum* capsules

A variety of combinations are commercially available. May be taken 3 capsules at a time, 3 times a day, reducing as blood lipids and glucose levels improve.

Formula for Type 1 Diabetes

Momordica (bitter melon) contains a mixture of steroidal glycosides known as charantin and has a hypoglycemic effect. *Momordica* also has an insulinomimetic effect.[77] *Gymnema sylvestre* is a traditional Ayurvedic herb for treating type 1 diabetes that may promote insulin production[78] and, possibly, β cell numbers and function.[79] Its name in Hindi translates as "sugar destroyer," because it has been used for thousands of years to treat diabetes and abolish sugar cravings.[80] Triterpenes are credited with some of the hypoglycemic and broad anti-inflammatory effects.[81] *Trigonella foenum-graecum* (fenugreek) seeds contain the alkaloid trigonelline, which is credited with a hypoglycemic effect, along with nicotinic acid, which can inhibit insulinase, the enzyme that breaks down insulin.[82] Human trials report *Trigonella* to improve fasting blood sugar levels, glucose tolerance, and postprandial glucose[83] and to reduce 24-hour urinary glucose excretion.[84]

Momordica charantia
Gymnema sylvestre
Trigonella foenum-graecum

Commercial capsules containing these ingredients may be purchased or the capsules can be made for one's self. Take 2 or 3 capsules at a time, 2 or 3 times a day. The ingredients can also be prepared as a tincture by combining in equal parts and taking a dropperful or 2 at a time, 3 times daily.

Simple Green Tea for Protecting β Cells

Epicatechin, a potent flavonoid found in *Pterocarpus marsupium* and *Camellia sinensis* may prevent toxin-induced damage of pancreatic β cells[85] and possibly promote regeneration of β cells.[86]

Camellia sinensis

Green tea can be consumed liberally for its many preventive, antioxidant effects. It can be combined with any of the other herbal therapies presented in this chapter.

Medicinal Foods for Diabetics

This chapter features more medicinal food recipes than perhaps any other in this set of formularies, because for those who have diabetes, every single bite of food should be medicinal, and every bite of a harmful food contributes to the disease and accelerates the development of end stage sequelae. While I often encourage people to steer away from traditional breakfast toast, bagels, muffins, and cereals, I do offer an oatmeal recipe in this section, with suggestions on making it as healthy as possible for diabetics. Otherwise, I encourage sautéed vegetables for breakfast, or a whole-food smoothie, or an atypical breakfast such as leftover soup. And throughout the day, I urge eating the more vegetables the better, especially green leafy vegetables and low carbohydrate options when aiming to improve insulin resistance and improve metabolic function.

Blueberry Dressing and Marinade

Blueberries and other berries offer protective effects, preventing damage to the blood vessels and the eyes. Both fresh or solid extracts can be used. *Nigella* (black cumin) seed oils have a strong thymelike flavor, limiting the quantity that is pleasant to consume at any one time, but *Nigella* is a powerful antioxidant with hypoglycemic, hepatoprotective, and neuroprotective effects. Because *Nigella* oil may reduce insulin resistance and help the vasculature resist formation of atheromatous plaques, including a teaspoon in foods such as this recipe can slow the long-term consequences of metabolic stress. *Nigella* oil may also exert hypotensive effects via promotion of nitric oxide and by acting as an angiotensin converting enzyme (ACE) inhibitor. The avocado in this recipe gives the dressing a creamy quality without the use of dairy. It can be omitted if a vinaigrette is desired, or the entire avocado may be used if a thick consistency is preferred. Luo han guo syrup is a noncaloric sweetener (see "Monk Fruit as a Noncaloric Sweetener" on page 85 for more information). This dressing may be sweet enough from the *Vaccinium* or fresh berries alone, so the luo han guo or other syrup is optional.

1 tablespoon balsamic vinegar

1 teaspoon *Vaccinium myrtillus* solid extract or
⅓ cup (50 g) fresh berries

1 teaspoon black cumin (*Nigella*) seed oil

1 small onion, coarsely chopped

2 to 5 cloves raw garlic

½ avocado, freshly peeled

½ to 1 teaspoon crudely chopped fresh ginger

½ to 1 teaspoon crudely chopped fresh turmeric

½ teaspoon luo han guo syrup (optional)

Place all the ingredients in a blender and liquefy. Sample and adjust the taste as desired. Use on raw salads or with steamed greens and vegetables. This recipe can be amended in numerous ways. For example, try omitting the avocado to use the dressing as a fish or poultry marinade. Add additional culinary spices such as ginger, salt, rosemary, or cayenne. Sugar-free hibiscus syrup (recipe on page 78) can be used in place of luo han guo syrup.

Metabolic Medicine Smoothie

An all-around goal of treating type 2 diabetes is improving insulin sensitivity and response at the insulin receptor. Magnesium, chromium, vitamin D, and inositol compounds are among the natural agents shown to improve insulin signal transduction. Many people are found to be low in vitamin D, and the degree of deficiency may correlate with the degree of insulin resistance.[87] Vitamin D is readily available in liquid form; one drop provides anywhere from 400 to 1,000 international units (IUs). Magnesium glycinate and inositol are both available in powders that dissolve readily in water-based or nut milk–based beverages. Medium-chain triglycerides (MCT) are a type of fat

Diabetes Diet Imperatives

Base the diet around whole foods and emphasize vegetables: vegetables for breakfast, vegetables for lunch, vegetables for dinner. And make a firm rule—NO sugar!—and stick with it.

Avoid

Table sugar, honey, syrup	Soda
Jam	Processed carbohydrates
Candy	Flour, bread, pasta, bagels
Pastry	

Enjoy

Lean meats, such as fish and poultry, in moderation	garlic, onions, and green leafy vegetables
8 to 10 vegetables per day, including baked squash, beans, beets,	Salads
	Stir-fries and soups
	Berries
	Mushrooms

that promotes weight loss by increasing fat metabolism, stimulating fat cells to take up glucose and burn it for energy.[88] MCT may also support beneficial intestinal flora,[89] promote the liver to burn fats and yield ketones, and offer neuroprotection.[90]

2 cups (480 ml) nut milk, chilled

2 to 3 teaspoons MCT oil

1 teaspoon carob powder

1 teaspoon cocoa powder

1 to 2 teaspoons lecithin granules or oil

1 teaspoon magnesium glycinate powder

1 teaspoon inositol powder

1 drop vitamin D (or more, to provide 1,000 IUs)

Place chilled nut milk, such as almond or hemp milk, in a blender, add the rest of the ingredients, and blend. Fiber such as psyllium powder or apple pectin can be added; 1 to 2 teaspoons of both as tolerated, as well as any other ingredients specific to the individual. Drink promptly and repeat each day.

Sugar-Free Hibiscus Syrup

Hibiscus sabdariffa contains more chromium per gram than any other plant. Its tart flavor and appealing pink-red color make a lovely syrup to use in making drinks or an ingredient in marinades. I find it delicious with salmon or in dressings to drizzle over a salad of apples, walnuts, blue cheese, chicken, and arugula. Mmmmm, right? Luo han guo, also known as monk fruit, is a noncaloric, noncarbohydrate sweetener available from herb shops or online. It is more expensive than *Stevia*, one of the only other noncaloric sweeteners, but it is so powerfully sweet that just a small amount may be sufficient, making it worth the investment. Luo han guo is also a good sweet option for people who do not care for the aftertaste of *Stevia*, and it is safe for use by diabetics. (See also "Monk Fruit as a Noncaloric Sweetener" on page 85.) The pectin in this recipe gives the syrup the thickness that sugar and natural pectin in fruits usually provide when making traditional syrups. Many brands and forms of natural pectin can be found in any hardware or grocery store that carries canning supplies.

4 cups (960 ml) water
1 ounce (30 g) *Hibiscus sabdariffa*, dried flowers
½ ounce (15 g) *Citrus aurantium*, granulated peel
¼ ounce (8 g) luo han guo powder
1 ounce (30 g) pectin gel or powder

Simmer the hibiscus flowers, orange peel, and luo han guo powder in water in a medium saucepan, covered, for 10 minutes. Remove from the heat and let stand 15 minutes more, strain, and return the liquid to the saucepan. Add the pectin gel or powder and bring to a hard boil, uncovered, for just 30 seconds, then immediately turn the heat down to the lowest possible setting. Stir for several minutes, removing any scum that forms on the top. Transfer to a canning jar or glass bottle and store in the refrigerator. Combine with oil and spices to create dressings and marinades. Or use a tablespoon or two combined with sparkling water to create beverages and mocktails.

Tea for Vascular Protection

Salvia miltiorrhiza (red sage) and *Paeonia* species are a common duo for treating a variety of vascular diseases including nephropathy, improving blood viscosity and lipid metabolism, and protecting endothelial cells from a variety of stressors.[91] *Astragalus* and *Angelica* are another common duo for vascular inflammation. Commercial TCM products exist in the form of pills or granules, and because these herbs are all mild-tasting, they can also be prepared as a tea.

Salvia miltiorrhiza	4 ounces (120 g)
Paeonia × *suffruticosa*	4 ounces (120 g)
Astragalus membranaceus	4 ounces (120 g)
Angelica sinensis	4 ounces (120 g)

This recipe will yield a pound of tea to store in a large glass jar in a dark cupboard. Gently simmer 1 heaping tablespoon in 4 cups (960 ml) of water for 10 minutes, remove from the heat, let stand 10 minutes more, and strain. Drink the entire amount each day, or at least 3 or 4 days a week, long term. For additional formulas containing vascular protectants, refer to Volume 2, chapter 2.

Magnesium to Improve Glycemic Control

Magnesium is the fourth most abundant mineral in the human body, half of which is found in the bones and the other half distributed about the cells and tissues. Magnesium is vital to health and is required for hundreds of biochemical reactions in the body. Magnesium is essential for carbohydrate metabolism, blood sugar regulation, and blood pressure regulation. Low magnesium levels are associated with heart disease, diabetes, high blood pressure, insulin resistance, metabolic syndrome, and PCOS, and taking magnesium supplements may help all of these conditions.

One study reported between 25 and 38 percent of people with type 2 diabetes also have low magnesium levels and that magnesium deficiency correlated with the worst degree of nerve damage and heart disease in diabetics. Magnesium supplementation is reported to correlate with improved glucose control and insulin resistance in type 2 diabetics. Dark green leafy vegetables and legumes are among the best dietary sources of magnesium.

Stevia for Sweetness without Calories

Stevia rebaudiana is a remarkable plant in that it is extremely sweet tasting, yet has no calories; its glycemic index is zero and studies show that it improves, rather than impairs, insulin resistance.[92] *Stevia* is high in chromium, a trace mineral of great importance to insulin reception and response, and animal studies show that *Stevia* may both improve insulin output from the pancreas (benefiting type 1 diabetes) and reduce insulin resistance in type 2 diabetes.[93] It may also offer protective effects on the insulin-producing β cells in the pancreas.[94] *Stevia* has not had extensive research but the steviosides are credited with hypotensive,[95] hypoglycemic,[96] and hypolipidemic[97] effects, as well as are shown to improve insulin production and sensitivity in cells.[98] See also the *Stevia* entry on page 98.

Stevia rebaudiana, stevia

Refreshing Flor de Jamaica Iced Tea

Hibiscus is so high in chromium as to rival a nutritional supplement. The calyces of *Hibiscus sabdariffa* are tart and delicious, especially in hot weather when lemonade-like beverages are appealing. Flor de Jamaica is a Central American standard *refresco*, meaning a cooling and refreshing beverage. Flor de Jamaica has also been recommended as a dieter's tea to support weight loss. This twist on the folkloric recipe keeps the traditional spices and substitutes xylitol for sugar. Sugar-Free Hibiscus Syrup (page 78) or stevia are alternative sweeteners.

2 quarts (2 L) water
1 cup (100 g) *Hibiscus sabdariffa*
1 to 2 tablespoons xylitol (optional)
2 teaspoons freshly grated ginger root
2 cinnamon sticks
3 or 4 hard, dried allspice berries
Fresh-squeezed juice of 1 lime, plus zest
Orange slices, for garnish

Heat the water in a saucepan and add all the ingredients, stirring occasionally to dissolve the xylitol. Bring the brew to a gentle simmer for 10 minutes, remove from the heat and let stand covered for 10 to 15 minutes more. Strain into a pitcher and refrigerate. Serve over ice with a slice of orange. Drink as much as desired. Makes about 8 cups (2 L).

Flax Crax

One of the most difficult aspects of adhering to a diet plan for improving diabetes is abstaining from breads, bagels, pastas, crackers, and other grain-based foods that are so prominent in the standard American diet. These crackers are a wonderful grain-free option and can be topped with vegetables or hummus as a sandwich substitute or spread with almond butter and fresh fruit slices as a satisfying light breakfast or a dessert. Try experimenting with variations on this recipe. I have added dark cacao powder to create dessert crackers and seaweed granules and smoked paprika to create an intensely flavorful cracker. I've also created a sweet cookie version of these crackers (see Yacon Cookies, page 80) These crackers can also be served with soup or can be broken into small pieces as a salad topping instead of croutons.

1 cup (170 g) organic flax seeds
⅓ cup (50 g) organic chia seeds
1 cup (240 ml) raw onions, puréed
⅓ cup (50 g) organic hemp seed meal
¼ cup (40 g) organic sesame seeds
¼ cup (35 g) dried garlic flakes
1 teaspoon caraway seeds
¾ teaspoon fine sea salt
¾ teaspoon organic black pepper

Preheat the oven to 200°F. Place parchment paper on a large cookie sheet. In a medium bowl, blend the flax and

Chromium to Improve Insulin Resistance

Chromium is a vital trace mineral for blood sugar disorders because it enhances the action of insulin. A little-understood compound named glucose tolerance factor (GTF) contains chromium and plays a role in glucose metabolism. Chromodulin is a similar molecule found in the human liver and kidneys that is known to bind chromium.[99] Like GTF, the activities of chromodulin in the body are not entirely understood but both compounds are involved in insulin binding and cellular responses.[100] A typical dosage range for chromium is from 200 to 1,000 micrograms (mcg) daily. Although the specific amount of chromium in any given herb may vary from sample to sample, the following herbs are the best tea and medicinal food sources of chromium, having the highest content per gram, according to the USDA Phytochemical and Ethnobotanical Databases created by the late Dr. James Duke.[101]

HERB	QUANTITY OF CHROMIUM PER GRAM DRIED PLANT MATERIAL
Hibiscus sabdariffa flowers and calyx	54 mcg
Taraxacum officinale leaves	50 mcg
Avena sativa bran	39 mcg
Stevia rebaudiana leaves	39 mcg
Cymbopogon citratus leaves	37 mcg

chia seeds and blend with the freshly puréed raw onion and let stand for 20 to 30 minutes. The seeds will become a sticky mass. Blend in the hemp seed meal, sesame seeds, garlic flakes, caraway seeds, salt, and pepper. Spread evenly over the parchment paper. Moistening a wooden spoon to spread the material is helpful. Place in the oven and leave to dehydrate for 90 minutes. Remove the pan from the oven and flip the crackers, return to the oven, and dehydrate for 30 to 60 minutes more (30 minutes may yield a chewier cracker, and 60 minutes a crispier cracker). Allow to cool and break into pieces to store in an airtight container. The recipe will yield roughly 20 crackers.

Yacon Cookies

This recipe uses beet (*Beta vulgaris*) and yacon (*Smallanthus sonchifolius*) powders as nutritious sweeteners, paired with high-fiber seeds. I learned of yacon when I was in the Andes, where the large juicy roots are used to make a syrup touted as a weight loss aid, despite its sweet flavor. The dry powder of yacon roots is available from dry herb vendors and can be used in many culinary ways.

1 cup (170 g) organic flax seeds
⅓ cup (50 g) organic chia seeds
1 cup (120 g) organic raw apple pieces, puréed
¼ cup (50 g) organic sesame seeds
1 tablespoon hemp seed meal
1 teaspoon anise seeds
¼ cup (30 g) beet powder
¼ cup (30 g) yacon powder
Carob syrup (optional)

Preheat the oven to 200°F. Place parchment paper on a large cookie sheet. In a medium bowl, blend the flax and chia seeds, mix in the freshly puréed raw apple, and let stand for 20 to 30 minutes. The seeds will become a sticky mass. Blend in the sesame seeds, hemp seed meal, and anise seeds. Add a tablespoon or two of water if the "dough" seems too difficult to work. Spread evenly over the parchment paper. Moistening a wooden spoon to spread the material is helpful. Place in the oven and leave to dehydrate for 90 minutes. Remove the pan from the oven and flip the cookies, return to the oven, and dehydrate for 30 to 60 minutes more (30 minutes may yield a chewier cookie, and 60 minutes a crunchier cookie). Score the cookies immediately upon removing from the oven to help break the flat mass into regular shapes. If desired, drip several drops of carob syrup onto each cookie before eating. The recipe will yield roughly 20 cookies.

Chia Chocolate Pudding

Chia (*Salvia hispanica*) seeds are high in alpha-linolenic acid and fiber, and animal studies suggest consumption to improve lipids and insulin response and reduce visceral adipose stores.[102] Similar studies show chia seed consumption to protect against the development of dyslipidemia and liver steatosis in animals fed a sucrose-rich diet;[103] to reverse impaired glycogen processing and insulin resistance;[104] to reduce the oxidation of fatty acids and protect the heart from collagen deposition;[105] and to improve glucose utilization by cardiac muscle fibers.[106] The basic success of this pudding relies on the tendency of chia seeds to absorb liquid and create

Yacon for Intestinal Health and Glycemic Control

Research on the microbiome and the impact of intestinal microbial species on a variety of disease processes is exploding. The importance of intestinal flora to diabetes is one such research arena, suggesting that types and ratios of bacterial populations in the gut can have a significant impact on body-wide metabolism. Probiotics (beneficial bacterial species) can be taken as a nutritional supplement to help populate the intestinal ecosystem with desirable organisms. Another therapeutic approach is to supplement or consume foods and molecules that feed the desirable gut bacteria; such supplements are referred to as *prebiotics*. Fructose polymers such as inulin occur naturally in a number of foods and are one such prebiotic. Yacon (*Smallanthus sonchifolius*) may offer modest improvements in serum lipids and glucose by acting as a prebiotic and improving the intestinal microbiome, among other mechanisms.

In the Andes, *Smallanthus* tubers are chopped and boiled down to produce a sweet delicious syrup used as a sweetener and medicinal food. The roots are high in fructooligosaccharides (FOS). Other plants high in these simple starches, such as chicory root, artichoke leaves, and Jerusalem artichokes, have long been claimed to improve diabetes. However, more than just a healthier carbohydrate option, FOS may improve glucose metabolism via beneficial effects on bacterial populations of the gut. Yacon consumption may also improve hepatic insulin sensitivity in cases of insulin resistance and improve blood glucose by this mechanism as well.[107] Animal studies show that eating yacon normalizes hyperglycemia by lowering hepatic glucose production and increasing whole-body insulin sensitivity.

Fructooligosaccharides escape enzymatic digestion in the upper gastrointestinal tract and reach the colon intact, where they are fermented by intestinal microbial flora and also promote the proliferation of *Bifidobacteria* species. FOS are fermented by intestinal bacteria into short-chain fatty acids (SCFAs), which also promote beneficial intestinal bacteria and inhibit pathogenic strains of *Escherichia coli* and *Clostridium* species. Animal studies show the consumption of yacon can change intestinal microbial communities within a week's time, doubling the presence of SCFAs.[108] The fermentation of FOS into SCFAs is a highly desirable effect because SCFAs act as substrates or signaling molecules in the regulation of the immune response, glucose homeostasis, and lipid metabolism.[109] As a result, glycemic levels, body weight, and colon cancer risk can be reduced. Regular consumption of yacon may improve the intestinal ecosystem and favor healthy populations of *Lactobacillus acidophilus*, *Bifidobacterium pseudolongum*, *Bifidobacterium animalis*, and *Barnesiella* species.

Inulin and FOS can be gas forming in the intestines as the starches readily ferment, but small doses are generally tolerated well. Those with small intestinal bacterial overgrowth (SIBO) may not tolerate inulin well, and they may need to treat their condition and improve the health of the intestines prior to implementing an FOS-containing prebiotic.

Smallanthus sonchifolius, yacon

a mucilaginous mass. The use of a nut or coconut milk as the liquid makes the "pudding" creamy, and approximately 3 tablespoons of seeds to a cup of the chosen liquid will usually be successful. Any desired fruits or flavoring can be used to provide sweetness and variety to the otherwise bland-tasting chia blend. Xylitol, stevia, monk fruit, or fresh fruit purees are good choices to help diabetics avoid sugar. This recipe uses stevia, but use whatever sweetener suits any individual's preferred tastes. Hundreds of similar recipes are available online, offering variations such as adding cinnamon, nutmeg, or spices. Avoid versions that call for chopped dates or other dried fruit because they would add a sugar load. And, of course, avoid recipes using honey, sugar, and other caloric sweeteners to best serve diabetic patients.

2 cups (480 ml) coconut milk or nut milk or a combination
1 tablespoon dark cocoa powder
1 teaspoon vanilla extract
10 drops liquid stevia extract
Fresh fruit (such as banana, peach, nectarine, grapes, as available), plus more for serving (optional)
Raw coconut flakes or toasted nuts, as desired, plus more for serving (optional)
½ cup (75 g) chia seeds

Place the milk in a blender and blend in the cocoa powder, vanilla, and stevia. If other amendments are to be added, such as fresh fruit or coconut flakes, add them to the milk at this time, rather than attempting to stir them into the prepared pudding. Transfer the liquid to a large canning jar, add the chia seeds, and shake well. Place the jar in the refrigerator and allow the pudding to "set" over 2 or 3 hours. To eat, transfer to serving dishes and

Herbs That Lower Fats and Cholesterol

Many of these plants can be used as culinary spices and condiments in the diet. Others can be used in herbal formulas and taken in capsule form.

Allium cepa	*Cynara scolymus*
Allium sativum	*Hibiscus rosa-sinensis*
Berberis aquifolium	*Lepidium meyenii*
Capsicum annuum	*Opuntia ficus-indica*
(also known as *C.*	*Silybum marianum*
frutescens)	*Trigonella*
Cinnamomum species	*foenum-graecum*
Commiphora mukul	*Zingiber officinale*

top with fresh fruit. If desired, add other toppings: coconut or toasted nuts, a drizzle of carob or yacon syrup, or a Yacon Cookie (see recipe on page 80). Makes roughly 4 servings of ½ cup each.

High-Fiber Truffles

High-fiber intake helps slow the glucose curve following meals, helps improve blood lipids, and supports weight loss. Eating a large number of vegetables each day provides a good amount of fiber, but until that can be accomplished, some patients may boost fiber intake with a high-fiber snack such as these truffles. For those who overeat or crave chips, candy, and snacks between meals, snacking on these truffles may reduce the intake of harmful foods by helping to fill the stomach and prevent "grazing." Some therapies might also employ modified fasts in which these truffles serve as a meal replacement, eaten in the morning along with one of the teas featured in this chapter, helping people to reduce caloric intake. Carob (*Ceratonia siliqua*) syrup serves as a sweetener in this recipe, having less harmful effects on blood sugar than other sweeteners due to the fact that it is a legume and a natural source of pinnitol and D-chiro inositol.[110] The best appetite-suppressant effect will be had by eating two truffles with a cup of hot black coffee.

¼ cup (30 g) apple pectin
¼ cup (50 g) cashew butter
2 tablespoons carob syrup
2 teaspoons cocoa powder
½ teaspoon cinnamon
Sesame seeds or coconut flakes, for coating

Combine apple pectin, cashew butter, carob syrup, cocoa powder, and cinnamon in a small bowl and blend together with a fork, adding more pectin or more carob syrup to achieve a cookie dough–like consistency. Roll into balls roughly 1 inch in diameter and roll in sesame seeds or coconut flakes until coated. Place the truffles in small candy papers, or simply store them in an airtight glass container. Try eating 2 or 3 truffles per day to boost fiber intake. The truffles can be amended in many ways, such as substituting almond butter instead of cashew butter, chopped dried fruit simmered into a gel instead of syrup, and other herbs instead of cocoa powder. Get creative! Makes 10 to 12 truffles.

Peanut Butter Blondies

I adore this recipe, legume fan that I am. These bars are highly satisfying when you are craving a treat, and they

Opuntia for Insulin Resistance

Opuntia fruits may be another tasty tool for improving insulin resistance because the fruits optimize basic metabolism[111] and normalize blood glucose and cholesterol in animal models of diabetes.[112] *Opuntia* juice lowers elevated blood sugar[113] and improves glucose control.[114] One human clinical trial on patients with metabolic syndrome found *Opuntia* to lower total cholesterol, HDL and LDL, and triglycerides in as little as 14 days,[115] and other studies confirm *Opuntia* to reduce blood fats[116] via increased glucose uptake and utilization by the liver and muscles.[117] Prickly pear seeds promote weight loss in animal models of diabetes and metabolic syndrome, a finding associated with a reduction in serum thyroxine and glucose and an increase in HDL and liver and muscle glycogen content.[118] Prickly pear fruits are high in betalains,[119] credited with many anti-inflammatory mechanisms and antioxidant effects.[120] Betalains exert protective effects on the vasculature,[121] have an ability to reduce cytokine-driven inflammation,[122] and are able to protect the liver from a variety of environment toxins that can impair insulin response. Because toxins, inflammation, and oxidative stress contribute to metabolic syndrome and endocrine disruption, *Opuntia* may be valuable both as a food and as a medicine in treating diabetes and metabolic dysfunction.[123]

Opuntia ficus-indica, prickly pear

contain *no* flour or grain. I have made them using maple syrup as a sweetener, but they are just as good with xylitol and monk fruit to reduce the carbs even further. My only complaint is that the batter is so thick that it becomes tedious to remove all the material from the blender. But the effort is worth it and a small price to pay, as garbanzo beans are one of the highest sources of d-chiro-inositol and are also high in fiber. I have prepared similar recipes that use black beans to make brownies, blending in unsweetened cocoa powder.

1 can (16 ounces) garbanzo beans, drained and rinsed
½ cup (180 g) peanut butter
⅓ cup (75 g) xylitol
½ teaspoon salt
¼ teaspoon baking powder
¼ teaspoon baking soda
¼ teaspoon coarse sea salt, for sprinkling
1 tablespoon cocoa nibs, for sprinkling

Preheat the oven to 350°F and coat an 8 × 8-inch glass baking dish with a thin layer of coconut oil. Place all ingredients except the sea salt and cocoa nibs in a blender or food processor and blend until smooth. Spread the batter evenly in the prepared pan and sprinkle the coarse salt and cocoa nibs over the top. Bake for 20 to 25 minutes or until a knife inserted into the center comes out clean. Cool on a wire rack. The blondies can be cut into 16 small squares; each serving contains very little xylitol. For a nonvegan version that has a more cakelike texture, blend in an egg at the start of the food processing.

Astragalus Oatmeal for Diabetes

Oats are a good source of fiber, and you can offer this recipe to patients as a good way to start the day—by switching from other breakfast foods to oatmeal with fresh fruit. Oat bran is lower in carbohydrates and higher in chromium than rolled oats. *Astragalus* is a legume family herb with many immunomodulating and blood sugar–balancing affects. (See "*Astragalus* Foods and Medicines for Glucose Regulation" on page 84.) *Pouteria lucuma* is a South American fruit that is becoming available commercially as a natural sweetener sold under the name of lucuma. *Lepidium meyenii* (maca powder) is another alternative that could be added to this recipe instead of *Astragalus*. Cooking the oats with some added raisins or topping the oatmeal with berries at the time of serving satisfactorily sweetens the porridge so no sugar or other harmful sweeteners are needed. This recipe is for a single serving and can be doubled or tripled when aiming to serve the whole family.

1 tablespoon *Astragalus membranaceus* powder
1 tablespoon lucuma powder (optional)
⅓ cup (30 g) oat bran
1 cup (240 ml) water
½ cup (120 ml) nut milk
½ cup (75 g) fresh berries

Blend the *Astragalus* and lucuma powders into the oat bran in the bottom of a small saucepan. Add the water, blend again, and bring to a gentle simmer until the oats are soft, roughly 5 minutes. Remove from the heat and stir in the nut milk and fresh berries. Transfer to a serving bowl. Additional toppings could include nuts, coconut flakes, or additional fruit.

Better Butter

Consuming quality seed oils leads to beneficial effects on cholesterol levels and anti-inflammatory effects because seed oils contain essential fatty acids. However, too much saturated fat in the diet can be problematic for those with diabetes, because both diabetes and metabolic syndrome can evolve into fatty liver, or into severe oxidative stress in which fats become oxidized and contribute to heart disease, atherosclerosis, and other issues. (While research on "good" fats and "bad" fats is controversial, with the ketogenic-diet folks on one side and low-fat advocates on the other, all authorities agree that hydrogenated fats such as margarine and vegetable shortening are harmful and must be avoided.) Because it can be difficult to obtain the beneficial omega-3 and omega-6 fatty acids provided by fish, nuts, and seeds on a daily basis, here is a recipe that combines healthy lecithin, medium-chain triglycerides, and flaxseed oil to create a fatty-acid-rich "butter." The recipe calls for MCT

Astragalus Foods and Medicines for Glucose Regulation

Astragalus is a legume-family herb that has been widely used in herbal medicine for centuries. *Astragalus* roots contain saponins that may help protect tissues and reduce inflammation caused by high blood sugar.[124] Furthermore, *Astragalus* may improve insulin response in cells.[125] *Astragalus* may also improve high blood sugar by helping the liver and the muscles take up glucose and use it for fuel or convert it to less harmful storage forms.[126] The polysaccharides in *Astragalus membranaceus* are credited with an ability to enhance insulin-signaling pathways in muscle cells,[127] and one small human clinical trial reported reduced insulin resistance in diabetic subjects.[128] *Astragalus* is available in herb shops as shredded roots and powders, as well as tinctures, pills, and formulas. The plant appears very safe to ingest and has numerous other benefits including immunomodulating benefits for chronic infections and allergic conditions. Many traditional Chinese medicines employ *Astragalus* in medicinal soups because it has a bland, starchy flavor. *Astragalus* roots can be boiled with bones, seaweeds, medicinal mushrooms, and spices to create nutritional broths. *Astragalus* powders can be stirred into oatmeal, nut butter, yogurt, applesauce, and smoothies.

Astragalus membranaceus, milk vetch

oil—MCT is short for medium-chain triglycerides—which is a fat found in butter and coconut oil shown to support weight loss, prevent fatty liver, and protect the brain and neurons. MCT oil is readily available as a dietary supplement.

1 cup (100 g) lecithin granules
½ cup (120 ml) flax oil
½ cup (120 ml) MCT oil

Place the lecithin in the bottom of a medium bowl and blend in the flax and MCT oils, a bit at a time, by whisking vigorously with a fork. Once all the oil has been added, blend a minute or two more to homogenize and transfer to a small ceramic crock or dish and store refrigerated. Because the oils are sensitive to heat, this "butter" is not used for baking or frying. The best way to use Better Butter is dotted on steamed vegetables. Dairy butter can also be added to the recipe if desired. Simply add ¼ cup (60 g) butter to the bowl along with the lecithin and then blend in the oils.

Agar-Agar "Jell-O" Squares

Agar-agar is a gelling agent derived from *Gelidium* or *Gracilaria* species of red algae (seaweed). Agar-agar contains more minerals than gelatin but has only 3 calories per gram and is vegan to boot. Agar-agar is flavorless, easy to work with, and useful in creating Jell-O–like desserts. Although there are health benefits to eating agar-agar, I include this recipe primarily as an alternative for diabetics to harmful sugar-sweetened gelatin desserts. Consuming agar may actually help diabetics and support weight loss, such as with the popular kanten diet in Japan (kanten being the Japanese word for agar-agar). This diet relies on the filling effects of agar to prevent overeating. As a general rule, use roughly 1 teaspoon of agar-agar powder per 1 cup of liquid. For the purpose of recipes for diabetics, high-flavonoid juices such as blueberry or pomegranate would be useful or a naturally sweet tea such as hibiscus tea sweetened with stevia would also work. Whisking an agar-infused hot liquid before it has time to set will result in a meringue-like

Monk Fruit as Noncaloric Sweetener

Monk fruit, known as luo han guo in Asia, is the spherical fruit of a gourd family vine, *Siraitia grosvenorii*, which can be used to prepare a noncaloric sweetener. The sweetener is available in liquid or powdered form. The sweet taste of luo han guo is due to a group of terpenes known as mogrosides that are estimated to be about 300 times as sweet as sugar by weight. Mogrol is the aglycone (nonsugar component) of several dozen mogrosides (glycosides in which mogrol is bound to a sugar) in luo han guo fruits, and it is absorbed from the digestive tract when digestive enzymes and acids, or the actions of intestinal microflora, hydrolyze the parent molecules. Extraction of the sweet compounds is a patented process that removes many other compounds that are astringent or bitter in nature, resulting in a purely sweet powder that is approved by the FDA as a safe food additive. The dry powder is so sweet-tasting that just a small amount is needed to sweeten a drink or a recipe. The fruits have not yet been extensively studied, but animal investigations have shown that the triterpene glycosides

may have broad immune-enhancing, antioxidant, antidiabetic, antitumor, and anti-inflammatory properties, explaining other traditional uses for hepatitis, epilepsy, and tumors. Animal studies show the plant to improve stamina associated with increases in liver and muscle glycogen, along with decreases in blood lactic acid and serum urea nitrogen levels.[129] Luo han guo fruit extracts may also suppress lipid accumulation due to various effects on adipogenesis.[130] Other research suggests that mogrosides may promote insulin secretion through direct anti-inflammatory effects on pancreatic β cells to exert an antidiabetic effect.[131] Several other species of *Siraitia* are also used medicinally.

Siraitia grosvenorii, monk fruit

consistency that works well for topping baked fruits and custards. These gelatin squares are the simplest way of working with agar-agar and akin to what is featured in the kanten diet.

1 cup (240 ml) pomegranate juice (may substitute
  prickly pear, blueberry, or tart cherry juice)
¼ to ½ cup (35 to 40 g) blueberries
1 teaspoon agar-agar powder

Puree the pomegranate juice with the berries in a blender and transfer to a saucepan. Gently heat the pomegranate-berry blend and add the agar-agar powder, stirring constantly to fully dissolve it. Slowly bring the mixture to a simmer, and pour into an ice cube tray or small candy molds. Allow to cool for 30 minutes and refrigerate until the liquid sets (2 to 3 hours). Then remove the gelled squares from the ice cube tray or molds and store in an airtight container in the refrigerator. Try serving these "Jell-O" squares with fresh berries, or grated carrots and thin apple slices. You can find many recipes on cooking with agar-agar online, and many websites offer beautiful inspiring photos of finished desserts that provide many options—from simple Jell-O–like desserts to cheesecake and vegan custards such as the Cashew Custard with a Nut Crust.

Cashew Custard with a Nut Crust

This recipe makes six individual custard tarts. Those with metabolic syndrome and diabetes can eat just one for a special occasion. Dates are not ideal for diabetics due to their dense simple carbohydrate content, but each of these tarts contains only one date, so one is not *too* bad as an occasional treat. The custard has a beautiful color due to the hibiscus tea (which also has medicinal value). The custard on its own is not very sweet and is perfect combined with the sweet crust.

FOR THE NUT CRUST

6 dates, finely chopped
2 tablespoons coconut oil or
  2 tablespoons salted butter, melted
1 tablespoon lecithin granules
1 cup (110 g) pecans
3 tablespoons coconut flakes

Preheat the oven to 350°F. Place the chopped dates in a small bowl and cover with 2 tablespoons of boiling water to soften, mashing with a fork to break them apart. Transfer to a small saucepan along with the coconut oil

and lecithin and melt together. Use a knife and a cutting board to chop the pecans into lentil-size pieces and add to the sauce blend along with the coconut flakes, stirring briefly to blend. Divide the mixture into six approximately equal portions and press each portion by hand into the bottom of an individual oven-proof tart pan. Bake for 10 to 15 minutes, removing the pans from the oven immediately once the crusts look crisp around the edges. Cool at room temperature.

FOR THE CUSTARD

2 tablespoons dried hibiscus flowers
2 cups (480 ml) water
1 cup (140 g) raw cashews
1 cup (240 ml) canned coconut milk
1 vanilla bean
1 tablespoon lucuma powder (may substitute
  1 or 2 teaspoons monk fruit powder, if available)
1 cup (120 to 150 g) fresh organic apple chunks
10 drops liquid stevia (optional)
1 scant teaspoon agar-agar powder
Fresh fruit slices or berries

Brew hibiscus tea by steeping the flowers in boiling water for 10 minutes. Strain and let cool for 10 more minutes. Combine the tea with all the ingredients except the agar-agar in a blender and liquify. Transfer the blended liquid to a medium saucepan. Add the scant teaspoon of agar-agar and stir the mixture continuously as you bring it to a gentle simmer. Promptly pour or ladle the custard mixture into the nut crusts and chill for several hours. Serve with fresh fruit slices or berries.

Winter Tea for Metabolic Support

This tea is flavorful and slightly warming and makes for a delicious winter brew, featuring blood sugar–balancing herbs. *Astragalus* is a legume, supplying inositol compounds that improve insulin resistance and offer immune support against winter colds and flu.

Astragalus membranaceus roots, shredded	5 ounces (150 g)
Glycyrrhiza glabra root, shredded	2 ounces (60 g)
Cinnamomum verum bark, small chunks	2 ounces (60 g)
Trigonella foenum-graecum seeds	½ ounce (15 g)

Combine the dry herbs in a jar or resealable plastic bag. Use 1 teaspoon of the mixture per cup of water. Simmer gently for 10 minutes, let stand 10 minutes more, and strain. Drink freely.

High Chromium Summer Tea for Metabolic Support

This tea provides chromium, a trace mineral important to insulin-receptor signaling. *Hibiscus sabdariffa* sepals are the highest known plant source of chromium, and *Stevia* is quite high in chromium as well. Chromium contributes to the metabolic effects of these herbs and offers blood sugar–balancing effects. *Medicago* is high in isoflavones and d-chiro-inositol compounds, both shown beneficial in supporting healthy blood sugar and lipid metabolism. The tea is especially nice in the summer due to its sour flavor and cooling effects.

Medicago sativa	6 ounces (180 g)
Stevia rebaudiana	2 ounces (60 g)
Hibiscus sabdariffa	2 ounces (60 g)

Combine the dry herbs in a jar or resealable plastic bag and mix well. Steep 1 tablespoon of the mixture per cup of boiling water for 10 minutes and then strain the brew. Drink as desired. May also chill and serve cold or over ice if desired.

Agua de Manzana (Apple Water)

Agua de Manzana means "apple water" in Spanish, and such waters are prepared by simply simmering apples or other fruits, sometimes along with quinoa, in water, like making a tea. I have learned to appreciate this refreshing drink from my time in Peru. The recipes that include quinoa, oats, or barley are popular as a breakfast drink, being similar to a very thin porridge. Berry water such as the Berry Water Spritzer might be even better for diabetics, and preparing a fruit beverage in this manner may provide an enjoyable alternative to those who crave

High-Flavone Fruits

Because of the bright red and purple pigments they contain, these fruits have protective and anti-inflammatory effects for blood vessels and are good choices for making herbal fruit waters, vinegars, dressings, and desserts. (One example is the Herbal Sipping-Vinegar Spritzers recipe on page 89.)

Prickly pear (*Opuntia* species)
Black cherry (*Prunus serotina*)
Pomegranate (*Punica granatum*)
Blueberry (*Vaccinium myrtillus*)
Grapes (*Vitis vinifera* and *Vitis* hybrids)

Medicago for Healthy Hormone Regulation

Medicago is a traditional remedy for diabetes, and modern research shows the plant to improve glucose metabolism[132] and its phytoestrogens to help treat heart disease and menopausal symptoms.[133] *Medicago* leaf teas and products are extremely nourishing and supportive to healthy hormonal regulation. *Medicago* contains trace amounts of vitamins K and D, which are rare in plants, as well as medicinally important saponins[134] and d-chiro-inositol. Steroidal saponins in *Medicago* lower cholesterol by decreasing its absorption and promoting intestinal excretion.[135] D-chiro-inositol may improve insulin sensitivity, and like most legumes, *Medicago* contains daidzein and genistein, which are shown to reduce hemoglobin glycosylation.[136] Isoflavones in *Medicago* are hormonally active and may support the bones and connective tissue in various cases of metabolic imbalances.[137]

Medicago sativa, alfalfa

soda pop, fruit juice, and other harmful beverages. These fruit waters are low in sugar and high in tissue-protective flavonoids. Using an herbal tea rather than water in these recipes supplies added medicinal effects.

2 medium organic apples
½ cup (90 g) quinoa
2 quarts (2 L) spring water or herbal tea of choice

Boil the apples and quinoa in the water or herbal tea in a large saucepan for 10 minutes, then let stand covered until cool. Transfer to a pitcher and refrigerate until cold. In Peru, people leave the particulate, seeds, and all in the water and just let it settle on the bottom of the pitcher. If you prefer a clear beverage, you can pour the apple water through a sieve into the pitcher to remove the largest pieces of apple, but if you're using the quinoa, it is intended to be consumed. Makes 10 servings.

Berry Water Spritzer

Similar to Agua de Manzana, this recipe provides flavonoids but contains very little sugar. This is also a healthy beverage to help wean people off sugary drinks or alcohol. For the herbal tea, try *Hibiscus*, alfalfa, *Astragalus*, licorice, and *Stevia*, individually or combined as desired, or any of the tea recipes featured in this chapter.

Berries, such as blueberries, raspberries, cranberries, or blackberries
Herbal tea of choice
Sparkling water
Ice cubes
Mint leaves, citrus peels, or whole berries frozen in ice cubes, for garnish

Place the berries in a blender and add enough herbal tea to just cover the fruit. Puree as finely as possible.

Xylitol as a Sugar Alternative

Xylitol is a naturally occurring sugar alcohol that may be purified from birch sap, oat straw, or some fruits and vegetables including plums, strawberries, raspberries, and cauliflower. It is probably never a good idea to purify single compounds out of whole food matrices, especially sugars, but xylitol is increasingly being used as an alternative sweetener. Xylitol tastes nearly identical to sucrose yet contains around 40 percent fewer calories. Xylitol is popular in sugar-free chewing gums because it has been found to deter bacteria associated with dental plaque and cavities. It may also deter ear infections, gingivitis, and periodontal disease. Xylitol is available as a white crystalline powder that can be substituted in most baking or other recipes 1:1 for cane sugar. I don't mean to propose that xylitol can be consumed with wild abandon by diabetic or metabolic syndrome patients, but for the occasional situation where a natural sweetener is desired, xylitol is certainly better than cane sugar and corn syrup. The ingestion of xylitol causes a smaller rise in plasma glucose and insulin concentrations than does the ingestion of glucose in healthy men and diabetics. Hence, it carries less risk of contributing to insulin resistance, even though it tastes like it should. Animal studies have shown xylitol to improve glucose metabolism in hyperinsulinemic rats, as well as to prevent the development of insulin resistance and visceral fat accumulation in animals given a high-fat diet. Esterified fatty acids in circulation are known to impair the uptake of glucose and increase inflammation that can lead to insulin resistance, and xylitol may improve blood sugar regulation via effects on fatty acid metabolism.[138] Xylitol may promote lipid-metabolizing genes in the liver, but it remains to be seen if promotion of lipogenic enzyme systems might cause excessive lipogenesis or carry a risk of promoting fatty liver.[139] Until more research is done, it is probably best to err on the side of caution and use only small quantities of purified xylitol in any given week and to avoid it on a daily basis in anything more than a piece of sugar-free chewing gum. I have included one recipe in this chapter, Peanut Butter Blondies (page 82), as an example of a grain-free dessert that substitutes xylitol for sugar. I recommend reserving it as a special occasion food, such as a birthday treat, for those with diabetes.

Transfer the puree to a pitcher or glass jar and refrigerate until cold. When ready to serve, place ½ cup (120 ml) of berry puree in a tall glass and dilute with 2 cups (480 ml) of sparkling water. Serve with ice cubes and a garnish of choice, such as a sprig of mint, twist of citrus peel, and/or berry-bearing ice cubes (ice cubes prepared with a whole berry frozen into each one).

Carob Milk

Ceratonia is a leguminous tree whose seedpods can be powdered and dried into carob powder or boiled into a carob syrup. Carob contains inositol compounds noted to improve insulin response, and this drink can be the basis for adding medicinal ingredients. This recipe is just an example, and the supplements and herbs added can be individualized to match an individual's presentation and personal taste preferences.

2 cups (480 ml) nut milk
2 teaspoons cod liver oil
2 teaspoons *Astragalus membranaceus* powder
1 teaspoon carob syrup
¼ teaspoon cinnamon
¼ teaspoon freshly grated nutmeg

Combine all ingredients except the nutmeg in a blender and blend well. Transfer to a small saucepan and heat gently. Transfer to a mug and sprinkle with grated nutmeg.

Herbal Sipping-Vinegar Spritzers

Sipping vinegars are prepared by macerating mangos, berries, papayas, citrus zest, high-mineral herbs, or other medicinal foods and herbs in basalmic, rice, wine, or apple cider vinegar for 4 to 6 weeks and then filtering out the fruit pulp. Macerating hot and spicy foods and herbs in vinegar yields a beverage popularly called Fire Cider (one recipe for fire cider is included in Volume 1, chapter 2). The resulting vinegar is then prepared into drinks by diluting with plain water or sparkling water and adding a dash of fruit juice or rose water. Vinegar enhances digestion and has a cooling effect on the body, and even if the notion of drinking vinegar is not immediately appealing, most people find the sour flavor enjoyable and refreshing, similar to lemonade, and surprisingly good tasting.

Ice cubes or frozen cranberries or blueberries
1 tablespoon fruit or herbal vinegar
1 tablespoon prickly pear juice
1 tablespoon rose water
2 cups (480 ml) sparkling mineral water

Place ice cubes or frozen cranberries or blueberries in the bottom of a tall glass. Add the vinegar, prickly pear juice, and rose water and top with sparkling water. Nice variations on these recipe ingredients include lemon juice, lemon peel twists, or chilled herbal tea. Makes 1 to 2 servings depending on the size of glass used.

Specific Indications: Herbs for Metabolic Conditions

The following herbs are used in the formulas throughout this chapter due to their ability to lower lipids and elevated glucose, improve insulin resistance, and protect the vasculature and organs from metabolic distress. There are many legumes that contain inositol compounds, many high-flavonoid herbs that protect the vasculature, and many high-chromium plants that improve insulin signal transduction intracellularly. Many herbs used in formulas for cardiovascular and peripheral vascular health are also useful in formulas for diabetes and metabolic syndrome, and you'll find additional valuable information about some of the herbs described here in Volume 2, chapter 2.

Allium cepa • Onions

While garlic is available in tincture and encapsulated form, onions are primarily used as a whole food. Onion consumption improves fasting and postprandial glucose levels in animal studies, lowering both plasma glucose and glycosylated hemoglobin,[140] and increasing insulin sensitivity in human clinical trials.[141] Some of the hypoglycemic effects are credited to α-glucosidase inhibition, whereby carbohydrate digestion in the intestines is partially inhibited, delaying postprandial glucose elevation. Like garlic, onions also have hypolipidemic and antimicrobial effects for diabetics prone to opportunistic infections. Include raw onions in salads and casseroles, sauté onions in stir-fries, and include onions and garlic in soups, stews, and wherever possible. I often puree two or three chopped onions in a blender to use as a soup base or to use as the liquid component when making salad dressings, Flax Crax (page 79), and marinades.

Allium sativum • Garlic

Include garlic bulbs in the diet and medicinal formulas for treating hyperlipidemia and diabetes, due to its ability to support metabolism and provide insulin-like effects. Garlic is also specific to include in formulas for diabetics with chronic fungal infections and as an expectorant antimicrobial for respiratory infections. *Allium* is useful to help protect the vasculature and may be an ingredient in formulas for retinopathy, hypertension, high cholesterol, and atherosclerosis.

Andrographis paniculata • King of Bitters

The leaves, which are extremely bitter, may improve metabolic functioning in diabetes and cardiovascular disease. Include *Andrographis* in formulas for infections, whether acute, chronic, or lingering in hypothyroid and diabetic patients. Consider *Andrographis* when acute viral infections cause a worsening of thyroid or metabolic symptoms. Andrographolide is a principal constituent, credited with hypoglycemic effects and an ability to slow carbohydrate absorption via α-glucosidase inhibitory activity. In Ayurvedic medicine, the herb is referred to as kalmegh or kalamegha, meaning "dark cloud."

Angelica sinensis • Dong Quai

Angelica roots are specific for vascular congestion, blood congestion associated with circulatory insufficiency, allergic hyperreactivity in the blood, inappropriate platelet activation and aggregation, and an increased risk of thrombi, atherosclerosis, and poor perfusion to the organs and limbs. Include *Angelica* in formulas for metabolic syndrome with hyperlipidemia, and for diabetics with neuropathy, nephropathy, and retinopathy. *Angelica*'s area of action is mainly on blood cells and cytokines, giving it "blood-moving" properties, anti-allergy effects, and an ability to enhance perfusion to various organs.

Arctium lappa • Burdock

Include *Arctium* roots in teas and tinctures to support hepatic clearance of hormones, lipids, and carbohydrates, to help optimize the intestinal microbial ecosystem, and to prevent fatty degeneration in the liver.

Astragalus membranaceus • Milk Vetch

Astragalus is available as shredded roots and powders, as well as tinctures, pills, and formulas and can be consumed on a daily or regular basis to improve insulin resistance and metabolic syndrome, as well as offer immunomodulating benefits for diabetics prone to chronic infections and allergic disorders. The roots are also available as large slices that can be decocted in soups and removed at the time of serving. *Astragalus* can improve metabolic function and can be included in formulas for frequent common infections that linger and are associated with fatigue and exhaustion. The saponins in *Astragalus* roots may help protect tissues and reduce inflammation caused by high blood sugar,[142] improve insulin response,[143] and have hypoglycemic effects by increasing liver and muscle uptake and utilization of glucose.[144] The polysaccharides in *Astragalus* enhance insulin-signaling pathways in muscle cells,[145] and one small human clinical trial reported reduced insulin resistance in diabetic subjects.[146]

Avena sativa • Oats

Oat bran is high in chromium and fiber and can be used as a medicinal food, or the fresh milky oat groats (referred to by herbalists as "milky oats") can be tinctured. Milky oat tincture can be used in formulas for nervous exhaustion and debility and to support recovery following exhausting illnesses, malnutrition, and addictive disorders. *Avena* is specific for fatigue following overwork and loss of vital fluids such as an acute hemorrhage and also as a supportive ingredient in formulas for adrenal dysfunction.

Berberis aquifolium • Oregon Grape

The bright yellow inner root bark of Oregon grape is the source of berberine. Isolated berberine has been the subject of more than 5,000 published papers, including thousands elucidating antidiabetic mechanisms of action. It is credited with benefits to metabolism that include improving hyperlipidemia and hyperglycemia and enhancing insulin sensitivity. Berberine is also credited with a thermogenic action due to its effects on brown adipose cells. Include *Berberis* in formulas for fungal infection in diabetes and as a general antimicrobial and hepatotonic in cases of intestinal dysbiosis related to hyperglycemia or sluggish elimination in hypothyroidism. *Berberis* may be included in formulas to help control lipids as one aspect of metabolic syndrome therapy. This plant is also known as *Mahonia aquifolium*.

Bupleurum falcatum • Chinese Thoroughwax

Also known as chai hu, *Bupleurum* roots have an anti-inflammatory effect on internal organs including the liver, kidneys, and spleen. *Bupleurum* has been used in

Traditional Chinese Medicine to help the liver "smooth the chi" and is appropriate for acute inflammatory reactions in the body. *Bupleurum* is often included in formulas for pathologies associated with diabetes including renal and hepatic insufficiency and inflammation. For more information on *Bupleurum*, see also the *Bupleurum falcatum* entry on page 55.

Camellia sinensis • Green Tea

Green tea leaves contain epicatechin, which is credited with broad antioxidant mechanisms, and animal studies have shown an ability to help pancreatic β cells survive oxidative stress and recover their insulin-producing function. Green tea beverages and pills can be a gentle metabolic stimulant for hypothyroidism and deficiency states. *Camellia* may benefit the heart, but present research suggests only unfermented or partially fermented green and oolong teas benefit hypertension—these benefits may not apply to "black" or fermented *Camellia* teas. The use of green tea epigallocatechin supplements improves lipid and glucose homeostasis and increases insulin sensitivity in part through increasing the expression of peroxisome proliferator-activated receptor proteins.

Ceanothus americanus • Red Root

The roots of *Ceanothus* are specifically indicated for thickening of the skin with doughy character, seen in long-term diabetics, and may improve lymphatic circulation and reduce the tendency to cyst formation and chronic fluid stagnation. *Ceanothus* is also specific for liver congestion, pelvic and portal congestion, splenomegaly, vascular congestion, and hypertension. It is traditionally asserted to have an affinity for the lymphatic system, alleviating vascular congestion via enhancing entry of interstitial fluid into the vasculature and enhancing venous return. There has been little to no modern scientific research on this plant.

Centella asiatica • Gotu Kola

The entire herbaceous plant is used medicinally or prepared into nourishing beverages. *Centella* may help protect tissues including the liver, nerves, heart, and blood vessels from fibrotic changes due to hyperglycemia and inflammation.

Ceratonia siliqua • Carob

Carob trees are in the legume family. Carob seeds are the source of locust bean gum, while the outer pods are processed in carob syrups and powders. Locust bean gum is high in galactomannan and may help treat gastrointestinal disorders, and the thick gum is also used as a carrier for other medicines and substances. Carob powder is one of the highest sources of inositol compounds, especially pinitol, noted to improve the sensitivity of insulin receptors to aid type 2 diabetics and those with metabolic syndrome. Pinitol may especially improve the uptake of blood glucose into skeletal muscle.[147] Carob flours and carob syrups can be used as medicinal foods; small amounts of the syrup are useful for replacing more harmful sugars and sweeteners. Carob is not typically used in tinctures or capsules. Many people point out, or even complain, that carob is *not* chocolate, but I have found that adding a tablespoon of ground coffee to a carob recipe makes the end result taste quite chocolate-like.

Chionanthus virginicus • Fringe Tree

The root bark of *Chionanthus* is specific for exhaustion, debility, and weight loss in diabetics according to traditional folklore. *Chionanthus* has not yet been widely researched but folkloric specific indications describe a presentation fitting a long-term diabetic with a fatty liver, as demonstrated by jaundiced skin and conjunctiva, pain in the right upper quadrant, steatorrhea, and glucosuria. Both tinctures and teas are available to include in formulas for diabetic and metabolic syndrome patients.

Cinnamomum verum • True or Ceylon Cinnamon

Cinnamon bark can be prepared into teas and tinctures to include in formulas for skin fungus and chronic skin infections in diabetics and those with poor peripheral circulation. Cinnamon essential oil may be diluted with water to use as a topical disinfectant for fungal infections. Cinnamon has hypoglycemic and hypolipidemic activities and may potentiate insulin; due to its pleasant flavor, it can be included in teas and used as a medicinal spice in foods and beverages. *Cinnamomum verum* is "true" cinnamon, also called Ceylon cinnamon, and an alternate species name is *C. zeylanicum*. *Cinnamomum cassia*, which is commonly called cassia cinnamon, is the species most commonly sold commercially as "cinnamon." Both species are used for culinary and medicinal purposes and are credited with hypotensive, antiatherogenic, and antidiabetic effects. Both of these species contain cinnamaldehyde and eugenol, which are credited with antimicrobial, anti-inflammatory, and other activities. Various species of cinnamon may regulate glucose transporters, improve glucose uptake and fasting blood glucose, inhibit α-glucosidase activity, and increase insulin sensitivity in those who have diabetes or are obese. Cinnamon bark is a warming blood mover for cold constitutions, and it reduces platelet aggregation, reducing the risk of clotting and thrombi. Cinnamon moves blood out of the pelvis and into the limbs. Cinnamon is a general circulatory effector with warming properties in the skin and gastrointestinal tract. Cinnamon may irritate the intestines in those with irritable bowel syndrome. Cinnamon may lower glycosylated hemoglobin due to increasing insulin sensitivity, and its use supports weight loss, lower lipid profiles, and improvements in various aspects of metabolic syndrome.

Cnicus benedictus • Blessed Thistle

Cnicus leaves are specific for sluggish liver function, and like its thistle family relatives Silybum and Cynara, Cnicus may improve the liver's processing of hormones and help protect the liver from fatty degeneration.

Cocos nucifera • Coconut

Botanists claim that coconut palms are the most naturally widespread fruit trees on earth.[148] Coconut sprouts, fiber, fruit pulp, fruit water, and other parts of the trees are all used medicinally. Coconut water is sourced from the inside of the unripe fruits. Coconut "oil" is actually a solid at typical room temperature and is an extract from ripe coconut meat. Although coconut oil is high in saturated fats, it is also a source of medium-chain triglycerides (MCT) that may support weight loss. Another benefit of coconut oil is its resistance to oxidation, making it a stable cooking oil. The popular opinion that coconut oil is healthy despite its substantial amount of saturated fat (a current trend at the time of this writing) is probably not entirely warranted, and some who claim this to be true may be mistaking the research on pure MCT oil and applying it to coconut oil. In fact, the consumption of coconut oil can raise total cholesterol levels and studies have not shown improvements in lipid profiles. There is more evidence that consuming unsaturated fats reduces the risk of heart disease than does consuming coconut oil.[149] I do *not* recommend that those with metabolic distress follow the popular fad of stirring spoonfuls of coconut oil into their morning coffee, for example. Coconut oil is a good choice as a frying oil.

Coleus forskohlii • Coleus

The roots of Coleus have been used for centuries in Ayurvedic medicine to improve poor metabolism, inflammation, and atony of organs and tissues. Research on Coleus has mainly focused on the diterpenoid constituent forskolin, which is shown to support fat metabolism. Coleus may reduce body fat percentage in overweight and obese human subjects,[150] and when combined with a low-calorie diet and exercise, Coleus may improve insulin resistance and metabolic risk factors. This species is also known as Plectranthus forskohlii. For more information on Coleus, see also the Coleus forskohlii entry on page 56.

Commiphora mukul • Guggul

Guggul is a traditional medicine made from the tree's resin, which is collected from the incised tree trunk. Guggul may help resist or reverse insulin resistance due to dietary stressors.[151] For more information on guggul, see the Commiphora mukul entry on page 56.

Coptis trifolia • Goldthread

Coptis roots are high in berberine and related isoquinoline alkaloids, useful for chronic infections in mucous membranes. Coptis may be included in formulas for diabetic and hypothyroid patients who have frequent sore throats or other common infections, with slow recovery time. Coptis can be used for fungal infections in the mouth, such as thrush, and topically/locally on the skin/vaginal infections common in patients with chronic hyperglycemia. Coptis is also useful internally for intestinal dysbiosis and sluggish digestion. Coptis has antimicrobial effects and may be included in formulas for infectious endocarditis and phlebitis. Berberine, found in Coptis roots, supports metabolism and liver function and is also appropriate to include in formulas for cardiovascular disease related to elevated lipids, diabetes, and hypertension. Due to its many metabolic effects, Coptis may help deter endothelial atherosclerotic plaque deposition and may help protect the brain against neurodegenerative effects due to metabolic syndrome. Coptis also supports liver metabolism, improving carbohydrate and lipid processing,[152] and offers cardioprotection.[153] Coptis chinensis is one of the 50 fundamental herbs used in TCM, where it is called duan e huang lian.

Crataegus species • Hawthorn

Various species of hawthorn trees are used medicinally; both the young flower buds with their leaves and the ripe berries are used. *Crataegus* is indicated for circulatory insufficiency and may be used in formulas to treat heart palpitations and hypertension related to hyperthyroidism or menopause. *Crataegus* can also help protect the blood vessels in cases of hypertension, hyperglycemia, and hyperlipidemia. *Crataegus* is appropriate in inflammatory processes associated with hyperglycemia and hyperlipidemia and can be a foundation herb upon which to rest more specific herbs to help prevent vascular damage and to use long term in treating chronic conditions. It is appropriate in cases of both hypertension and hypotension, bradycardia and tachycardia, and sluggish circulation and tendency to clots, and it is not contraindicated during Coumadin therapy or for diseases associated with a tendency to hemorrhage.

Curcuma longa • Turmeric

The bright yellow tuberous roots of *Curcuma* have antioxidant and anti-inflammatory properties that may help protect the liver from fatty degeneration in cases of hyperlipidemia and metabolic syndrome. *Curcuma* can improve lipid and carbohydrate metabolism in cases of diabetes and hyperlipidemia and may improve intestinal dysbiosis and propensity to fungal and other infections. For more information on this herb, refer to the *Curcuma longa* entry on page 57.

Cynara scolymus • Artichoke

Cynara leaves are edible, but often overlooked for their food value when growing artichokes. *Cynara* acts as a liver tonic and may improve cholesterol and hormonal imbalances by supporting the hepatic processing of these substances. *Cynara* leaves have been used since ancient times in North Africa and the Mediterranean as choleretic and hepatoprotective agents, and optimization of fat digestion and cholesterol processing may be useful for those with metabolic syndrome. Phenolic compounds in *Cynara* extracts may inhibit cholesterol biosynthesis and improve glucose and lipid processing.[154] One human clinical trial on patients with nonalcoholic steatohepatitis (NASH) showed the use of 2,700 milligrams of *Cynara* to improve liver enzymes and blood lipid profiles compared to control medication.[155] *Cynara* can be recommended for diabetics and metabolic syndrome patients for protecting the liver from hyperlipidemia.

Echinacea angustifolia, E. purpurea • Coneflower

Echinacea root medicines may be included in formulas for severe and long-standing diabetic ulceration that threatens to become gangrenous. *Echinacea* is also a folkloric classic to reduce the chronicity of simple common infections. It can also enhance immune strength for exhausted, hypothyroid, or diabetic subjects who suffer from frequent infections or for those who recover slowly or have simple colds that progress to the sinuses or bronchi. *Echinacea* can stabilize the hyaluronic acid matrix of the skin's connective tissue and support healing, as well as prevent dissemination of the infection in cases of cellulitis.

Eleutherococcus senticosus • Siberian Ginseng

Eleutherococcus roots act as an adrenal tonic in both hypo- and hyperfunction of the adrenal gland. *Eleutherococcus* is specifically indicated for long-term stress resulting in nervous symptoms, fatigue, poor sleep, and poor stamina, with easy exhaustion following minor exertion. *Eleutherococcus* may be a supportive ingredient in formulas for endocrine imbalances. For more information on this herb, see the *Eleutherococcus senticosus* entry on page 57.

Equisetum arvense • Horsetail

Equisetum aerial parts are an ancient medicine for wound healing and trauma recovery and to support the excretory function of the urinary system. *Equisetum* is high in silica and other minerals that contribute to its healing reputation. Include *Equisetum* in formulas to support connective tissue; it may help wound healing in chronic diabetic ulcers and other inflammatory conditions of the skin or connective tissue in the body. Anti-inflammatory effects may also be due to an ability to limit T cell proliferation in various states of active inflammation[156] and may contribute to the plant's folkloric reputation to limit fibrosis in arthritic, traumatic, and possibly metabolic-induced inflammatory states. Furthermore, the plant has been shown to have antidiabetic properties and beneficial effects on reducing urinary creatinine in rats.[157] *Equisetum* is known to bind heavy metals in the soil and is used in bioremediation to cleanse toxic sites. *Phytochelation* is the term for the process by which a compound in plants binds heavy metals and toxins, but no studies have been done to determine whether such compounds may have chelating effects in vivo. Toxins can contribute to oxidative stress that contributes to metabolic syndrome, and thus any agent that reduces toxicity in the tissue would be an important initial therapy. Silica in *Equisetum* is partly responsible for the plant's ability to hold lead and other heavy metals and detoxify them within the plant tissue. It is plausible that ingestion of *Equisetum* may help the human body to process and detoxify heavy metals, contributing to the plant's anti-inflammatory and connective tissue–strengthening effects.

Foeniculum vulgare • Fennel

Fennel seeds are a traditional remedy for polyuria, tachycardia, and hypertension, which are symptoms that may accompany advanced diabetes and heart disease. Folkloric texts report fennel seeds to be specific for rapid pulse. Fennel has many pharmacological properties that may be valuable in the treatment of metabolic disorders; these properties include antimicrobial, antiviral, anti-inflammatory, antispasmodic, antithrombotic, hepatoprotective, hypoglycemic, and hypolipidemic effects.[158] Because of the pleasant flavor, fennel seeds can be used in cooking as well as featured in teas and tinctures. Fennel seeds have also been used to suppress the appetite, and one old herbal in my collection recommends keeping a few fennels seeds tied up in one's handkerchief when in church, to nibble on discreetly to prevent the stomach from growling during long sermons. Fennel teas and seeds can be featured in weight loss protocols because they may help control appetite and possibly have antidiabetic effects. Trans-anethole, a volatile oil in fennel, has a molecular structure similar to catecholamine and, like amphetamine, may suppress the appetite. Even inhaling fennel volatile oil may suppress the appetite.[159]

Galega officinalis • Goat's Rue

Galega leaves are a traditional remedy for diabetes and to help protect, and possibly regenerate, pancreatic β cells. A search for the blood sugar–lowering compounds in *Galega* identified the biguanide substances galegine and guanidine and led to the development of antidiabetic pharmaceuticals used to reduce insulin resistance to this day. *Galega* has also been shown to improve white blood cell numbers and function, which is often impaired in metabolic disorders, a condition that contributes to infection susceptibility and disrupted tissue barriers.[160] *Galega* has also been used folklorically to treat obesity and shown to reduce body weight in animal research, with effects on adipocytes credited to galegine. Galegine may support weight loss by improving blood glucose metabolism and reducing food intake, as well as by exerting antilipolytic effects in adipocytes.[161] Improved glucose and lipid metabolism occurs in part

due to increased glucose uptake by fat and muscle cells, without signs of cytotoxicity.

Ganoderma lucidum • Reishi

The entire woody reishi mushroom is used medicinally. It can be processed into tinctures or powdered and encapsulated, and reishi slices can also be decocted in teas. Reishi and other medicinal and edible mushrooms contain β-glucans credited with immunomodulating activity. *Ganoderma* may help protect the nerves from oxidative stress and support the immune system in situations of poor immunity, including in adrenal and thyroid disease, and in diabetic patients.

Ginkgo biloba • Ginkgo

Ginkgo leaves are specifically indicated for oxidative stress in the blood, clotting, and circulatory insufficiency. *Ginkgo* may support circulation in people with heart disease and protect the heart in those suffering from metabolic syndrome and hyperlipidemia. Include *Ginkgo*, also called maidenhair tree, in formulas for diabetic retinopathy, nephropathy, and neuropathy, as well to treat impotence due to diabetes and poor circulation. The research on *Ginkgo* is extensive and shows numerous antioxidant, antiallergy, vasodilating, vascular-protective, and other tissue-protective anti-inflammatory effects.

Glycyrrhiza glabra • Licorice

Licorice is specifically indicated for exhaustion, loss of strength, weight loss, and debility. Glycyrrhizinic acid is a component of the roots shown to have hepatoprotective effects, preventing changes in cell membrane permeability by inhibiting phospholipase.[162] Licorice extracts may also improve dyslipidemia. Licorice may improve insulin resistance, and licorice flavonoids are shown to suppress abdominal fat accumulation in obese mice and rats.[163] Peroxisome proliferator-activated receptor γ-2 (PPARγ2) has been identified as a contributor to obesity, and licorice may improve fat cell response to insulin via PPARγ2 -mediated effects and thus support weight loss.[164] The triterpenes in licorice roots are particularly credited with metabolic effects.[165] Due to the extremely sweet flavor of licorice, it is highly useful and versatile to flavor medicinal teas and wean diabetic patients from harmful sodas and fruit juices. Licorice powder can also be used to sweeten medicinal truffles, sauces, and other foods. For more information, see "Licorice: A Model Adaptogen" on page 53.

Gymnema sylvestre • Gurmar

The common name gurmar means "sugar destroyer." The leaves of *Gymnema* are indicated for arteriosclerosis and atherosclerosis associated with diabetes and have been used in India for such complaints for thousands of years. The plant has earned its name because chewing the leaves destroys the ability to taste sweetness in foods for an hour or two afterward. This explains its traditional use to curb sweet cravings and support diabetics attempting to change their diet and lose weight. The plant has also been used for vascular disease and to treat both high and low blood pressure. Large doses of the plant may be contraindicated in those with rapid hypoglycemic reactivity, because it may cause blood sugar to drop quickly. A group of triterpenoid saponins called gymnemic acids have been identified and are thought to be responsible for the observed antidiabetic activity. Another group of saponins, the gymnemasaponins, are credited with the antisweet phenomenon. *Gymnema* may be used by both type 1 and type 2 diabetics. *Gymnema* stimulates insulin secretion, and human studies have suggested that patients on insulin may require less when *Gymnema* is used in tandem. *Gymnema* may be able to protect the pancreatic islets, possibly even regenerate or repair them, and increase their number.

Gynostemma pentaphyllum • Jiaogulan

Gynostemma pentaphyllum, commonly called jiaogulan or makino, is a climbing, perennial vine that grows in several parts of Asia, where it has been used as a food, a beverage, and a medicine. The plant is in the cucurbit family and can yield a small inedible purple gourd. The leaves have been used as an adaptogen-like medicine that is purported to be a longevity tonic. Antidiabetic effects have been demonstrated in animal models of type 2 diabetes, in which *Gynostemma* saponins exert hypoglycemic, hypolipidemic, immunomodulating, and antioxidant activities. Small human clinical trials have shown *Gynostemma* to improve insulin sensitivity in type 2 diabetics[166] and to complement antidiabetic drugs as an add-on therapy.[167] Animal studies show *Gynostemma* to be a possible substitute for statin drugs, exerting a hypolipidemic action via supporting liver metabolism and supporting phosphatidylcholine levels.[168] *Gynostemma* may also benefit type 1 diabetics, reducing hepatic glucose output and potently stimulating insulin secretion when insufficient. *Gynostemma* may promote pancreatic β cells, in part due to effects on calcium ion channels.[169]

Hibiscus sabdariffa • Roselle

Also called flor de Jamaica, this herb is a Caribbean, South American, and African traditional medicine for cardiovascular disease, obesity, and metabolic disorders. Rich in flavonoids, *Hibiscus* sepals are best used as a sour tea taken as often as possible to prevent the occurrence or at least slow the progression of heart disease. *Hibiscus* may protect the heart, blood vessels, and other tissues from the long-term consequences of oxidative distress due to metabolic disorders. Roselle teas and medicinal foods can be used in tandem with other herbs and pharmaceuticals. *Hibiscus* is the richest known plant source of chromium and may improve insulin response and metabolism of glucose. Therefore, *Hibiscus* might be specifically indicated for hypertension, atherosclerosis, type 2 diabetes, and metabolic syndrome. While the sepals of roselle are especially valuable as medicine and medicinal food, the dried petals of *Hibiscus rosa-sinensis* are also used medicinally in Asia and go by the name Chinese hibiscus. The brightly colored flavonoids in the sepals and petals of several species may also protect the vasculature from high blood sugar and lipids.

Hypericum perforatum • St. Johnswort

St. Johnswort's flower buds can be tinctured, prepared into oils for topical use, dried to use in teas, or encapsulated for oral use. St. Johnswort may be used as an ingredient in formulas to strengthen and reduce inflammation in all types of vascular tissue and as a circulatory and neural anti-inflammatory for retinopathy, retinitis, and vascular trauma with bruising and hematomas, a common problem for end-stage diabetics. *Hypericum* is a traditional medicine for bruising, strains, sprains, and trauma and may also support healing of stasis ulcers in patients with diabetes and circulatory insufficiency. Animal studies suggest that *Hypericum* has the ability to enhance fibroblast proliferation and collagen synthesis and to support revascularization and wound healing. *Hypericum* has also been used folklorically for neuropathy and trauma to highly innervated areas. Internal and topical medication may also be useful for diabetics with peripheral neuropathy, and flavonoids in *Hypericum* may offer neuroprotective action. *Hypericum* is specifically indicated for neuralgic pains and may be included in formulas for polyneuropathies in diabetic patients. *Hypericum* is traditionally emphasized for shooting, lancinating pain, or buzzing and tingling sensations.

Its use may help reduce neuronal inflammation and may reduce irritable bladder and bladder weakness in diabetics, especially when taken as a tea.

Lepidium meyenii • Maca

Maca grows exclusively at high elevations in the Andes, where the small turniplike tubers are an important part of the diet, believed to offer strength and stamina into old age and to act as a reproductive and circulatory tonic. Modern research is validating some of these claims, and maca powder is now sold around the world for infertility, menopausal symptoms, erectile dysfunction, and cardiovascular disease. *Lepidium* tubers contain linoleic and linolenic essential fatty acids, which may improve blood lipids[170] and reduce high blood pressure by enhancing renal processing of sodium, potassium, and chloride.[171] Animal studies show *Lepidium* to help treat metabolic syndrome[172] and offer hypoglycemic effects.[173] Maca has been shown to increase the urinary excretion of glucose, thereby reducing blood glucose levels.[174] Because maca can be ground into a flourlike powder, it may be used as a flour substitute in many baked goods and stirred into juices, milks, and fruit purees to enhance the medicinal value of the daily diet.

Mahonia • See Berberis

Medicago sativa • Alfalfa

Medicago is a highly nutritious plant that contains isoflavones, inositol compounds, and traces of vitamins D and K. *Medicago* is traditionally indicated for nutritional deficiencies and mineral imbalances and has also been used in formulas for impotence in men. Alfalfa sprouts may also offer some hormonal regulatory benefits. *Medicago* phytoestrogens are credited with preventing heart disease, cancer, and menopausal symptoms[175] due to the many hormone-regulating effects. Fresh *Medicago* leaves may be juiced in a wheatgrass juicer to consume immediately or frozen into ice cubes and stored in resealable plastic bags to include in future soups, drinks, teas, and daily drinking water.

Momordica charantia • Bitter Melon

The fruit of the bitter melon is an insulin-enhancing agent, traditionally used for diabetes to promote the entry of glucose into cells, where it can be stored by the liver and muscles. The fruit and rind are also used as a digestive bitter that may improve digestive insufficiency. A scant amount of animal research has suggested *Momordica charantia* to have a remarkable ability to protect and possibly renovate β cells in the pancreas[176] or to support the repair of partially destroyed ones, making bitter melon especially useful for type 1 diabetes.

Nigella sativa • Black Cumin

The small seeds are commonly used in Middle Eastern and North African traditional medicine for many purposes, including diabetes. The seed oil has a strong thymelike flavor due to the presence of thymoquinone, and consumption may improve glucose metabolism by enhancing insulin sensitivity. The oil is insulinotropic and has insulin-like activities in pancreatic β cells, skeletal muscle cells, and adipocytes. The plant also goes by the name black seed, and the oil is most commonly sold as "black seed oil," which may be used in cooking or to prepare various medicines for oral medicines.

Opuntia species • Prickly Pear

Prickly pear fruits are used traditionally for obesity and diabetes, as well as for hangovers due to the ability to stabilize blood sugar. *Opuntia* supports glucose and oxygen utilization in muscles and can be used as a medicinal food or beverage ingredient in the treatment of diabetes and metabolic syndrome.

Panax ginseng • Ginseng

Ginseng is traditional to support recovery for exhausting illnesses and to treat endocrine imbalances following long-term overwork and stress. *Panax* can be included in adrenal support formulas and may improve energy in hypothyroid, diabetic, and adrenal insufficient patients; it may also help normalize cortisol levels, whether elevated or deficient. *Panax* is a warming herb best for those with cold or neutral constitutions and is specifically indicated for exhaustion, weight loss, and debility. *Panax ginseng*, also known as Korean red ginseng, has multiple modes of action to improve diabetes and metabolic insufficiency with many such mechanisms credited to ginsenosides. *Panax* may upregulate insulin- and noninsulin-stimulated glucose transport and peroxisome proliferator-activated receptors (PPAR), a type of receptor involved with the regulation of fatty acid storage and glucose metabolism. *Panax quinquefolius*, North American ginseng, appears to act in similar ways.

Plantago species • Plantain

Plantago products, also called psyllium, are readily available as fiber supplements that can be supportive to weight loss, hyperlipidemia, and hyperglycemia. The rough textured seed powder can be blended into water or other beverages to improve bowel function and support a healthy microbiome, stimulating peristalsis and encouraging beneficial probiotic species. Psyllium powder can also be part of a weight-loss program, filling the stomach with low-calorie bulk and helping to curb hunger. I teach patients to make psyllium powder truffles by blending the powder with nut butter, carob syrup, and any other appropriate herbs and rolling into small balls that I call "truffles." The balls can then be rolled in dark cocoa powder or unsweetened coconut flakes to consume with hot coffee or tea in place of breakfast or lunch. See the High-Fiber Truffles recipe on page 82.

Pterocarpus marsupium • Kino Tree

Pterocarpus trees are used in traditional Ayurvedic medicine for diabetes, and research shows bark extracts to have supportive effects on pancreatic β-cell survival, repair, and output of insulin. Isolated marsupin, pterosupin, and liquiritigenin compounds have all shown hypolipidemic activity. Like insulin, the flavonoid epicatechin also stimulates oxygen uptake into fat cells and increases glycogen content of muscles, suggesting an insulinogenic action. Another flavonoid, pterostilbene, is also reported to be insulinotropic. The tree also goes by the names Malabar kino and Indian kino tree.

Pueraria montana var. *lobata* • Kudzu

This species is native to Asia and used in TCM for hypertension, menopause, bone density, hormonal cancer, renal failure, and many other ailments. The large starchy roots can be processed to eat or used in herbal teas, powders, and tinctures. The plant is in the legume family and may offer the insulin-signal enhancing effects of d-chiro-inositol seen with many other legumes. *Pueraria* reduces serum lipids and improves hepatic lipid metabolism. *Pueraria* roots have many cardiovascular and circulatory-enhancing effects, contributing to *Pueraria*'s ability to improve renal cellular function in cases of renal failure and insufficiency, a common end-stage sequelae of diabetes. Some other species are also used medicinally.

Punica granatum • Pomegranate

Pomegranates are a source of flavanols useful in medicinal drinks to help protect the heart and blood vessels from hyperglycemia and hyperlipidemia. The rind, leaves, stem bark, and fruit are all used in traditional medicine, with the fruit being the most delicious for use in the diet and the rind being studied for hypoglycemic actions in current research for treating metabolic disorder and protecting the tissues from oxidative stress. Traditional uses include many conditions associated with long-standing metabolic distress, including cancer, obesity, Alzheimer's disease, diabetes, and erectile dysfunction. The fruits and the rind are rich in a variety of phenolic compounds including the anthocyanidins, cyanidin, pelargonidin, and delphinidin as well as flavonoids such as kaempferol, luteolin, and quercetin, all credited with anti-inflammatory, antioxidant, and tissue-protective effects. Animal studies show the juice and the rind to lower blood sugar level and increase the number and insulin-producing capacity of pancreatic β cells.[177] *Punica* extracts may also improve the sensitivity and response of insulin receptors to benefit type 2 diabetics. Hepatoprotective and hypolipidemic effects have also been demonstrated.

Salvia hispanica • Chia

Also known as salba, chia seeds are a rich source of linolenic fatty acid. Chia is used as a medicinal food and fiber and is not used as a tea or a tincture herb. The seeds are used in the Flax Crax recipe on page 79 and the Chia Chocolate Pudding recipe on page 80; these recipes are good examples of how to best consume chia seeds, which form a mucilaginous mass when mixed with any liquid. Animal studies show benefits to glucose regulation, insulin response, and resistance to oxidation of fatty acids and deposition of fats in tissues.

Salvia miltiorrhiza • Red Sage

Also known as dan shen, red sage roots can be used medicinally to protect the blood vessels in patients with elevated levels of lipids and glucose. Include *Salvia miltiorrhiza* in formulas for diabetes, metabolic syndrome, vascular reactivity, and Grave's exophthalmia. It can also be used to improve circulation to the eyes in treating retinopathy, to protect the kidneys against and to treat nephropathy, and to help protect the heart in cases of high lipids and oxidative stress.

Silybum marianum • Milk Thistle

Milk thistle is specific for liver inflammation and can protect the liver from fatty degeneration. *Silybum* can improve hepatic metabolism of fats and sugars and can help protect the kidneys and liver in states of metabolic stress and improve cellular function in the organs.

Siraitia grosvenorii • Monk Fruit

The fruit of this gourd family vine also goes by the name luo han guo and was formerly classified as *Momordica grosvenorii*. Monk fruit has traditionally been used for cough, sore throat, and constipation. It is credited with cooling and moistening properties and is becoming popular as a noncaloric sweetener. Several companies in China produce compressed powders that are used to prepare beverages, and the bulk powder is becoming increasingly available from a variety of vendors. The sweet taste of the fruit is due to a group of terpenes known as mogrosides, estimated to be about 300 times as sweet as sugar by weight, making the plant suitable to use in baking, cooking, and drinks.

Smallanthus sonchifolius • Yacon

The large clusters of juicy roots are processed to produce syrups used as natural sweeteners, or dried into flourlike powders to use in baking and cooking as a sugar alternative. The fructooligosaccharides in the roots may help to shift the intestinal ecosystem in a manner that improves metabolic function. For detailed information about yacon, see "Yacon for Intestinal Health and Glycemic Control" on page 81.

Smilax ornata • Sarsaparilla

Smilax is an aromatic alterative agent with hormonal and adrenal supportive properties, and steroidal saponins in the roots may also support metabolic activity. *Smilax* may stimulate lipolysis promoting β-adrenergic activity.[178] *Smilax* may be included in protocols and formulas for weight loss as the herb may promote lipid metabolism and reduce lipid accumulation.[179] See also the *Smilax* entry on page 62.

Stevia rebaudiana • Sweet Leaf

Stevia is native to the highlands of Bolivia and Paraguay where the leaves are commonly used to sweeten teas and as a folk medicine for diabetes. *Stevia* is high in chromium, which may improve insulin signaling, and is indicated in teas and medicinal beverages for elevated lipid and glucose levels. While most of the research has concentrated on antidiabetic actions, *Stevia* may also be cardioprotective, protecting cardiac muscle mitochondria from ischemic and reperfusion injury. The glycoside stevioside is credited with the sweet flavor as well as hypoglycemic activity and has demonstrated insulinotropic and glucagonostatic actions. Diterpenes such as isosteviol are also shown to have hypotensive and hypoglycemic actions. Include *Stevia* in teas and medicinal foods to protect the vasculature in cases of diabetes, coronary artery disease, stroke recovery, and cardiomyopathy. *Stevia* is an excellent choice as a sweetener for use in weight-loss efforts as it is noncaloric, does not elevate blood glucose, and does not provoke an insulin response. The dried leaf powder is the least processed sweetener and may offer the broadest range of nutrients and chemical constituents, but many people prefer sugarlike isolates and liquids offering the purified steviosides—the sweet compounds.

Syzygium aromaticum • Cloves

Cloves are the unopened flower buds of a tropical tree, which are dried and used for culinary purposes due to their aromatic and warm, flavorful properties. Of the common culinary spices, cloves are reported to be a more powerful antioxidant than cinnamon, pepper, ginger, or garlic.[180] Antioxidant compounds in cloves include polyphenols such as gallic acid, ellagic acid, quercetin glucoside, and ellagic acid derivatives, all of which may contribute to clove's ability to protect the pancreas from lipid peroxidation. Cloves are also a traditional herb for diabetes, hyperglycemia, and glycosuria, specifically indicated for weakness, excessive thirst, and emaciation. Insulin-like actions have been reported from clove buds, and the spice may lower blood sugar by improving liver uptake of glucose.[181] Triterpenes are credited with hypoglycemic effects accomplished through inhibition of glucose transport through the small intestine. Triterpenes in cloves include oleanolic acid and maslinic acid. Animal research shows cloves to reduce the postprandial rise in blood glucose following the ingestion of a carbohydrate load.[182]

Syzygium cumini • Java Plum

Also known as the Malabar plum or jambolan, seeds from the Java plum have been used in Ayurvedic medicine for thousands of years, including in formulas for obesity, liver disease, diabetes, and heart disease. The seed powder may significantly reduce body weight, white adipose tissue accumulation, blood glucose, serum insulin, and the concentration of plasma lipids such as total cholesterol, triglyceride, LDL, and HDL. *Syzygium cumini* seed powder also increases glutathione concentration in the liver and prevents the infiltration of inflammatory cells, fibrosis, and fatty droplet deposition and fibrosis in rats fed a high-fat diet.[183] The plant is also known as *Syzygium jambolanum*.

Tamarindus indica • Tamarind

The sour seeds are processed into pastes and powder and used in cooking but are also a traditional remedy for diabetes. The plant may increase insulin output when insufficient due to restorative effects on β cells in pancreatic islets and improve glycogen uptake in type 1 diabetes. Tamarind may also benefit type 2 diabetes via improving glucose storage as glycogen in the liver and muscles.

Taraxacum officinale • Dandelion

Dandelion leaves are used as a general diuretic and mineral tonic in cases of diabetic renal insufficiency. Dandelion root may be used in formulas to support the liver and protect against fatty and cirrhotic degeneration and is specifically indicated for gastric headaches, biliary insufficiency, jaundice, coated tongue, pain in the right upper quadrant of the abdomen, hepatic torpor and constipation, and night sweats due to liver disease.

Trigonella foenum-graecum • Fenugreek

Fenugreek is a legume family plant, but rather than producing edible beans, it produces aromatic seeds, emphasized folklorically for type 2 diabetics. Fenugreek seeds are ground into a spice with a slightly pungent aromatic quality that many people liken to maple syrup, but the bitter flavor limits its use as a culinary spice. Fenugreek seeds have significant hypoglycemic, antiatherosclerotic, anti-inflammatory, and other actions that benefit diabetics and those with metabolic syndrome. Research reports hypoglycemic mechanisms to include inhibition of intestinal glucosidase enzymes that reduce the absorption of carbohydrates and positive effects on glycolytic, gluconeogenic, and lipogenic enzymes to restore glucose homeostasis in various animal models of diabetes and metabolic distress.

Vaccinium myrtillus • Blueberry

Vaccinium fruits and leaf teas help to protect the blood vessels in cases of hyperglycemia and hyperlipidemia. Animal studies have shown compounds in blueberries to reduce blood glucose and to enhance insulin sensitivity in type 2 diabetic mice, with one mechanism of action being promotion of the intracellular "second messenger" cyclic AMP[184] and associated protein kinases. Human studies on glucose regulation are sparse, but one study in overweight and obese women with metabolic syndrome reported the consumption of blueberries to have a slight but statistically significant effect on total weight and waist circumference compared to other berries and isolated chemical concentrates from berries.[185] *Vaccinium* is one of richest sources of anthocyanosides, flavonoids shown to protect blood vessels and capillaries from oxidative stressors that occur with metabolic dysfunction.[186] *Vaccinium* preparations are specifically indicated for diabetic retinopathy and can be included in formulas for paresthesias and polyneuropathy. Research models report that 160 milligrams of anthocyanins per day is helpful for retinopathy. Consume blueberries regularly, take ½ teaspoon of solid extract daily, or use *Vaccinium* leaves in teas and tinctures; the anthocyanosides in all of these forms of *Vaccinium* can be helpful for bruising and skin trauma and to improve microcirculation in the skin.

Vitis vinifera • Grapes

Grapes are high in resveratrol, which has a protective effect on the heart and blood vessels and can be taken as a supplement in cases of metabolic syndrome and elevated blood lipids and glucose. Resveratrol is well studied for its numerous antioxidant and anti-inflammatory effects shown to protect the tissues from fibrotic, inflammatory, and allergic processes.

Zingiber officinale • Ginger

Ginger can help improve digestion, reduce elevated lipid levels, and reduce inflammation in the blood and tissues. Ginger can be included in teas and tinctures for elevated lipids and diabetes, as well as used as a medicinal food. Ginger is specifically indicated for cold and deficient constitutions to stimulate circulation and bring more heat to the digestive system. Ginger's antihyperglycemic effects are due to enhancing insulin release and improving carbohydrate and lipid metabolism.[187] Among the active ingredients in ginger are the pungent gingerols and shogaol. The regular consumption of ginger may also offer protective effects on the liver, kidneys, eyes, and neurons, all commonly injured with diabetes. One human clinical trial on adult diabetes mellitus subjects showed 2 grams of ginger powder per day to significantly reduce fasting blood sugar, hemoglobin A1c, and inflammatory markers compared to baseline measurements.[188]

— CHAPTER FOUR —

Creating Herbal Formulas for
Reproductive Endocrine Conditions

In my early years of practice, I quickly became skilled in treating menstrual cycle disorders and menopausal difficulties due to the sheer number of patients seeking help. I learned that a few core herbs worked well for common complaints among my patients: heavy menstrual bleeding often responds to *Achillea millefolium* and *Cinnamomum* species; uterine pain and cramping respond to *Viburnum opulus* or *Dioscorea villosa*; menopausal hot flashes often respond to *Salvia officinalis*; and irregular cycles and midcycle spotting often respond to *Vitex agnus-castus*. I address herbal treatment for all of these conditions in this chapter. This chapter also details some of the hormone-regulating phenomena being elucidated at the time of this writing, including endocrine disruptors, selective estrogen response modifiers, amphotericism, and estrogen dominance. *Endocrine disruptors* are environmental compounds that interfere with hormone receptors and signaling; *selective estrogen response modifiers* (SERMs) are medicines capable of having precise and targeted effects on the various estrogen receptors in different tissues; *amphotericism* is the ability of a phytoestrogen to increase or decrease hormonal activity depending on the physiologic situation; and *estrogen dominance* is a common clinical scenario where estrogen levels are excessive with respect to the progesterone level, also known as *unopposed estrogen*.

These four broad physiologic situations are of great clinical relevance and are also highly relevant to herbal research. I discuss many herbal molecular mechanisms of action in this chapter, weaving together some of the complex details of estrogen receptor subtypes, cross talk between the endocrine organs, and herbal constituents acting as agonists and antagonists at the various estrogen receptors.

Taraxacum officinale, Berberis aquifolium, Curcuma longa, and other alterative herbs can act as foundational herbs in formulas for numerous female and male reproductive issues when toxicity and heavy hormonal loads in the body contribute to hormonal acne, breast tenderness and cysts, and proliferative disorders of

Unopposed Estrogen and Estrogen Dominance

The physiologic situation of unopposed estrogen or estrogen dominance is extremely common in the general population and contributes to mastalgia and breast cysts, endometriosis, ovarian cysts, and prostatic hypertrophy. Herbal formulas for all these conditions might emphasize alterative herbs to assist the liver in conjugating estrogen and to help clear the tissues of excessive hormonal stimulation. High estrogen levels in the body can be excessively stimulating to estrogen-sensitive tissues. Endometriosis, uterine fibroids, ovarian cysts, and breast cysts are all fairly common conditions, and all are associated with a high estrogen level, especially when that estrogen is not matched or "opposed" by comparable levels of progesterone. Therefore, one clinical approach for treating these conditions would be to reduce estrogen levels by avoiding exogenous sources in the diet or exposure to chemical estrogens (xenoestrogens) and by supporting the liver to clear estrogen from the bloodstream.

Alterative herbs such as *Berberis, Curcuma, Silybum marianum*, and *Taraxacum* can be included in herbal formulas for this purpose. Therapy might also aim to raise progesterone by supporting the adrenal glands' production of precursor hormones. The adrenal glands synthesize DHEA, the "mother hormone," which is then converted to other reproductive hormones. Thus, supporting adrenal gland health with adaptogenic herbs such as *Withania somnifera, Eleutherococcus senticosus,* or *Glycyrrhiza glabra* may be another prong in herbal formulas for estrogen dominance. Stress can shunt more DHEA into the production of stress hormones, leaving relatively less for producing progesterone, so any stress-reducing lifestyle practices such as meditation, laughter, or gentle exercise may complement the use of adaptogens and alteratives as a foundational approach to correcting estrogen dominance. See also "Improving the Clearance of Estrogen" on page 103.

the reproductive organs. These herbs help reduce hyperestrogenism, reduce the proliferative effects of hormones on hormone-sensitive tissues, or reduce exogenous estrogens by supporting liver clearance and conjugation of hormones. See the "Unopposed Estrogen and Estrogen Dominance" sidebar for more about this condition.

Herbs that affect neuroendocrine regulation of the reproductive system are important prongs of therapy for many reproductive disorders. Dopamine, γ-aminobutyric acid (GABA), and other neurotransmitters are increasingly being shown to control the release of prolactin, follicle-stimulating hormone (FSH), and luteinizing hormone (LH) from the anterior pituitary. Disorders such as polycystic ovarian syndrome (PCOS) may involve elevated prolactin, or male infertility may involve LH insufficiency, and herbal medicines that optimize the pituitary's release of gonadotropins are often effective. Adaptogenic herbs that strengthen the hypothalamus-pituitary-adrenal (HPA) axis can be trophorestorative foundations in formulas for complex

endocrine cases. Adaptogenic or other supportive herbs may be chosen based on specific symptoms and presentations as exemplified throughout this chapter. Due to synergism and cross talk among the brain, thyroid, adrenal glands, and gonads, many reproductive disorders are complex and involve multiple endocrine imbalances. The HPA axis embraces the synergism between the brain and its feedback loops with the adrenal glands, and the reproductive organs are involved in numerous such complex feedback loops as well. Chapters 2 and 3 offer further guidance on choosing herbs for reproductive conditions that also involve metabolic or thyroid and adrenal imbalances.

The more the mechanisms of steroid regulation of cells have been elucidated over the last century, the more researchers have learned about complex receptor agonism and antagonism, up- and downregulatory phenomenon, enzymatic influences, and other mechanisms. For example, there are α estrogen receptors and β estrogen receptors, and subtypes of each, with the α most active during embryologic development and puberty, and the β type most involved with maintaining tissues. There are estrogen receptors on the vasculature, on the bones, on the endometrium, in the brain, and on the endocrine organs, and agents that bind the receptors may have different binding affinities and different effects given the tissue and the situation. A myriad of endogenous and exogenous substances are capable of binding to estrogen receptors and act as agonists, partial agonists, or antagonists with various electromagnetic strengths, which researchers refer to as ligand affinities. Other hormones, vitamin D, and vitamin A derivatives can coregulate estrogen's cellular influence at the level of the nuclear membrane, rendering it stronger or weaker. And many enzymes and the hormone receptors themselves may be up- and downregulated by diet, stress, thyroid and adrenal hormones, and other factors. Therefore, any one study on a single hormone, in isolation of the holistic system, is only a single data point and will not apply to all tissues, situations, or individuals. Herbal medicines have been found to affect all aspects of hormone regulation, as I discuss throughout this chapter.

Because genes that control cellular and hormonal receptor quantities and response signals are shown to be unfavorably influenced by exogenous chemicals[1] and favorably influenced by phytoestrogens,[2] dietary influences can both promote and protect against various hormonal cancers. Liver-supportive herbs that improve detoxification of exogenous pollutants can be valuable in both treating and preventing hormonal diseases. The maternal diet is shown to affect gene expression and hormonal regulation in offspring, even across several generations. Embryologic development of the breasts, prostate, and reproductive organs involves both estrogens and androgens. Steroidal metabolism and signaling may be "disrupted" early in life through exposure to hormonally active synthetic chemicals. Exposure to some exogenous estrogens—such as hormonally active synthetic chemicals—during the neonatal period may increase the risk of hormone-related cancers later in life. Natural plant-based compounds such as the phytoestrogens may reduce the chances of hormone-related reproductive cancers later in life, however, mitigating the severity of endocrine disruptors and offering positive hormonal influences on the breast and prostate.

This chapter explores all such molecular details on hormonal regulation and herbs that affect them, but I would be remiss to not steer the conversation back to women as whole persons. Over the decades I have met with hundreds of women seeking support and solutions for menstrual symptoms, menopause, emotional and mood challenges, and general hormonal imbalances. I have witnessed that many women are very discerning and seek out options in thoughtful and intelligent ways. They want information and answers. Sadly, though, I have heard many a story about women who asked intelligent questions only to have their curiosity about alternative therapies belittled as unscientific. Dare I say, some clinicians do pressure women into pharmaceutical treatment by instilling fear that their bones will become brittle or their skin and vaginal mucosa will

Improving the Clearance of Estrogen

Some herbs and foods are noted to improve the metabolism and clearance of estrogen, progesterone, and testosterone. Sulfur-containing indoles found in broccoli and Brassica family vegetables are known to improve the clearance of estrogen[3] and should be liberally consumed in the diet as prevention against many diseases including reproductive cancers. For those with breast cancer, endometriosis, or other hyperproliferative disorders, these indole compounds can be taken in concentrated capsules as part of an overall treatment plan. Diindolylmethane (DIM) is one such compound commercially available.

Other herbs can be used to affect enzymes involved with estrogen synthesis. *Urtica*,[4] *Prunus africana*, and *Serenoa*[5] are noted as aromatase inhibitors, and they diminish the conversion of progesterone and testosterone to estrogen in peripheral adipose cells. These herbs, classically pigeonholed as male reproductive herbs, are also appropriate for women due to these actions and can be considered as support and adjuvant herbs in formulas for complex hormonal imbalances involving estrogen dominance or low progesterone.

dry up if they don't accept a prescription for synthetic hormones. I, on the other hand, think that it is very possible to restore hormonal balance without prescription hormones and believe that herbal medicines are extremely valuable for this purpose. In some cases, therapy may be as simple as removing the obstacles to hormonal balance—be they poor sleep, poor diet, lack of exercise, exposure to exogenous estrogens, or stress and poor adrenal function. In other cases, improving estrogen clearance and offering symptom-specific protocols using herbal formulas such as those in this chapter are of great value. Helping women navigate the medical empire is an enormously valuable way of helping them make their own informed choices and thereby claim their own power. I am passionate about empowering women to be a healing force on the planet. Even in the twenty-first century, women are the main cooks and caregivers for their families, and so the more balanced and heathy the women, the healthier the family. And the healthier the family, the healthier the community, and it snowballs from there toward planetary wellness. Empowered women supported by a healthy community are a force to be reckoned with.

Historically, the *wise women* were those mothers, grandmothers, beloved aunts, sisters, and neighbors who had herbal and other healing skills and were vital to the health of their communities. I feel that I have become part of that wise woman lineage in the work that I do in my own medical practice. I also believe the concept can be expanded to those women who are the wise and patient teachers of the community's children; those female business owners who create vibrant towns; those women who become politicians, civic leaders, advocates for the environment, and midwives of change. I feel a sense of urgency that more women assume leadership and help to guide us toward a new era of honoring Mother Earth. After all, healthy, loving women will raise sons and daughters who become aware, respectful, and gentle men and women. I believe that the lack of honoring the sacred feminine extends to a willingness to do harm to Mother Earth. Healing can begin with a single woman, and I believe the effort of nurturing and restoring balance becomes amplified through the family, community, and workplace out through wider and wider circles of humanity.

An Overview of Reproductive and Genital Infections

The most common reproductive infections treated in a general family practice are simple yeast or dysbiotic vaginitis, along with fungal infections of the external genitalia. Herpes simplex, chlamydia, and human papillomavirus (HPV) are also common sexually transmitted infections and are each treated specifically. Human immunodeficiency virus (HIV) is a devastating sexually transmitted disease, but I do not cover HIV in this chapter, largely because I am not very experienced in the management of the infection, and to do the subject justice, one should be an expert.

Herpes simplex virus (HSV) causes painful genital ulcerations, and the herpes virus family of DNA viruses all cause chronic infections that are highly resistant to "cure." Many people who suffer from HSV learn what sort of stressors weaken their immune system and result in an outbreak, and manage the disease by avoiding those stressors. Common stressors include emotional and mental stress, ingesting sugar or alcohol, lack of sleep, or in the case of herpes simplex cold sores, too much sun exposure. HSV infects epithelial cells and progresses to local nerve endings, where it persists in sacral ganglia. HSV has a pattern of becoming repeatedly reactivated, where it travels down nerve axons to the genital mucosa, leading to new outbreaks of ulcers.[6]

Chlamydia trachomatis is a sexually transmitted bacterial infection that spreads from the lower genital tract to the upper genital tract where it may infect the fallopian tubes and lead to scarring and infertility or increase the risk of ectopic pregnancies.[7] Human papillomavirus (HPV) infection is a leading cause of cervical cancer in women and of anal, penile, and oropharyngeal cancer among young men and is reported to be increasing in incidence worldwide.[8]

In addition to herbs noted to be antiviral, antibacterial, or antifungal against certain organisms, the formulas for infections included in this chapter may also include general immune support, as well as adjunctive herbs that address underlying contributors to outbreaks such as stress, poor sleep, or poor digestion. These infections can also cause highly uncomfortable symptoms, and some of the formulas also offer palliative ideas to address itching, burning, and other discomfort.

Formulas for Vaginitis

The majority of vaginitis cases are infectious due to yeast, dysbiotic bacterial strains, and other pathogens. Some cases may be atrophic, rather than infectious, occurring in older women when falling hormone levels fail to adequately maintain the vaginal mucosa and ecosystem. Cases of atrophic vaginitis are best treated with systemic hormonal support and local lubricating and nourishing agents by means of suppositories, hormone or other creams, or soothing, moistening sitz baths. When there is severe thinning of the vaginal wall postmenopausally, with dyspareunia and chronic atrophic vaginitis, some women will benefit from the use of various forms of estrogen creams, such as the popular Bi-Est (estradiol with estriol) or pure estriol applied directly to the vagina.

Infectious vaginitis requires the use of local antimicrobial agents. Vaginal infections are usually more of a discomfort than a serious health threat however, when severe or occurring in immunocompromised individuals, infections can lead to serious complications such as pelvic inflammatory disease, infertility, chronic pelvic pain, and premature birth. *Hydrastis canadensis, Calendula officinalis, Melaleuca alternifolia, Berberis* (also known as *Mahonia*), *Allium sativum, Hamamelis virginiana, Thuja* species, *Achillea, Usnea barbata*, and *Coptis chinensis* are helpful in treating a variety of common vaginal pathogens and are among the best antimicrobial choices to treat vaginitis when put in suppository and vaginal douche recipes. All types of infectious vaginitis may respond to boric acid suppositories, which serve to acidify the vagina and support desirable acidophilus bacteria and other beneficial vaginal flora. Clinical trials suggest that garlic (*Allium sativum*) creams used vaginally may improve vaginal discharge and vulvar erythema and alleviate secretions, itching, irritation, dyspareunia, and painful urination in cases of yeast vaginitis.[9] *Berberis* has been shown to boost the efficacy of metronidazole gel in treating bacterial vaginosis.[10] The alkaloid berberine is noted for broad antimicrobial and metabolic-enhancing activity and also has antifungal activity against common yeast infections due to *Candida albicans*.[11]

Trichomoniasis is the most prevalent nonviral sexually transmitted infection, with 170 million new cases reported each year. In addition to general discomfort and irritation, trichomoniasis is associated with low birth weight and negative birth outcomes. The topical use of *Commiphora molmol* (myrrh) improves *Trichomonas vaginalis* infections in both men and women.[12] The vaginal use of *Punica granatum*[13] bark douches and various *Eucalyptus globulus* extracts also inhibit the growth of *Trichomonas*.[14] The fact there are published studies on *Punica* and *Eucalyptus* for trichomonas should in no way suggest that they are the best two herbs for vaginal infections—it simply means that there are published studies to cite. Numerous essential oils deter vaginal infections but must be diluted to avoid irritating the tissue. Tea tree, cedar, and lavender essential oils may be available in various commercial suppositories.[15] Numerous other antimicrobial herbs, including *Achillea, Berberis*, and *Hydrastis*, may also be effective in sitz baths and vaginal douches.

It is also important to consider whether there are any underlying contributors to vaginitis. For example, diabetics may be prone to vaginal infections because constant high blood sugar levels allow pathogens to thrive in the tissues and create a hospitable environment for fungi and bacteria, shifting the flora of the vaginal ecosystem to an unhealthy state. Those who have used antibiotics frequently or recently may have inadequate levels of desirable vaginal bacteria and may benefit from instilling lactobacillus and other probiotic strains directly into the vagina. It may also be important to optimize digestive health and the microbiotic ecosystem to help women recover from chronic vaginal infections. Thus, alterative herbs might be one prong of an herbal protocol. High estrogen levels may support the growth of yeast and undesirable bacteria, so herbal formulas for vaginitis might also include alterative and liver herbs to enhance estrogen conjugation and clearance.

Tea Tree Oil Suppositories

When a strong antimicrobial agent is desired, tea tree (*Melaleuca alternifolia*) can be powerful against fungus, bacteria, and *Trichomonas*, yet gentle on the tissues. The suppositories can be purchased commercially as well as easily made for yourself if you have suppository molds.

Cocoa butter	2 ounces (60 g)
Melaleuca alternifolia (tea tree) essential oil	½ ounce (15 ml)
Suppository molds and wide-bore droppers	

Melt the cocoa butter in a double boiler. Pour into suppository molds as explained in the instructions for Suppositories for Vaginitis. This recipe will make around 25 to 30 suppositories. Add 4 or 5 drops of tea tree essential oil to each individual mold, then cover with a tiny piece of plastic wrap to help prevent the essential oil from volatilizing, and refrigerate. The finished suppositories may be stored in the molds, placing 7, or 10, or 25 suppositories in a resealable plastic bag or other container suitable to dispense to patients for a week's or month's supply. Patients may tear a single suppository from the roll along the perforation and peel the suppository from the plastic mold at the time of use. Use vaginally, 1 each night for chronic vaginitis, 5 days on and 2 days off, for several weeks to a month.

Suppositories for Vaginitis

This is an all-purpose formula for infectious vaginitis. These herbs can be helpful for infections due to yeast, *Trichomonas*, and all types of bacteria. *Commiphora, Usnea*, and *Achillea* are broad-spectrum antimicrobials; *Hydrastis* and *Geranium* are astringent antimicrobials that help to allay swelling and vaginal discharge. *Althaea* provides a soothing mucilaginous base to soothe pain and balance the other herbs in the formula. For cases of atrophic vaginitis, these suppositories could be made with *Althaea* alone.

Commiphora myrrha	½ ounce (15 g)
Usnea barbata	½ ounce (15 g)
Hydrastis canadensis	½ ounce (15 g)
Althaea officinalis	½ ounce (15 g)
Geranium maculatum	½ ounce (15 g)
Achillea millefolium	½ ounce (15 g)
Cocoa butter	1 ounce (30 g) or more, as desired
Suppository molds and wide-bore droppers	
Melaleuca alternifolia (tea tree) oil (optional)	
Essential oils (optional)	

Blend the herb powders together thoroughly in a resealable plastic bag or glass jar. Melt 30 grams of cocoa butter in a double boiler. Stir 30 grams of the powdered herb blend into the cocoa butter (you may eyeball equal volumes if desired). Use a wide-bore dropper to transfer the melted mixture into suppository molds. This recipe will make around 20 to 30 suppositories depending on the size of the molds. The herbs will continually sink to the bottom of the pan, so the mixture must be briskly stirred every minute or so. Work quickly or the cocoa butter will begin to harden. Tea tree oil or other antimicrobial volatile oils may also be added to this formula, but bear in mind that the heat of the melted cocoa butter will tend to dissipate the volatile oils. Immediately after pouring the melted mixture into the suppository molds, add 1 or 2 drops of essential oils to each suppository. It may help to work with another person while preparing this recipe,

one to transfer the mixture into the molds, and another to keep stirring the mixture and add the essential oils to individual suppositories. Cover the suppositories with plastic wrap after adding essential oils to help prevent the volatile compounds from escaping into the air. Once the suppositories have hardened, they may be stored in the refrigerator in resealable plastic bags of 6 to 12 suppositories. The suppositories will retain their potency for several years in refrigerated storage. Patients may tear a single suppository from the roll along the perforation, and peel the suppository from the plastic mold at the time of use. Insert 1 suppository each night before bed for 5 nights in a row, break for 2 days and repeat for another 5 days. This is often effective in treating vaginal infections, but the routine may be repeated for several more courses if necessary.

Sitz Bath or Douche for Vaginitis

Powders or tinctures may be used in preparing this douche formula, but powders would be the more economical. Similar in concept to the Suppositories for Vaginitis, this formula contains herbs that are very gentle yet still antimicrobial. For women with acute pain and red, swollen labial or internal tissues due to bacterial or yeast vaginitis, a douche or a sitz bath such as this may be more soothing and provide faster pain relief than suppositories.

Berberis aquifolium	4 ounces (120 g)
Calendula officinalis	2 ounces (60 g)
Hamamelis virginiana	2 ounces (60 g)

Combine the dry herbs and mix well. Steep 6 to 8 heaping tablespoons of the mixture in 8 cups (1,920 ml) of hot water for 15 minutes. Strain and use as a douche or sitz bath. Adding 2 teaspoons of boric acid powder and 10 to 20 drops of tea tree oil will enhance the antimicrobial effects of this formula. For best results, prepare and use this sitz bath or douche at least twice daily for several days in a row.

To prepare a sitz bath, the finished brew is placed in a rubber tub, large enough to comfortably sit in, immersing the buttocks and pelvis. It is usually convenient to place this tub in the bathtub and soak in the sitz bath for 15 to 20 minutes, sitting with the knees folded up. The sitz bath water may be poured down the bathtub drain when finished, as long as it doesn't have a large amount of plant material or particulate remaining. If that is the case, dispose of the used liquid on a lawn or garden area outdoors. When preparing a douche, use a commercial douche bag. Strain the tea and fill the bag. Hang the bag from the bathroom shower or curtain rod above the bathtub to allow gravity to create a strong flow of liquid. Stand in the shower or tub and insert the small tube at the end of the hose attached to the elevated douche bag into the vagina. The flow of the fluid serves to lavage the vaginal vault and is simply allowed to flow out into the shower stall or bathtub. As with the sitz bath, the douche may be performed daily for a series of days.

Suppository for Atrophic Vaginitis

Antimicrobial herbs are not needed for atrophic types of vaginitis. Rather, formulas should emphasize herbs noted to soothe, nourish, repair, or regenerate vaginal mucosal cells. *Althaea* acts as a soothing, moistening demulcent herb in this formula. *Centella* supports healthy connective tissue growth, and *Calendula* supports microvascular circulation. This combination would also be appropriate for dyspareunia.

Althaea officinalis powder	1 ounce (30 g)
Centella asiatica powder	1 ounce (30 g)
Calendula officinalis powder	1 ounce (30 g)
Cocoa butter	4 ounces (120 g)
Liquid vitamin E supplement	2 tablespoons

Combine the powders. Melt the cocoa butter, add the vitamin E, and stir in the herb powder mixture. Pour into vaginal suppository molds, as explained in the instructions for the Suppositories for Vaginitis on page 106. The suppositories may be used Monday through Friday with weekends off, and continued for a month or more until the situation is improved.

Vaginal Douche for Vaginitis

Douches are indicated for irritation of external genitals as well as for active yeast or bacterial vaginitis. They can be used acutely to treat infections as well as to help relieve discomfort. Douches should not be necessary when the vaginal and intestinal ecosystems are healthy.

Berberis aquifolium powder
Calendula officinalis powder
Geranium maculatum powder
Symphytum officinale powder

Combine the powders in equal parts and mix well. Pour 5 cups (1,200 ml) of hot water over 3 to 4 heaping tablespoons of the powdered herb mixture. Steep for 10 minutes, strain, and use as a douche, sitz bath, or herbal wash for vaginal yeast or bacteria. Add boric acid powder and/or tea tree oil as needed. Use ¼ cup (30 g) of boric acid powder per 5 cups of water. Ten drops of tea tree or other essential oil can be added to the finished product at the time of using a wash, douche, or sitz bath, taking care to fully blend by stirring or agitating to disperse the essential oil to prevent it from floating on the top of the water and having potentially irritating effects. Double the quantity, as 10 cups (2,400 ml) may be needed if preparing a sitz bath to ensure that the quantity of liquid is sufficient to immerse the genitals. Refer to the Sitz Bath or Douche for Vaginitis for instructions if using this formula as a sitz bath.

Decoction for Atrophic Vaginitis

Asparagus racemosus roots, known in Ayurvedic traditions as shatavari, is a purported female rejuvenating herb that provides demulcent effects with an affinity for the genitourinary mucosa. *Pueraria* offers hormonal support. Drinking this decoction daily might be sufficient to provide vaginal comfort for women who wish to avoid prescription hormones or for those for whom hormone replacement therapy is contraindicated.

Asparagus racemosus	1 pound (450 g)
Pueraria lobata	1 pound (450 g)

Mix the dry herbs together. This creates enough of the herb blend to consume daily for many months. Decoct 1 teaspoon mixed herbs per cup of hot water and aim to drink 3 (720 ml) or more cups daily, continuing for several months. Additional herbs such as cinnamon, ginger, or orange peel may be added if desired to vary the flavor. Evaluate efficacy following 6 months of daily use.

Tincture for Vaginitis Due to Obesity or Diabetes

Both obesity and diabetes can lead to fluid and circulatory stagnation in the tissues and a shift in beneficial intestinal and vaginal flora. Antimicrobial douches and suppositories such as those described previously may help temporarily, but patients should also be offered systemic therapies to help treat the underlying stasis. The herbs in this formula are chosen due to folkloric emphasis for pelvic stagnation with profuse discharges, as can occur with severe yeast vaginitis. This formula is *not* for vaginal douching. It should be taken orally for many months to treat underlying stasis in the pelvis. This formula could be amended to better address diabetes or any other contributor.

Ceanothus americanus	20 ml
Hamamelis virginiana	15 ml
Angelica sinensis	15 ml
Zingiber officinale	10 ml

For oral use as a preventive and restorative, take 1 teaspoon of the combined tincture, 2 to 4 times daily. Use in conjunction with a local topical treatment as needed.

Formulas for Pelvic Inflammatory Disease

Pelvic inflammatory disease (PID) is an inflammatory and/or infectious disorder of the upper female genital tract, including the uterus, fallopian tubes, and adjacent pelvic structures. PID may spread upward to the peritoneum, and the condition may be infectious, inflammatory, or (commonly) both. Infection and inflammation of the uterus (endometritis) and fallopian tubes (salpingitis) occurs most often in young sexually active women, and *Chlamydia* (and other bacteria) and viruses are possible causative pathogens introduced via sexual intercourse. Sexually transmitted diseases (STDs) that may cause PID, especially chlamydia and gonorrhea, are typically harbored in both sexual partners, so all partners of women with PID should be examined and treated accordingly. The fallopian tubes can be scarred and obstructed resulting in infertility, even with one episode of PID, so all cases are to be treated immediately and aggressively, typically with antibiotics. PID may also result from the use of IUDs or following vaginal delivery or abortion. Severe abdominal pain, fever, and copious vaginal discharge are the typical symptoms. Excruciating pain is experienced with palpation of the adnexal regions and bimanual pelvic examination. Frequently, infectious PID becomes chronic as adhesions and obstructions cause scar tissue pockets where bacteria may flourish.

Although PID is probably best managed with antibiotics, herbal and other therapies are useful complements. Castor oil packs over the pelvis, herbal antimicrobials and immune tonics, hydrotherapy, vaginal steams, or sitz baths are all appropriate. Herbs that may reduce fibrosis or excessive inflammatory activity include *Curcuma*, *Astragalus membranaceus*, *Equisetum arvense* or *E. hyemale*, and *Centella*. PID infections can also involve a large amount of exudate that may spread into the abdominal cavity and cause peritonitis or result in the formation of large abscesses. *Berberis*, *Coptis*, and *Hydrastis* are some of the best herbs for reducing abundant discharges from mucous membranes. *Andrographis paniculata* is used widely in China for infectious disease and immune support and is a traditional ingredient in formulas for PID. Animal studies show the herb to exert significant anti-inflammatory effects in a manner that may reduce fallopian scarring or adhesions.[16] *Smilax ornata* root preparations are also widely used in China to treat PID, and animal models have shown *Smilax* to inhibit hyperplasia of inflamed endometrial tissues and promote recovery.[17]

Antimicrobial Tincture for Acute PID

This formula is based on herbs mentioned in current literature for PID and complements the use of antibiotics because it is aimed more at preventing scarring, obstruction, and adhesions than for a strong antimicrobial effect. *Centella* has wound-healing effects and an affinity for connective tissue. *Salvia miltiorrhiza* was noted to reduce the sequelae of tubal occlusion when given concomitantly with antibiotics in cases of acute chlamydial PID in mice[18] and is included here because of that possible beneficial effect. Both *Salvia miltiorrhiza* and *Angelica sinensis* are blood-moving herbs that have many anti-inflammatory effects on blood cells and pelvic organs. *Astragalus* provides immune support, and *Thuja* is an antimicrobial with an affinity for the reproductive organs and is emphasized in the folkloric literature for STD symptoms.

Smilax ornata	10 ml
Andrographis paniculata	10 ml
Salvia miltiorrhiza	10 ml
Astragalus membranaceus	10 ml
Centella asiatica	10 ml
Angelica sinensis	5 ml
Thuja plicata	5 ml

This 2-ounce (60 ml) formula may doubled or quadrupled, as it is best to employ it aggressively, taking 1/2 to 1 teaspoon of the combined tincture hourly, reducing as symptoms improve. Consider combining the use of this formula with encapsulated *Andrographis* or another strong antimicrobial, to complement prescription antibiotics.

Tea for PID

This tea blend is aimed at reducing inflammation and the likelihood of adhesions or scarring of the fallopian tubes and would complement the Antimicrobial Tincture for Acute PID.

Smilax ornata
Centella asiatica
Berberis aquifolium
Astragalus membranaceus

Combine equal parts of the dry herbs, blend, and store in an airtight container. Gently decoct 1 to 2 teaspoons of the mixture per cup of hot water. Drink as much as possible throughout the day and continue for 2 to 4 weeks, depending on the severity or chronicity of the infection.

Andrographis Capsules for Chlamydia-Induced PID

Chlamydia trachomatis is the causal organism of the most common sexually transmitted bacterial disease worldwide. If left untreated, chlamydia may ultimately damage tissue and scar the fallopian tubes, leading to infertility. Andrographolide, a diterpenoid lactone in *Andrographis paniculata*, is shown to inhibit *C. trachomatis* organisms and significantly reduce the secretion of interleukins, a type of inflammatory mediator, and cellular damage induced by interferon-γ.[19]

Andrographis paniculata capsules

Take 2 capsules every 3 or 4 hours for 3 or 4 days, then reduce to 3 times per day, continuing until symptoms have fully resolved.

Berberine Capsules for PID

In TCM, berberine-containing plants are thought to clear damp and toxic heat. Herbs that contain berberine include *Berberis aquifolia*, *B. nervosa*, *Phellodendron amurense*, and *Coptis chinensis*. *Phellodendron* contains the isoquinoline alkaloids berberine, palmitine, and phellodendrine, which are credited with many pharmacological effects including antimicrobial and anti-inflammatory actions. Research has shown *Phellodendron* to reduce inflammation and deter microbes in PID.[20] *Coptis* roots contain berberine alkaloids and are available in the marketplace as small vials of tablets packaged in small boxes under the name of huang lian su. Berberine is also available purified from various plants, and berberine capsules are readily available on the market.

Berberine capsules
Huang lian su tablets

Take either 2 berberine capsules or 2 huang lian su tablets, 3 or 4 times daily for acute PID. Continue dosing 1 to 2 times daily for an additional 1 to 2 weeks after the acute symptoms have cleared.

Formulas for Genital Herpes

The herpesvirus family includes more than 70 viral members, including the herpes simplex virus responsible for recurrent cold sores and genital herpes. The herpesviridae are an extremely successful family of DNA viruses, capable of infecting all known vertebrate species.[21] It is estimated that about half a million new cases of genital herpes arise worldwide each year and that there are about 30 million people with genital herpes in the United States alone.[22] The herpesvirus family includes varicella-zoster virus, the virus responsible for causing chicken pox and shingles, and several viruses associated with chronic fatigue syndrome, including cytomegalovirus and the Epstein-Barr virus. Cytomegalovirus[23] is capable of damaging the brain and nervous system in susceptible individuals such as infants and AIDS patients. Many or even most patients infected with genital herpes have no symptoms and shed the virus intermittently through the genital tract. The herpes virus can be transmitted to neonates during birth, and cesarean section is recommended for woman with active genital herpes outbreaks at the time of delivery.[24] Genital herpes is also believed to increase the risk of HIV infection.[25]

The herpes family of viruses tends to affect nerves and is able to lie dormant in the body for many years, becoming active during times of stress or immune challenges. As the virus becomes active, tingling, burning, or crawling sensations are often produced as the virus inflames local nerves. *Hypericum perforatum* has an affinity for the nerves, in addition to antiviral affects, and so may be a foundational herb in many formulas for treating genital herpes. Another characteristic of herpes simplex and relatives is to emerge with tiny painful blisters, which quickly ulcerate and create red, raw, inflamed lesions that take several weeks to scab over and heal. Like most viruses, outbreaks are accompanied by enlargement of the lymph nodes. Rosmarinic acid has activity against the herpes virus and is found in *Melissa officinalis*, *Teucrium polium*, *Ziziphora clinopodioides*, *Salvia rhytidea*, and other species.[26] Vitamin C may enhance white blood cell activity and interferon production[27] and be a complementary therapy in the treatment of condyloma and herpes infections.[28] Zinc enhances cell-mediated immunity and benefits herpes patients by aiding the immune system in identifying and fighting off the virus.[29] Zinc is best taken in divided doses of 15 or 30 milligrams at a time, due to potential stomach upset, for a total of 60 to 90 milligrams of zinc per day. Lysine is an essential amino acid that may benefit those with genital herpes[30] and may be taken 500 milligrams at a time, 3 to 5 times a day, with food, immediately at the onset of new herpes outbreaks, or in cases of frequent outbreaks, as a daily supplement. In addition to being a source of

rosmarinic acid, *Melissa officinalis* (lemon balm) contains many antiviral constituents. *Melissa* is shown to inhibit the attachment of herpes virus particles to host cells.[31] *Melissa* may also improve the efficacy of antiviral pharmaceuticals, overcoming drug resistance.[32] Quality *Melissa* essential oil is expensive, but low concentrations of 0.025 to 1.5 microgram/milliliter have been shown effective,[33] justifying the investment in an office stock bottle to use in formulations.

Tincture for Chronic Genital Herpes Outbreaks

Herbs demonstrated to have activity against herpes viral strains include *Hyssopus*, *Melissa*, and *Quercus brantii* (also known as *Quercus persica*).[38] Other herbs, including *Panax ginseng*,[39] may boost innate immune responsiveness and improve the body's ability to eradicate viruses.

Hyssopus officinalis
Melissa officinalis
Panax ginseng
Hypericum perforatum
Andrographis paniculata

Combine in equal parts and take 1 dropperful of the combined tincture, 3 times daily. Use as often as hourly for new eruptions.

Licorice Simple for Chronic Herpes Outbreaks

Licorice root has antiviral activity[40] and may alleviate pain and speed healing of herpes lesions.[41]

Glycyrrhiza glabra root

Prepare a tea by boiling 1 teaspoon of shredded root per cup of water for 5 to 10 minutes. Drink 4 or 5 cups (960 to 1,200 ml) of the tea daily at the onset of a herpes outbreak; continue drinking several cups per day during convalescence. Soaking a cloth in the tea water and applying it to ulcers may also speed healing.

Licorice-Lemon Balm Topical Cream for Herpes

Clinical trials show that topical application of *Melissa* extracts to initial herpes outbreaks can prevent recurrence[42] and speed healing. Although the essential oil is expensive, as little as a few drops added to a topical cream may be effective.[43] Aqueous extracts of *Melissa* may also be useful[44] as sitz baths, douches, or topical

Aloe to Treat Viral Infections

The well-known *Aloe* plant has long been used as a topical treatment for burns and wound healing.[34] Immunostimulating compounds in aloe gel,[35] including acemannan, a long-chain polysaccharide, have been noted to have immune activity useful in combating viral infections. Acemannan, like other immune polysaccharides, has been shown to promote white blood cell numbers, their phagocytic abilities, their production of interferon, and their ability to fight viruses.[36] Since *Aloe vera* is an excellent antiulcer and wound-healing agent, it may be applied topically to herpes lesions as well as taken internally to have an antiviral effect.[37] One or 2 ounces of the liquid may be combined with other fruit juices, teas, or blended drinks.

Aloe barbadensis, aloe

compresses. Tea tree oil (*Melaleuca*) is much less expensive and has also shown activity against the herpes virus.[45] *Glycyrrhiza* may be applied topically, as well as taken internally for best results.[46] Salves containing glycyrrhizinic acid, one of the antiulcer compounds in licorice, are commercially available.

Aloe vera gel	2 ounces (60 ml)
Glycyrrhiza glabra solid extract	2 ounces (60 ml)
Melaleuca alternifolia essential oil	100 drops
Melissa officinalis essential oil	10 drops

Place the aloe gel in a shallow bowl and whisk in the *Glycyrrhiza* solid extract by blending rapidly with a fork. When thoroughly combined, blend in the essential oils and transfer to a 4-ounce salve jar. Apply topically every half hour at the onset of an outbreak. Reduce the frequency of application as the lesions heal and pain subsides.

Herbal Sitz Bath for Genital Herpes

Azadirachta indica potently inhibits the entry of HSV-1 into cells and may be included in formulas to deter outbreaks of genital herpes lesions as well as treat initial infections.[47]

Azadirachta indica	8 ounces (240 g)
Quercus alba	8 ounces (240 g)
Hamamelis virginiana	8 ounces (240 g)
Calendula officinalis	8 ounces (240 g)
Aloe vera gel	2 to 4 tablespoons

Combine equal parts of the powdered herbs and add 1 to 2 cups (120 to 240 g) of the blend to 15 cups (3,600 ml) of water in a sitz bath. Add the *Aloe vera* gel to the bath. Refer to the Sitz Bath or Douche for Vaginitis on page 106 for instructions on preparing and using a sitz bath. At the close of each bath, apply additional aloe gel directly to the lesions.

Formulas for Human Papillomavirus

The human papillomavirus (HPV), one of the causes of genital warts, is among the most common sexually transmitted diseases, thought to affect around 25 million people in the United States alone.[48] There are more than 60 types of HPV: Some infect the skin, causing common warts, and around one-third of HPV types may be spread sexually and cause venereal warts. These types are associated with an increased risk of reproductive cancer.[49] In many cases, no visible warts or lesions can be observed, but the presence of the HPV virus can be detected with a routine Pap smear. Some types of HPV are more virulent and carry a greater cancer risk so pathologists will analyze to identify the precise strains of the virus in order to flag those types of the greatest concern. Colposcopy and cervical biopsies may be employed to better determine cellular changes. HPV is also associated with oral and throat cancers and lesions, transmitted by having oral sex with an infected person. No screening method such as the Pap smear yet exists for screening HPV-caused oral and throat cancer.

Younger people may clear the virus more readily than older people, typically within a few months to a few years. However, in older people or those with immune deficiency, the virus is often tenacious. Supporting the strength of the immune system with nutrition and lifestyle changes would be appropriate. There is some evidence that smoking may increase the tendency to cancerous potential of HPV.

Modern medicine offers laser surgery, cryosurgery, and electrocautery (LEEP procedures) to destroy visible lesions or to ablate the entire cervical os; however, this may be problematic for future pregnancies in younger women. Imiquimod ointments may be effective on external genital lesions, but they are too irritating to use in the vagina, and therefore, on the cervix. While it does not exert antiviral effects on its own, imiquimod primes the microenvironment and supports the body's own immune-mediated clearance of HPV lesions. It has long been used topically on actinic keratosis and is useful on the external genitalia or superficial skin lesions.[50] Retinoids are vitamin A derivatives and have also long been known to suppress cancer development, especially epithelial cancers. Topical application of retinoids will promote regression and healing of cervical cancer in some women.[51] Retinoid compounds are sometimes included in commercial suppositories and are often part of naturopathic physicians' protocols for eradicating HPV. Lopimune, a pharmaceutical used orally for HIV, has been shown to deter HPV when the capsules are used vaginally, twice a day for 2 weeks.[52] Alpha interferon is expensive, only minimally effective, and does not reduce the rate of recurrence. For severe or resistant cases, interferon has been injected directly into the warty growths with some reported success.

Many natural products are in use or under investigation for the treatment of HPV-induced cervical dysplasia

Human Papillomavirus

and cervical cancer. *Podophyllum peltatum*, *Sanguinaria canadensis*, and *Thuja* have antimicrobial ability against sexually transmitted viruses, such as those that cause venereal warts. *Podophyllum* fruits yield a caustic resin that may be applied topically to destroy HPV lesions, but systemic therapies may still be required to fully eradicate the virus. Due to potential irritating and even caustic effects, these herbs are used in only small

Topical Options for Venereal Warts

Pharmaceutical and natural therapies for topical application or vaginal use are as follows:

Azadirachta indica. This traditional Ayurvedic antimicrobial herb has been developed into a medication sold under the trade name Praneem as intravaginal tablets and creams to treat sexually transmitted viruses, including HIV and HPV. Although the product is not available in the United States, it would not be difficult for an herbalist to prepare a vaginal suppository or ointment from neem oil, which is readily available.

Berberine. This isoquinoline alkaloid can be extracted from various plant species and is noted to deter a broad variety of microbes, including HPV.

Cidofovir. This antiviral drug is used in topical creams and as intralesional injections and reported effective in treating papillomavirus-induced skin warts and lesions.

Curcuma longa. *Curcuma* is shown to boost the efficacy of paclitaxel,[53] making smaller doses of paclitaxel possible. *Curcuma* has anticancer effects on its own,[54] and extracts are also included in proprietary ointments used in Asia against venereal warts and to help clear HPV-induced cervical cancer.

Green tea polyphenols. Sold under the trade names Polyphenon and Veregen, this ointment concentrates the polyphenols from green tea to create a powerful antioxidant for topical and local application. This ointment is unreasonably expensive, and many clinicians use simple green tea capsules vaginally in lieu.

Imiquimod. This pharmaceutical cream has been used for actinic keratoses and is shown to help treat basal cell skin carcinomas when topically applied. The cream is also reported to help clear warts and lesions due to HPV.

Larrea tridentata. The resinous leaves are high in nordihydroguaiaretic acid (NDGA), credited with potent antioxidant and anticancer effects. The leaves can be used to prepare oils for making suppositories and ointments or can be prepared into aqueous douches, sitz baths, and compresses.

Lopimune. This prescription AIDS antiviral medication has had only one published clinical trial, but the results were so impressive in clearing HPV with vaginal use that the drug requires further investigation.

Paclitaxel. Isolated from *Taxus brevifolia*, paclitaxel is effective against a wide spectrum of human cancers, including human cervical cancer. However, dose-limiting toxicity and high cost limit its clinical application.

Podophyllotoxin. Also known as podofilox (trade name Condylox), these gels are prepared from a toxic lignan derived from *Podophyllum peltatum* or *Juniperus communis* and used topically as antimitotic agents on anogenital warts.

Thuja plicata. *Thuja* has antiviral properties and is traditionally included in wart and cervical cancer protocols in traditional herbal medicine. Labdane terpenoids may have antimutagenic effects, and many *Thuja* species also contain podophyllotoxin.[55]

Trichloroacetic acid. A strong acid, this caustic agent is applied daily to visible warts.

Vitamin A. Vitamin A and retinoid metabolites are essential for epithelial cell health and barrier function, and carotenoid and retinoid deficiency may increase the risk of malignant transformation. Naturopathic physicians often include vitamin A suppositories and oral supplementation in protocols for venereal warts and cervical cancer.

amounts, combined with other less irritating herbs, to treat condyloma and HPV infections.

Alternative practitioners report success in eradicating HPV, often with rather extensive protocols involving the use of herbal and nutritional systemic therapies and local therapies known as escharotic treatments, which are described in detail in the Escharotic Treatments formula on page 114. One clinic has reported success in reversing high-grade cervical intraepithelial lesions using two escharotic treatments per week for 5 weeks. *Thuja* suppositories are employed during the month of escharotic treatments, followed by green tea suppositories for another 2 months, complemented by systemic antiviral and immune support, mentioned below.[56] I, too, have successfully eradicated HPV using escharotic treatments.

The use of antiviral and immune-supportive herbs and nutrients are part of most alternative protocols. *Thuja* is a traditional herb used for genitourinary tract infections and warts, and modern research shows the plant to have activity against HPV.[57] *Andrographis paniculata* is widely used for infectious diseases, and activity against HPV[58] is credited to suppression of viral transcription by andrographolide.[59] *Camellia sinensis* is credited with numerous antioxidant and anticancer properties, including antiviral activity,[60] largely credited to epigallocatechins. *Larrea tridentata* (chaparral) contains nordihydroguaiaretic acid (NDGA) credited with potent antioxidant and anticancer effects.[61] *Berberis* is a source of berberine with numerous antimicrobial actions, including an ability to suppress HPV transcription.[62] See also Volume 1, chapter 5, for a discussion of HPV and formulas and protocols for warts.

Topical Compress for Genital Warts

Genital or venereal warts are highly contagious and may take forms ranging from a miniscule pimple to massive, tissue-destroying, tumorlike growths. Genital warts are also referred to as condyloma acuminata because they often accumulate in tiny clusters. In women, condyloma warts may occur on the labia, vagina, and cervix. In men, condyloma appears on the shaft or more commonly the glans of the penis. In both sexes, condyloma warts may also occur around the anus. Genital warts may simply go away on their own without treatment or may induce cancerous changes, without obvious warts or lesions initially. *Thuja* is listed in many historical herbals as specific for sexually transmitted diseases and genital lesions. *Larrea* may be difficult to find as an oil and need to be made for one's self. Vitamin A (retinyl palmitate) can be used from a commercial liquid or may be squeezed from

several gel capsules. A standard dose provided by a gel cap, dropper, or liquid is 25,000 international units (IUs).

Podophyllum peltatum tincture	1 ml
Larrea tridentata oil	1 ml
Vitamin A oil	1 ml
Thuja plicata essential oil	30 drops

Combine all in a small dish or measuring cup and blend well. Saturate a cotton ball with the mixture and apply to the wart(s), and lie down for 20 minutes to help keep the application in place. Rinse off. Repeat twice daily and continue for at least 2 weeks.

Topical Application of Podophyllin for Warts

Podophyllin is a caustic resin derived from the mayapple plant, *Podophyllum peltatum*. It has been used topically to treat warts for at least 150 years and is on the World Health Organization's list of essential medicines. Podophyllin is typically prepared as a 20 percent solution, diluting the plant resin with tincture of benzoin. This solution is applied topically to the labia, penis, and other external genital tissues and washed off 4 or 5 hours later. Though the application is not usually painful, podophyllin will destroy the tissues it comes into contact with and will cause the wart to slowly dry up and simply flake away. Occasionally, some people may experience redness, irritation, and itchy inflammation at the site of application. To protect the skin surrounding the wart from the caustic podophyllin resin, apply a greasy ointment such as *Calendula* salve. Podophyllin is not appropriate to use intravaginally, on the cervix, or in the anus, because the medicine won't stay put on these moist mucous membranes and will harm surrounding healthy tissues. Several synthetic versions of *Podophyllum* resin are available on the market, including podofilox (Condylox). Podophyllin resin is very expensive, but usually highly effective. Podophyllin contains podophyllotoxin, a molecule known to inhibit cell division in warty growths, where condyloma or papillomaviruses stimulate excessive cell proliferation and increase the occurrence of cervical, penile, and other genital lesions.

Podophyllin resin fixed in tincture of benzoin

A single application of a single drop of podophyllin may be sufficient to destroy a wart in many cases. For a large, fleshy growth or unresponsive cases, the podophyllin resin may be applied every 2 to 4 weeks as needed. Due to the potential to cause birth defects, podophyllin should not be used, even topically, by pregnant women.

Douche for HPV and Abnormal Paps

This formula is a variation of the traditional "vag pack" and escharotic treatment. This formula may be less powerful, but it is much safer and is more easily prepared by those who don't have fast or easy access to the ingredients needed to prepare the traditional remedies described in the Vag Pack formula that follows and the Escharotic Treatments formula on page 115. *Sanguinaria* is potentially caustic when applied topically as a black salve. A tincture is less caustic, but even a tincture should be diluted in other fluids, as with the other tinctures in this formula. (See also the *Sanguinaria canadensis* entry on page 177). *Podophyllum* resin prepared from the mayapple fruit is highly caustic, but leaf powder is less concentrated and gentle enough to be prepared into a vaginal douche. *Hydrastis* is an excellent douche herb, but it is expensive; *Berberis* or *Coptis* powder are possible substitutes.

Podophyllum peltatum dried leaf	4 ounces (120 g)
Thuja plicata dried leaf	4 ounces (120 g)
Hydrastis canadensis dry powder	2 ounces (60 g)
Sanguinaria canadensis tincture	4 ounces (120 ml)

Combine the dry herbs and powder and blend. This quantity of ingredients allows for the preparation of 3 douche treatments per week for many months. Decoct 1 heaping tablespoon of the blended powders in 2 cups (480 ml) of hot water, let stand until tepid, strain, and add 1 teaspoon of *Sanguinaria* tincture. Add enough additional warm water to fill a douche bag or prepare into a compress to use as a poultice on external genitalia. Label the dried herb bag well as NOT for oral use. Discontinue use if any pain, burning, or skin irritation occurs. Follow each treatment with a vitamin A suppository.

The Vag Pack: A Traditional Eclectic Formula

This formula is for intravaginal use for uterine fibroids, cervical dysplasia, and chronic vaginitis. Anhydrous magnesium sulfate is available from chemical supply houses. Be sure to specify USP (US pharmacopoeia) or food-grade product for this use and not an industrial-grade chemical.

Vegetable glycerine	2 cups (480 ml)
Anhydrous magnesium sulfate	1 cup (200 g)
Hydrastis canadensis powder	3 ounces (90 g)
Citrus aurantium (bitter orange) essential oil	1 ounce (30 ml)
Thuja plicata essential oil	1 ounce (30 ml)

Place all the ingredients in a quart jar and shake vigorously to create a thick, soupy blend. Form tamponlike structures by spreading a liberal tablespoon on lamb's wool (available from pharmacies) and wrapping the wool to house the vag pack medicine inside. Use dental floss or strong string to secure the pack, leaving a long "tail" to assist in removal. Placement of the pack against the cervix or area to be treated should be done by a physician or clinician with the aid of a speculum. This is usually done at the close of an escharotic treatment. Remove the pack in about 24 hours. Repeat on a weekly or biweekly basis for several months.

Escharotic Treatments (Traditional Naturopathic Formula)

Eschar means to throw off or to scar, and escharotic treatments involve the use of caustic herbs topically on the cervix to "throw off" dysplastic cells and scar over lesions. This technique is best learned under the tutelage of an experienced physician because the herbs are caustic and potentially harmful if the procedure is not performed safely. *Sanguinaria* has numerous antimicrobial and anticancer mechanisms.[63] The fresh juice or poultices can be caustic and, when mixed with zinc chloride, markedly so. Bromelain helps to loosen the most superficial cells and helps to allow the medicine to penetrate more deeply. The FDA has warned that escharotic treatments in the form of "black salves" that contain these ingredients are "fake" cancer cures, and indeed, such salves are extremely dangerous and are not proven to cure cancer. Yet I have seen this escharotic therapy return dysplastic Pap smears to normal and

---

Nutraceutical Agents for HPV

In addition to herbal options for HPV, many nutraceutical agents may be complementary, supporting the immune system, protecting the cervix from oncogenic changes, and helping to clear the HPV virus. Many naturopathic physicians suggest supplementing with any or all of the following agents for 6 months or possibly for as long as a year or two.

Green tea extract	500 mg/day
Diindolylmethane (DIM)	300 mg/day
Folic acid	10 mg/day
Vitamin C	6 g/day
β-carotene and/or natural mixed carotenoids	150,000 IU/day
Vitamin E	400 IU/day

cure carcinoma in situ, and so have my colleagues.[64] The website Quackwatch issued a warning against the use of escharotics in 2008, and the American Academy of Dermatology has issued a press release to warn of the potential danger inherent in using such caustic substances on skin lesions. I concur that these agents are not safe for public consumption, but argue that in skilled hands, such agents may be powerful herbal tools.

Bromelain capsules, to open for topical application
Calendula officinalis succus
Zinc chloride liquid
Sanguinaria canadensis tincture

Caution: I describe this procedure here in detail, but I do not recommend attempting it without first receiving one-on-one instruction in the technique from an experienced practitioner. Bromelain powder is applied to the cervix by a physician with the aid of a speculum and a large cotton swab and followed with the use of a heat lamp to help activate the powder. After 20 minutes, the powder is rinsed off with *Calendula* succus. Then 5 to 10 milliliters each of zinc chloride liquid and *Sanguinaria* tincture are combined and applied to the cervix, taking care not to touch the vaginal walls or the external genitalia in the process. After 5 minutes, the blend is rinsed off with the *Calendula* succus. Any fluid that pools in the vagina from this procedure is immediately mopped up using large cotton swabs. A vaginal pack (vag pack), as described on page 114, is inserted and left in place for 24 hours. The escharotic series involves 2 treatments per week for 4 to 5 weeks. Allow the cervix to heal for at least several months before repeating a Pap smear.

Formulas for Treating Menstrual Cycle Disorders

Menstrual irregularities are a common complaint in a general practice and may accompany premenstrual syndrome (PMS) or occur independently, as in the case of amenorrhea, irregular menses, menorrhagia, and dysmenorrhea. Abnormal uterine bleeding includes metrorrhagia, menorrhagia, midcycle spotting, prepubertal bleeding, and postmenopausal bleeding. Such cases that are not caused by organic lesions are termed dysfunctional uterine bleeding (DUB). Polycystic ovarian syndrome (PCOS) may also involve menstrual irregularities, especially amenorrhea. (See "Formulas for Polcystic Ovarian Syndrome [PCOS]" on page 144.)

Serenoa and *Vitex* are appropriate foundational herbs in formulas for amenorrhea. *Vitex* is also a good foundational herb for irregular menses of all types, including both long and short cycles. For more about *Vitex*'s many hormonal actions, see "*Vitex agnus-castus* for Reproductive Complaints" on page 148. Heavy menses can often be improved by lowering elevated estrogen with alterative herbs such as *Taraxacum* and increasing progesterone with *Vitex*, while including hemostatic agents such as *Achillea* and/or *Cinnamomum*. Dysmenorrhea, meaning painful or difficult menses, often responds to the aptly named cramp bark, *Viburnum opulus*.

Before detailing all the presentations that appear to pathologize the menstrual cycle, I wish to reflect on the importance of honoring the sanctity of hormonal rhythms. There are inherent rhythms throughout the planet that reflect cosmic forces: day followed by night, waxing and waning moon cycles, inspiration and expiration, the beating of the heart, birth and death, and so on. The cyclical nature of women is a reflection of the universal yin, a feminine cosmic force. In this perspective, the ebbs and flows of a woman's energy, patience, strength, and creativity are not a pathology, but rather the natural and healthy state of being. During a woman's monthly cycle, there are cyclical rhythms in her brain wave patterns, blood protein content, heart rate, vitamin C content of blood cells, olfactory acuity, and a myriad of other biochemical patterns. To *not* experience ebbs and flows during the month in physical symptoms as well as mental-emotional tone is not only difficult to accomplish, but it is also physiologically impossible. But too many aspects of our culture do not honor these cycles as sacred and, in fact, pathologize them. Not all symptoms related to the menstrual cycle are medical problems, but may in fact result from our cultural and societal inability to honor the sacred rhythms with appropriate actions (staying home, having someone else cook and do the chores, spend time on art and dreams instead of work and activities that are challenging to the power of the physiologic state).

If a woman doesn't honor her own power, she certainly can't ask her husband or significant other or children to solve this for her. If a couple is fighting against the natural cycles and rhythms, then the relationship cannot

Types and Causes of Amenorrhea

Primary amenorrhea. Failure of initial menarche is typically due to hormonal abnormalities, but less commonly may also be due to genetic abnormalities, such as Turner syndrome. Signs of hypogonadism include the lack of development of breasts or immature vaginal mucosa. An imperforate hymen is another cause of primary amenorrhea.

Secondary amenorrhea. The cessation of menses at any point in time after initial menstruation has begun is typically due to hormonal imbalances, sometimes simple as in the case of hypothyroidism,

and sometimes quite complex as in the case of polycystic ovarian syndrome (PCOS). Gynecologic tumors, ovarian cysts, or other pathology may also cause secondary amenorrhea.

Unsuspected pregnancy. This possibility should always be considered and ruled out in any case of amenorrhea in sexually active women.

Low body weight; insufficient body fat. If a physical exam reveals very low body weight and insufficient body fat along with amenorrhea, the cause may be anorexia or excessive physical exercise (such as the exercise regimen of endurance athletes).

---

thrive. Honoring a woman's rhythms has a tremendous ripple effect. It is not just for the benefit of women; it is significant for the balance of the entire planet. Many ancient societies recognized the importance of working *with*, not *against*, lunar cycles and women's rhythms. Native Americans had the moon lodge; the Middle East had the menstrual tent.

Any attempt or expectation to "not" have any mood swings, ups and downs, or display monthly rhythms is not reasonable. Some women are crippled by these swings, and I offer herbs and a variety of approaches to make this process as comfortable as possible for them. I also engage them in conversations aimed at understanding the beauty of the cycle. Irritable outbursts and tearful breakdowns are usually not a sign of mental illness in an otherwise healthy woman. They are likely signs of a woman allowing herself to be tossed about on the stormy seas of hormonal currents. A woman may have kindly and patiently picked up the dirty dishes strewn about the house, have taken on unfinished tasks of her coworkers' duties, ignored the neighbors' barking dog, or shrugged off errors made by an incompetent business partner, and then one day vents in a less-than-grounded manner about the burdens. Whether it's high stress in the workplace or fatigue from caring for young children or worry due to financial insecurity, a woman may maintain a certain stoicism for weeks. But then at a certain time in the cycle, some little thing breaks through her armor and she lashes out or breaks down in

tears. If a woman understands exactly what is going on in her body in these moments, she can better have a bit of humor about them and aim for healthy and balanced ways to vent, to shed a tear, and to claim some personal space without alienating her family, friends, neighbors, and coworkers. With awareness of and appreciation of the earth's and her own rhythms, she can anticipate the high hormone days of the month, and learn to master the power of these days, rather than be controlled by them. And herbs can help!

Tincture for Dysfunctional Uterine Bleeding (DUB)

Formulas for DUB might include hormonal balancers, alteratives, and general hemostatics, when needed. *Vitex* may often serve as a lead herb in formulas for many types of DUB because it improves gonadotropin control of the menstrual cycle at the level of the hypothalamus and pituitary. *Vitex* may even help control bleeding irregularities due to the use of an IUD.[65] Alteratives and liver herbs are useful if the bleeding irregularity is suspected to be associated with unopposed estrogen or the liver's inability to conjugate and eliminate estrogen. *Arctium lappa, Taraxacum, Berberis, Curcuma*, and other liver herbs are appropriate alterative choices. *Viburnum* is a good addition if the bleeding is associated with uterine muscle tone problems, both atonic and hyperspastic. Unlike the Tincture for Menstrual Cramps (page 117), this formula contains no strong styptics, so is best used as a foundational tonic attempting to remedy the underlying causes of DUB.

Vitex agnus-castus	30 ml
Taraxacum officinale	15 ml
Viburnum opulus	15 ml

Take 1 dropperful of the combined tincture, 3 to 6 times daily for several months to allow for the formula to take effect. If no results have occurred in 3 months, reevaluate the case.

Tincture for Menstrual Cramps

Dysmenorrhea is known to involve increased production of proinflammatory prostaglandins that lead to contraction of uterine vasculature and the myometrium.[66] *Boswellia* and *Angelica*[67] in this formula may reduce inflammatory prostaglandins. Bromelain and *Oenothora* (evening primrose) oil may also produce this effect and may be complementary to this herbal tincture. Uterine contraction also involves the influx of calcium ions into myometrial cells, and isoliquiritigenin from *Glycyrrhiza* and tetrahydropalmatine from *Corydalis yanhusuo* can reduce uterine spasms by significantly decreasing the level of intracellular calcium ions.

Boswellia species
Angelica sinensis
Corydalis yanhusuo
Glycyrrhiza glabra
Viburnum opulus

Combine in equal parts. Take 1 dropperful, 3 times daily, aiming to reduce uterine hyperexcitability over time. The formulas may also be used acutely by taking a dropperful every 10 to 15 minutes at the time of menstrual cramps.

Tea for Menstrual Cramps

This formula contains a combination of leading pelvic pain and fluid stagnation herbs from both Western and Eastern traditions.

Bupleurum chinense
Paeonia lactiflora
Viburnum opulus
Angelica sinensis
Glycyrrhiza glabra
Cinnamomum verum

Combine equal parts of the dry herbs. Simmer 1 teaspoon of the mixture per cup of hot water and then strain. For severe cramps, prepare 6 or more cups (1,440 ml) of the brew and sip constantly through the day.

Dysfunctional Uterine Bleeding Versus Organic Lesions

About 25 percent of abnormal bleeding cases are due to organic lesions including uterine fibroids, endometrial lesions, and cancers. The cancerous lesions are most commonly seen in postmenopausal bleeding disorders and necessitate an endometrial biopsy, as do any other suspicious cases. The other 75 percent of bleeding abnormalities are entirely due to faulty hormonal rhythms and controls. When organic lesions have been ruled out, DUB (dysfunctional uterine bleeding) may be diagnosed.

Hypothyroidism and/or PCOS are both associated with abnormal uterine bleeding due to hormonal abnormalities. Spotting that occurs characteristically only at midcycle is usually due to high estrogen and relatively low progesterone and is a common side effect of birth control pill (BCP) use. Midcycle spotting in women not using BCPs indicates either that there is a high estrogen load in the body—such as with exposure to exogenous estrogens or faulty liver conjugation and clearance—or low progesterone. Spotting that occurs sporadically, not at midcycle, is more likely due to fibroids, polyps, or other organic lesions.

Hemostatic Tincture for Menorrhagia

Heavy menstrual flow can be due to hyperestrogenism and can be addressed with alterative herbs, liver support, and avoidance of exogenous estrogens. Heavy uterine bleeding can also occur due to uterine fibroids and is difficult to control. (See "Formulas for Uterine Fibroids" on page 150.) *Vitex* promotes progesterone to help address hyperestrogenism, *Achillea* helps reduce excessive flow, *Berberis* helps conjugate and excrete estrogen to reduce excessive stimulation, and *Viburnum* supports a healthy uterine tone to reduce menstrual pain. This formula could be easily amended or simplified when only some of the symptoms are present. The hemostatic herbs *Achillea* and *Cinnamomum* make this formula more appropriate for acute hemorrhage and prolonged menses than the Tincture for Dysfunctional Uterine Bleeding (page 116). Additional hemostatics may be useful and include *Capsella bursa-pastoris* and *Erigeron canadensis*.

Vitex agnus-castus	20 ml
Achillea millefolium	10 ml
Berberis aquifolium	10 ml
Viburnum opulus	10 ml
Cinnamomum verum	10 ml

Take 1 teaspoon of the combined tincture, 3 to 6 times daily, continuing until symptoms subside.

Tincture for Acute Menstrual Hemorrhage

Whatever the underlying cause, there are times when menstrual bleeding is so heavy as to be hemorrhagic, and the most powerful herbal hemostatics that can be formulated are needed. This formula combines three of our most effective hemostatics for aggressive acute treatment but does nothing to address what is causing the complaint. Other long-term therapies will be necessary after the bleeding is controlled to address fibroids, uterine polyps, endometrial hyperplasia, or other causes of menorrhagia. *Cinnamomum* species are shown to improve dysmenorrhea including menorrhagia.[68] Ergotamine is a natural product derived from fungus that is also useful; it is available by prescription. When oral therapies are not effective at controlling the blood loss, the next course of action is often a hysterectomy.

Achillea millefolium	30 ml
Erigeron canadensis	15 ml
Cinnamomum verum	15 ml

Take ½ teaspoon of the combined tincture as often as every 15 minutes for heavy bleeding. If associated with uterine cramps, add *Viburnum* to the formula. This formula is most effective for hormonal imbalance causing excessive endometrial buildup. It is less effective for heavy bleeding due to uterine fibroids, but nonetheless helpful.

Tincture for Short Menstrual Cycles

Short cycles are often associated with low progesterone levels and/or hyperestrogenism. *Vitex* can promote progesterone, and alterative herbs (*Achillea* in this formula)

Herbs for Menorrhagia

Excessively heavy or prolonged menstrual flow is referred to as menorrhagia. The following herbs are listed in herbal and homeopathic folkloric texts as being specific for helping to control menorrhagia.

CONDITION	HERBS
Menorrhagia due to endometrial hyperplasia	*Achillea millefolium* *Alchemilla vulgaris* *Hamamelis virginiana* *Vitex agnus-castus*
Menorrhagia due to fibroids or polyps	*Achillea millefolium* *Cinnamomum verum* *Hamamelis virginiana* *Hydrastis canadensis*
Menorrhagia with PMS	*Achillea millefolium* *Actaea racemosa* *Vitex agnus-castus*
Menorrhagia with pain and cramps	*Achillea millefolium* *Actaea racemosa* *Dioscorea* spp. *Leonurus cardiaca* *Viburnum opulus*, *V. prunifolium*
Menorrhagia that is worse standing up or with jarring	*Achillea millefolium* *Cinnamomum verum*
Menorrhagia beginning at climacteric	*Viburnum opulus*, *V. prunifolium* *Vitex agnus-castus*

can help reduce hyperestrogenism. The formula may be made more complex as individual symptoms suggest.

Vitex agnus-castus	30 ml
Achillea millefolium	30 ml

Take 1 dropperful of the combined tincture, 3 to 6 times per day. Therapy may begin with 3 times per day dosing, and if ineffective after 2 or 3 months, one may increase to 2 dropperfuls, 3 times a day, or 1 dropperful, 6 times a day for a few more months before moving to another therapy. This frequently is helpful in several months' time, however, at which time the dose can be cut to 1 to 2 times a day for a few more months, and then discontinued altogether.

Uterine Tonics

Uterine tonics are agents able to normalize tone, be it atrophic or hypertrophic. Uterine tonics may be able to allay uterine pain and spasm or limit hyperexcitability of the uterine muscle, or they may increase tone in cases of flaccid, prolapsed, or fibrotic uterine tissue.

Angelica sinensis
Caulophyllum
   thalictroides
Actaea racemosa
Mitchella repens

Rubus idaeus
Senecio aureus
Viburnum prunifolium
Vitex agnus-castus

Tincture for Menorrhagia with Emotional PMS

This formula uses *Achillea* and *Vitex* to help normalize cycle length and flow and adds *Actaea* to help provide a calming antianxiety and antidepressant effect.

Achillea millefolium
Vitex agnus-castus
Actaea racemosa

Combine in equal parts. Take 1 dropperful of the combined tincture, 3 to 6 times per day. This formula should help improve the menstrual cycle over the course of 3 to 6 months and may be gradually reduced as symptoms improve.

Tincture for Menorrhagia with Pain and Cramps

In this formula, *Achillea* and *Vitex* help normalize cycle length and flow, and *Viburnum* acts as a uterine antispasmodic. Scopoletin in *Viburnum* is an effective smooth muscle antispasmodic and can often be fast acting for pain relief.[69]

Vitex agnus-castus
Achillea millefolium
Viburnum opulus

Combine in equal parts. Take 1 dropperful, 3 times daily. Increase intake at times of cramping or switch to a *Viburnum* simple.

Tincture for Menorrhagia at Climacteric

Some women have normal menses their entire lives until perimenopause, when hormones shift, which can cause menorrhagia. Supporting progesterone is often effective in such cases, and if this fails, add adaptogenic and alterative herbs. The goal is not to have perfect 28-day cycles in perimenopause, but to avoid prolonged bleeding, heavy bleeding, or extremely short cycles, such as menses that are just 10 or 14 days apart.

Vitex agnus-castus
Viburnum prunifolium
Glycyrrhiza glabra

Combine in equal parts. Take ½ to 1 teaspoon, 3 to 4 times daily for 3 to 6 months. Once the cycles are more acceptable, the dose may be reduced to just twice a day.

Tincture for Menorrhagia Due to Uterine Atony

Excessive menstrual flow due to atony of the uterine muscle is most likely to occur in older nulliparous women. *Viburnum* is capable of improving uterine muscle tone in cases of menstrual cramps but is also capable of improving excessive flow in cases of uterine atony. The Eclectic authors of the early 1900s believed that the herb could improve tone such that over time, the herb would no longer be needed. *Viburnum* and the other herbs in this formula improve both atonic and hypertonic issues in the uterine muscle. *Angelica* is a well-known "blood mover" in TCM, useful for improving pelvic circulation, and also has spasmolytic action on the uterine muscle.

Viburnum opulus	30 ml
Caulophyllum thalictroides	15 ml
Angelica sinensis	15 ml

Take 1 dropperful of the combined tincture, 3 to 4 times daily. Continue for 3 to 6 months, aiming to improve uterine muscle tone over time.

Tincture for Menorrahgia Due to Endometrial Hyperplasia

In contrast to menorrhagia from uterine atony, heavy menstrual bleeding may also occur with excessive proliferation of the endometrium due to hyperestrogenism. This formula aims to promote progesterone and to oppose estrogen with *Vitex*, while assisting the liver to conjugate and excrete excess estrogen with the alterative properties of *Achillea* and *Curcuma*. *Achillea* also has hemostatic properties for heavy bleeding, and when combined with cinnamon, rounds out this formula for excessive flow.

Vitex agnus-castus	20 ml
Achillea millefolium	15 ml
Curcuma longa	15 ml
Cinnamomum verum	10 ml

Take 1 teaspoon of the combined tincture, 3 to 6 times daily, reducing as bleeding subsides.

Tea for Menorrhagia

For women who tend to have heavy menses, this tea formula may support any of the tinctures for menorrhagia in this chapter. These herbs are chosen for their palatable flavors. The name *Matricaria* is derived from *matrix*, meaning womb, and it is a classic uterine tonic herb, as well as having stress-relieving and balancing effects on the HPA axis. *Alchemilla* is thought to promote

Aromatase Inhibitors

Aromatase is an enzyme that converts circulating testosterone and progesterone into estrogen in peripheral tissues. Aromatase inhibitors can help block excessive estrogen synthesis, which can be a helpful part of therapy for breast and prostate disorders.[70] Aromatase enzymes are abundant in peripheral adipose cells and occur in high amounts in prostate cells as well.[71] Many coumarin-type phytoestrogens, common in legume plants, have been shown to inhibit aromatase,[72] as have many flavonoids. Aromatase inhibitors such as those listed below may help treat conditions related to hyperestrogenism, including uterine fibroids, ovarian cysts, fibrocystic breast disease, endometriosis, and hormonal cancers such as endometrial, breast, and prostate.

Agaricus spp. mushrooms	Dihydroxyflavone
Coix lacryma-jobi	Galangin
Euonymus alatus	Genistein
Glycyrrhiza glabra[73]	Grape seed extract
Scutellaria barbata	Honokiol
α-napthoflavone	Indole-3-carbinol
Apigenin	Isoflavones
Baicalein	Naringenin
Biochanin	Red and purple
Chrysin	plant pigments

progesterone, and cinnamon is one of the most effective herbs for menorrhagia and has a better flavor than the other very effective choice, *Achillea*.

Alchemilla vulgaris leaf	2 ounces (60 g)
Matricaria chamomilla flower	3 ounces (90 g)
Cinnamomum verum powder	1 ounce (30 g)

Steep 1 tablespoon of the combined herbs per cup of hot water. Drink as much as possible, at least 3 cups (720 ml) per day.

Tea for Menstrual Nausea and Digestive Upset

I see this condition less commonly than painful or heavy menses, but menses that occur with nausea, vomiting, diarrhea, and digestive disturbances often respond to *Foeniculum*, *Mentha piperita*, *Zingiber*, and *Matricaria chamomilla* as a tea or tincture. For those too queasy to keep anything down, even the essential oils of mint or fennel applied topically may be helpful. Clinical trials have shown *Foeniculum* to be valuable for menstrual nausea.[74] *Dioscorea* is useful for menstrual cramps and digestive contraction and bloating.

Foeniculum vulgare
Dioscorea villosa
Zingiber officinale

Combine equal parts of the dry herbs. Simmer 1 teaspoon of the mixture per cup of water for 10 minutes and then strain. Drink in small sips for nausea. Carbonated beverages are sometimes better tolerated in cases of nausea, and the prepared tea can also be chilled and combined in equal parts with ginger ale or sparkling water.

Tincture for Frequent Spotting

Spotting refers to small amounts of menstrual bleeding at inappropriate times. Frequent spotting can be due to uterine polyps or fibroids but can also be due to hyperestrogenism. Estrogen causes a buildup of the endometrial tissue, but when progesterone elevates midcycle, the endometrium develops many crypts and invaginations instead of proliferating more thickly. When estrogen is excessive, or when progesterone is lacking, the endometrium may build excessively and start to shed at times other than with the normal menses. Agents that reduce excessive estrogen, such as alterative herbs, may be one prong of a formula for spotting. *Achillea* is chosen as an alterative in this formula for its additional hemostatic effects. Agents that increase progesterone, such as *Vitex* or *Glycyrrhiza*, would be the other important prong of

Dioscorea and "Wild Yam" Creams

Progesterone creams available as over-the-counter products may do more harm than good in the treatment of reproductive disorders. Hormones such as progesterone, even if described as "natural" progesterone, may put stress on the liver and will further compromise estrogen clearance. The use of progesterone cream, while it may help for a short period of time, is rarely a significant help for many pathologies in the long run. For many women, their progesterone may actually be in the normal range, but their estrogen is so excessive that the small amount of supplemental progesterone supplied in a cream is insufficient to oppose it. The use of progesterone cream will temporarily oppose the elevated estrogen but do nothing to lower estrogen and give the liver even more hormones to process. Furthermore, use of progesterone cream on the skin could actually lead to the production of more estrogen because the aromatase enzymes in the subcutaneous fat cells convert progesterone into estrogen. Thus, herbal, nutritional, and dietary therapies that reduce exogenous estrogens, correct hormonal imbalances in the body, and improve the liver's clearance of estrogen may be a better place to start than using progesterone cream without clear laboratory testing to suggest that it is indicated.

"Wild yam" creams are popular in the commercial marketplace to treat menopause. Wild yam (*Dioscorea* species) contains steroidal saponins such as diosgenin that may offer some adrenal supportive effects. Diosgenin can be manipulated to turn

Dioscorea villosa, wild yam

into progesterone and has been used as a raw material to produce steroid products ranging from birth control pills to progesterone creams. However, I do not believe that wild yam diosgenin converts into progesterone *in the body* when orally ingested or topically applied, despite promotional claims, and I think herbalists have many more-effective tools than wild yam cream for treating menopausal symptoms. Furthermore, if diosgenin has been turned into progesterone, such creams should be called "progesterone" and not wild yam products. Hence, there is much confusion about wild yam in the marketplace.

Diosgenin also occurs in fenugreek (*Trigonella foenum-graecum*) and fennel seeds (*Foeniculum vulgare*), both of which are thought to be galactogogues and are known to have hypolipidemic and hypoglycemic effects. Diosgenin inhibits proliferation and induces apoptosis in a wide variety of human cancer cells, including breast cancer.[75] Diosgenin also inhibits the activity of a protein complex referred to as NF-κB and the gene expression it regulates. Diosgenin also abolishes cyclooxygenase and lipoxygenase activity, and all of these mechanisms may offer chemotherapeutic effects.[76] *Dioscorea* may support fertility, and animal studies have shown diosgenin to enhance ovarian follicles in models of aging.[77] Therefore the oral use of *Dioscorea* in herbal teas and tinctures can be helpful for many health conditions including menstrual and digestive cramps, infertility, lactation support, adrenal insufficiency, low progesterone, and possibly reproductive inflammation or cancer, and oral consumption of quality products may be more effective than wild yam creams with unknown content.

therapy. *Viburnum* is included here to improve the tone of the uterine muscle underlying the endometrium, as yet a third approach to the treatment of spotting.

Vitex agnus-castus	15 ml
Achillea millefolium	15 ml
Glycyrrhiza glabra	15 ml
Viburnum opulus	15 ml

Take ½ to 1 teaspoon of the combined tincture, 3 or more times daily for several months, reducing as spotting subsides. A full 3 to 6 months may sometimes be needed to tone the uterus and correct underlying hormonal imbalances.

Tincture for Primary Amenorrhea

Primary amenorrhea is the lack of onset of initial menses (menarche) in adolescents. This formula attempts to support pituitary function with *Vitex* and *Leonurus*, adrenal function with *Glycyrrhiza*, and circulation to the pelvis with *Angelica sinensis*. *Glycyrrhiza* has been widely used in Asian countries for amenorrhea of all types.

Vitex agnus-castus	20 ml
Angelica sinensis	20 ml
Glycyrrhiza glabra	10 ml
Leonurus cardiaca	10 ml

Take 1 teaspoon of the combined tincture, 3 to 6 times daily. This formula may be taken by a 16- or 17-year-old woman who has not begun menses and continued for 3 to 6 months, or even 1 year, to support hormonal regulation. The formula may be taken just 1 or 2 times a day once menarche occurs, and then discontinued.

Herbs for Amenorrhea Based on Lab Findings

Lack of monthly menses can occur due to a number of pituitary and complex hormonal irregularities. The following lab markers can be useful to assess the origins of amenorrhea in order to address it as directly as possible.

High prolactin. Prolactin may be elevated in around one-third of women with amenorrhea. The cause may be a prolactin-secreting pituitary adenoma or simply inappropriate pituitary output due to complex endocrine abnormalities. L-dopa, phenothiazines, and reserpine may stimulate prolactin secretion, as will elevated thyrotropin-releasing hormone (TRH). *Salvia officinalis* may suppress elevated prolactin, and I have used it in cases of excessive lactation, and several cases of prolactinemia. Dopamine has an inhibitory effect on pituitary prolactin output, and dopaminergic herbs such as *Vitex* and *Melissa* may affect prolactin via this mechanism.

Low LH and FSH. Low levels of these hormones may occur with hypogonadism and hypopituitarism. Elevated LH and FSH may occur when the ovaries are unresponsive to the pituitary gonadotropins and may represent ovarian failure or a chromosomal abnormality. Consider *Panax* and *Glycyrrhiza* in panhypopituitarism and hormonal insufficiency states.

Normal hormonal levels. Hormone panels may be normal in many amenorrheic women whose menstrual irregularity is due to dysfunction of hormonal rhythms and feedback loops. Some women with PCOS may have abnormal hormonal levels, but many may have normal levels. Consider *Vitex*, *Serenoa*, and *Glycyrrhiza* as key foundational herbs to support gonadotropins, adrenal function, and general hormonal balance.

Elevated testosterone. Elevated testosterone is seen in women with PCOS. *Glycyrrhiza glabra* has been shown to have an estrogenic action in some tissues[78] and to reduce elevated testosterone such as occurs with PCOS,[79] and clinical trials have shown restoration of menses in amenorrheic women.[80] *Serenoa repens* may reduce α reductase activity and thereby reduce the synthesis of dihydrotestosterone. Spearmint (*Mentha spicata*) may also reduce elevated testosterone; as little as 2 cups (480 ml) of tea per day may produce an effect.[81]

Tea for Amenorrhea

Rubus is a classic uterine tonic, and what little modern research exists indicates it may contain constituents with estrogenic effects. *Paeonia* is widely used in TCM as a blood and yin tonic and likely also promotes estrogen. *Angelica* and *Reynoutria multiflora* (also known as *Polygonum multiflorum*) are also traditional fertility herbs and may enhance uterine circulation and provide hormonal support.

Rubus spp.
Paeonia lactiflora
Angelica sinensis
Reynoutria multiflora

Combine equal parts of the dry herbs. Simmer 1 teaspoon of the mixture per cup of water and drink 3 or more cups (720 ml) per day. Use the tea daily for many months, and once menses are restored, continue consuming several cups every few days for another month or two to restore pelvic circulation and support healthy menstruation.

Tincture for Anovulatory Cycles

This condition is noted or diagnosed only when a woman is charting her basal body temperature and fails to observe a midcycle rise in temperature, or when a woman struggles with infertility and urinary hormone assays reveal that she is not ovulating. Anovulatory cycles may occur with complex endocrine imbalances such as polycystic ovarian syndrome (PCOS). This formula uses *Vitex* and *Glycyrrhiza*, as in the Tincture for Primary Amenorrhea, and includes *Panax*, a long-standing fertility and chi tonic herb, and *Serenoa*, a genitourinary tonic and agent noted to help reduce elevated testosterone in women.

Vitex agnus-castus	15 ml
Glycyrrhiza glabra	15 ml
Panax ginseng	15 ml
Serenoa repens	15 ml

Take 1 teaspoon of the combined tincture, 3 to 6 times daily, continuing for 3 to 6 months. Discontinue should pregnancy occur or once regular ovulatory cycles are restored.

Tincture for Amenorrhea Related to PCOS

Women with PCOS typically have elevated androgens that interfere with reproductive functions, and *Serenoa* may help reduce elevated testosterone. *Vitex* is a favorite herb for supporting healthy progesterone levels and regular menses. *Glycyrrhiza* is noted to treat amenorrhea,

and *Leonurus* supports healthy dopamine signaling and thereby pituitary hormone levels.

Serenoa repens	15 ml
Vitex agnus-castus	15 ml
Glycyrrhiza glabra	15 ml
Leonurus cardiaca	15 ml

Take 1 teaspoon of the combined tincture, 3 to 6 times daily. This combination can be effective for promoting menses in several weeks to several months' time.

Viburnum Simple for Menstrual Cramps

Dysmenorrhea may involve just a few hours of cramping and discomfort at the onset of menses, cramping over the first day or two, or in some of the worst cases, may involve a full week or even more of painful cramping. *Viburnum* is often so effective that I use it as a simple. If there are other clues in a person's case—from allergies to anxiety to pelvic stagnation—*Viburnum* could be combined with other herbs as specifically indicated. See "Uterine Sedatives" on page 124 for other herbs that may have an antispasmodic effect on the uterine muscle. *Viburnum* tends to work quickly, so in some cases it may be taken just when needed. In the worst cases, however, it may be best to use *Viburnum* daily for 3 months to help improve the tone of muscle and prevent uterine cramping. *Viburnum* tones the uterine muscle in cases of both atonic spasm and hypertonic spasm, making it appropriate in

---

Vitex for Reducing Elevated Prolactin

Vitex species are often effective for restoring menses in cases of amenorrhea related to elevated prolactin. Casticin, a flavonoid found in various species of *Vitex*, reduces elevated prolactin in animals as effectively as the prescription drug Bromocriptine.[82] Human clinical investigations report that women with mastalgia and other PMS symptoms respond to *Vitex*[83] due to normalizing effects on elevated prolactin.[84] Another clinical study found *Vitex* to improve irregular menses related to elevated prolactin.[85]

nearly all cases. If it does not help in the first 24 hours, the formula should be amended. In women with severe menstrual cramps, *Viburnum* may tone the uterus over time such that the herb is no longer needed.

Viburnum opulus tincture

Take as often as every 10 minutes for an hour, reducing as symptoms improve in the case of acute menstrual cramps. *Viburnum* as a simple can be highly effective and need not be taken all month long.

Tincture for Menstrual Cramps

This formula combines three of the most mentioned uterine antispasmodic herbs. See "Menstrual Cramp Specifics from Traditional Literature" for other suggestions of herbs that may benefit menstrual cramps of a specific character.

Viburnum opulus
Dioscorea villosa
Angelica sinensis

Combine in equal parts. Take 3 or 4 dropperfuls per day for several months, then just premenstrually and at times of cramping, as needed. In some cases, this may improve the uterine tone enough that the complaint resolves and the medication is no longer needed.

Tea for Menstrual Cramps

Tinctures are more practical for menstrual cramps, as the tincture bottle can be kept in the purse or desk drawer. However, this tea can be complementary to such tinctures for use in severe cases to take a break from the tincture, such as at home in the evenings, and/

or to create a multipronged approach when necessary. Cinnamon can help with excessive bleeding and a small clinical trial showed moderate pain-relieving effects in women with dysmenorrhea.[86]

Cinnamomum verum bark
Glycyrrhiza glabra bark
Viburnum opulus bark
Achillea millefolium leaf and flower
Matricaria chamomilla flowers

Decoct the mixture of the tree bark ingredients using ½ to 1 teaspoon per cup of water. Add the *Achillea* and *Matricaria*, using 1 to 2 teaspoons of the mixed herbs for every cup of water, letting them steep for 5 to 10 minutes. Strain and drink several cups promptly at the onset of pelvic cramps (uterine, intestinal), repeating as necessary.

Topical *Lobelia* Compress for Menstrual Cramps

A thin cotton cloth soaked with *Lobelia* tea or vinegar may be applied to the oil-prepared abdomen and pelvis and covered immediately with heat. For best results, castor oil (*Ricinus communis*) may be rubbed into the lower abdomen just prior to the application of the *Lobelia*

Uterine Sedatives

Uterine sedative herbs relax the uterine muscle and are indicated for simple menstrual cramps or any other situation associated with excessive tone or irritability of the uterine muscle.

Achillea millefolium	*Humulus lupulus*
Actaea racemosa	*Lobelia inflata*
Anemone pulsatilla	*Matricaria chamomilla*
Angelica sinensis	*Mentha piperita*
Caulophyllum thalictroides	*Piscidia piscipula*
Dioscorea villosa	*Valeriana officinalis*
Foeniculum vulgare	*Viburnum opulus*

Menstrual Cramp Specifics from Traditional Literature

Women may experience menstrual cramps as having a particular quality. Folkloric texts describe the following herbs as being specific for different sensations or qualities of pain.

Cramps that are tight squeezing: *Viburnum, Actaea, Glycyrrhiza, Salix alba, Dioscorea*

Cramps of a laborlike bearing-down sensation: *Viburnum, Lobelia inflata*

Cramps with hemorrhoids, heavy aching sensation: *Angelica, Hamamelis, Aesculus hippocastanum, Caulophyllum*

Cramps, general: *Viburnum, Matricaria, Mentha*

to help wick the medicine into the body. The castor oil is helpful alone, or *Mentha piperita* essential oil may be added in various strengths. A basic starting preparation is around ⅛ cup castor oil and ⅛ teaspoon *Mentha* oil. The heat source may be a warm moist pack or a heating pad.

Lobelia inflata dried leaves	4 ounces (120 g)
Castor oil	8 ounces (240 ml)
Mentha piperita essential oil	½ ounce (15 ml)

Prepare *Lobelia* tea by steeping 1 tablespoon per cup of water. Apply castor oil or a castor oil/mint oil blend to the abdomen/pelvis area. Strain and soak a cloth in the tea liquid to prepare a compress to lay on the abdomen/pelvis area. Cover with a heating pad.

Topical Oil for Dysmenorrhea

I have seen women with such severe menstrual cramps that they become nauseous with debilitating pain and where they may miss several days of work each month to stay home in bed with a heating pad over the abdomen. Having a topical pain reliever can be a welcome complement to a tincture or tea.

Wintergreen essential oil	¼ ounce (7.5 ml)
Fennel essential oil	¼ ounce (7.5 ml)
Peppermint essential oil	½ ounce (15 ml)
Hypericum perforatum oil	1 ounce (30 ml)

Combine the essential oils into the *Hypericum* oil and shake well. To ensure safe application, place a single drop of essential oil on the inner wrist, and wait 5 minutes before applying the essential oils to larger skin surfaces. If the wrist does not display any redness, itching, or irritation, it can be assumed that it is safe to rub into the abdomen or pelvis. Dilute with more *Hypericum* oil for sensitive individuals. Massage into the lower abdomen and cover with heat for an analgesic, antispasmodic effect.

Formulas for Premenstrual Syndrome

Premenstrual syndrome (PMS) is sometimes classified into various categories based on when cramps, breast pain, anxiety and irritability, or acne are the dominant symptoms. In most cases, however, there is an overlap of the various fluid-retentive, mental-emotional, breast, and digestive symptoms. One foundational, all-purpose approach to treating general PMS might be the use of alterative herbs to improve liver and bowel clearance of hormones, because helping the body to metabolize hormones will be beneficial in most cases. Other herbs can be chosen as complementary to address specific symptoms, constitutions, and presentations, such as breast tenderness, bloating and fluid retention, digestive upset, emotional irritability, headaches, and other complaints including pelvic pain and cramping, premenstrual acne, premenstrual vascular congestion causing backache, pelvic heaviness, flareups of hemorrhoids or varicose veins, carbohydrate and sugar cravings, and dysglycemic symptoms. I offer many such formulas in this section of the chapter. Most of these formulas combine alterative and cholagogue herbs such as *Taraxacum*, *Arctium*, and *Berberis* with other supportive herbs specific to the situation. Complementary nutraceutical therapies for PMS include B vitamins and "lipotropics"—combinations of alterative herbs with choline, inositol, and other nutrients that promote the liver's ability to process lipids (and thereby the fat-soluble hormones). Coach women to avoid ingestion of exogenous estrogens in animal products and to avoid exposure to environmental and household chemicals, which may have hormonal effects in the tissues and contribute to the hormonal and liver burden in the body.

Many nervine herbs have a calming effect due to dopaminergic activity, and because dopamine acts as a regulator of FSH and LH, dopaminergic herbs such as *Melissa*, *Actaea*, *Verbena hastata*, *Leonurus*, *Vitex*, and *Hypericum* can alleviate mood swings and the mental-emotional aspects of PMS via effects on the neuroendocrine system. One clinical study showed *Melissa* to improve PMS symptoms in high school students.[87] I have found each of these dopaminergic herbs invaluable in treating both mental-emotional and hormonal symptoms that are commonly seen in my daily practice.

Herbs may also improve PMS via many other mechanisms of action. *Zingiber* may improve mood and physical symptoms of PMS when taken consistently for several months' time.[88] *Humulus lupulus*, *Angelica sinensis*, *Panax ginseng*, and numerous other herbs may help treat women with PMS as specifically indicated. Essential fatty acids from wheat germ oil, evening primrose oil, flax seed oil, and fish oils exert anti-inflammatory and prostaglandin-balancing effects.

Alterative Tincture for PMS

This is an example of a general formula using one alterative (*Taraxacum*), one adrenal supportive and HPA-balancing adaptogen (*Glycyrrhiza*), and one dopaminergic pituitary agent noted to increase progesterone and benefit hyperestrogenism (*Vitex*). This is a general starting formula for PMS to add to, or subtract from, to treat more specific symptoms.

Taraxacum officinale root	15 ml
Taraxacum officinale leaf	15 ml
Glycyrrhiza glabra	15 ml
Vitex agnus-castus	15 ml

Take ½ to 1 teaspoon of the combined tincture, 3 to 5 times daily.

Tincture for Agitation and Depression with PMS

Dysphoric mood swings, occurring in a cyclical manner in relation to the menstrual cycle, are a very common chief complaint in general practice. These herbs are chosen for their specificity for anxiety and depression concomitant with hormonal imbalances. *Vitex*, *Melissa*, *Hypericum*, and *Actaea* are known dopaminergic agents, which may be one mechanism explaining their historical use for female mood swings and emotional disturbances. *Berberis* is added as an alterative ingredient to assist the liver in processing hormones.

Hypericum perforatum
Melissa officinalis
Actaea racemosa
Vitex agnus-castus
Berberis aquifolium

Combine in equal parts. Take 1 teaspoon, 3 times daily for 3 months or more.

Formula for Premenstrual Fluid Retention

Dependent fluids may accumulate due to direct hormonal effects on estrogen receptors on the vasculature but may also result from vascular congestion and/or constipation, both common in some women premenstrually. As the endometrial lining builds to its fullest state, uterine congestion may lead to congestion in nearby blood vessels and extend to interstitial tissue congestion from there. "Blood movers" and diuretic herbs are often helpful. *Angelica sinensis* acts as blood mover in this formula, although other herbs from *Aesculus* to *Hamamelis* to *Leonurus* may be appropriate substitutes. *Petroselinum*

and *Urtica* are used here as diuretics, with *Taraxacum* leaf and *Equisetum* being appropriate alternatives. This formula would work as either a tincture or a tea and might be started midcycle to help prevent premenstrual fluid retention.

Angelica sinensis	20 ml
Petroselinum crispum	20 ml
Urtica species	20 ml

Take 1 teaspoon of the combined tincture, 3 or 4 times daily, long term. The frequency of dosing may be increased to every hour or two at time of acute fluid retention. To formulate as a tea, combine 3 or more ounces of each dry herb and blend. Gently simmer 1 teaspoon of the mixture per cup of water for 10 minutes and let stand in a covered pan for 10 minutes more. Strain and drink 3 or more cups (720 ml) per day, beginning roughly 1 week prior to the onset of menses, the time in the cycle when fluid retention typically occurs.

Tea for PMS Mood Swings

This is a general nervine and adrenal support tea, featuring palatable herbs with effects on neurotransmitters and the HPA axis. Alteratives may be complementary to this mood-stabilizing tea when clinically indicated by symptoms. This nervous system trophorestorative formula may have lasting effects when consumed regularly over time in those with PMS related to stress, exhaustion, and adrenal downregulation, but may be less effective in those whose symptoms are related to liver congestion, toxemia, or hyperestrogenism.

Matricaria chamomilla
Verbena hastata
Melissa officinalis
Scutellaria lateriflora
Avena sativa
Glycyrrhiza glabra
Zingiber officinale

Combine the dry herbs in equal parts. Steep 1 tablespoon of the mixture per cup of hot water and drink freely, at least 3 cups (720 ml) per day at times of anxiety or irritability.

Tincture for Premenstrual Breast Tenderness

Hormonal stimulation may result in fluid retention and breast tenderness for several days to several weeks prior to the menses. Alteratives, diuretics, and herbs that promote progesterone to oppose estrogen are all helpful, and an effective formula might combine one herb acting by each mechanism. In this formula *Taraxacum* root acts as an alterative agent, *Taraxacum* leaf as a diuretic agent, and *Vitex* as a progesterone agent. *Ceanothus* and *Phytolacca* are herbs mentioned as being specific for breast pain, fluid congestion, and breast cysts in historic herbal literature. *Phytolacca* is most indicated for discrete cysts and breast infections and thus will not be indicated in all cases and may be omitted where unwarranted.

Taraxacum officinale root and leaf	15 ml
Vitex agnus-castus	15 ml
Ceanothus americanus	15 ml
Phytolacca americana	15 ml

Take 1 to 2 dropperfuls of the combined tincture 3 or 4 times per day, and use all month long for several months, then only premenstrually, as needed.

Tincture for Premenstrual Headaches

Fluid retention or hormonal influences on vascular blood vessels or specific blood cells can cause both congestive headaches and migraines. This formula may help reduce hormonally promoted migraines. It is based on *Actaea* and *Vitex*, both well-known hormonal balancers and emphasized historically for menstrual headaches. *Actaea* is also appropriate for musculoskeletal pain, and *Salix* and *Tanacetum* may have effects on platelet aggregation and the release of histamine, which may initiate some types of vascular headaches. They can also help with inflammatory pain in general. This formula may be amended when the cause of headaches is suspected to be mainly muscle contraction or mainly vascular by varying the ratios of the herbs or adding the herbs most specifically indicated and subtracting those not appropriate. Alteratives may be added where indicated or prepared separately in teas and capsules.

Actaea racemosa	20 ml
Vitex agnus-castus	20 ml
Salix alba	10 ml
Tanacetum parthenium	10 ml

Take 1 teaspoon of the combined tincture, 3 times daily for prevention, and as often as every 15 to 30 minutes for acute pain.

Tincture for Premenstrual Backache and Vascular Congestion

The buildup of the endometrium may lead to vascular congestion and pelvic congestion, especially in those

Phytoestrogens and Amphoterism

Phytoestrogens bind estrogen receptors acting as either agonists or antagonists, depending on the physiologic situation, the tissue, and the receptor type, making absolute statements about their physiologic activity difficult. Phytoestrogens have such varying activity and ligand affinity for the various tissues and subreceptor types of estrogen, that they could be considered to be selective estrogen response modifiers (SERMs). For example, isoflavones might act as weak agonists of estrogen receptors in situations of low estrogen in the body and exert an estrogenic effect, yet that weak agonism can compete with a body's high estrogen load and reduce estrogenic stimulation in a situation of hyperestrogenism. This dual action of phytoestrogens, to both offer estrogenic support in some situations and reduce excessive estrogen stimulation in others, is referred to by herbalists as *amphoterism*. The inclusion of isoflavones and other phytoestrogens in the diet or in the form of herbal supplements is thought to offer many health benefits, including protection against hormone-related cancers.[89] Synthetic phytoestrogen-like compounds have been tested and, interestingly, display no such amphoteric action.[90]

with liver congestion, constipation, or a constitutional tendency toward fluid stasis. Blood-moving herbs and liver support can be very helpful. *Taraxacum* root serves as an alterative foundation and the leaf as a nutritive diuretic helping to excrete tissue fluid. *Achillea* is a peripheral vasodilator that may affect hemodynamics and shunt blood from the pelvis into the limbs. *Aesculus* is specific for portal congestion in the liver and vascular congestion in the lower body, including varicosities. *Angelica* is a general blood mover and platelet anti-aggregator and folkloric literature often suggests it for improving circulation in the pelvis. *Caulophyllum* is also mentioned folklorically for aching in the limbs and may be added when such symptoms are present.

Taraxacum officinale root and leaf	15 ml
Achillea millefolium	15 ml
Aesculus hippocastanum	15 ml
Angelica sinensis	15 ml

Take 1 teaspoon of the combined tincture, 3 times daily for prevention, and as often as every 15 to 30 minutes for acute pain.

Uterine Emmenagogues

Emmenagogues are agents capable of stimulating menstrual flow and can be used when menses are scanty, sluggish, or slow to start. **Caution:** At high and repetitive doses, these herbs may induce an abortion. There may be considerable toxicity to the body when such herbs are overdosed; should an abortion be the goal, a surgical dilation and curettage (D&C) is probably safer, having less risk when performed by experienced physicians.

Artemisia abrotanum	*Gratiola officinalis*
Artemisia absinthium	*Hedeoma patens*
Artemisia vulgaris	*Juniperus sabina*
Caulophyllum	*Mentha pulegium*
thalictroides	*Mitchella repens*
Gossypium hirsutum	*Ruta graveolens*

Uterine Vascular Decongestants

Congestion of the pelvic vasculature may occur due to obesity, portal congestion, biliary disease, or other challenges to the circulation of the pelvis. Pelvic congestion is often worsened by the premenstrual state when vascular congestion occurs as part of the normal buildup of the endometrium. These herbs may be included in formulas for pelvic pain due to vascular congestion, a different quality of discomfort from spastic menstrual cramps.

Achillea millefolium	*Angelica sinensis*
Aesculus	*Collinsonia canadensis*
hippocastanum	*Leonurus cardiaca*

Tea for Vascular Congestion

Those prone to vascular congestion in the pelvis and uterus may also suffer from fluid congestion in the pelvic tissues and lower limbs premenstrually. Even normal

Support for Premenstrual Acne

Acne often begins at puberty as the reproductive hormone levels rise. Testosterone can make the skin more oily and cause the sebaceous glands to become more active, and estrogen may also affect the texture of the skin and cellular activities. Androgens activate sebaceous cells, keratinocytes, and ductal lining cells and lead to proliferation. Ductal lining cells can become plugged and sebum accumulates, and the acne bacteria, *Propionibacterium acnes*, thrives as the overproduction of sebum provides fatty acids on which the bacteria feeds. Many women note that breakouts occur more premenstrually, with lesions present during the menses, and slowly heal over the midcycle weeks, until the cycle repeats again. Acne is sometimes worse in women with a high androgen level, such as those with PCOS. The following nutrients and practices may help optimize epithelial cell growth and sloughing, reducing proliferation and helping resist a *Propionibacterium* infection. Although not strictly a nutrient, alpha hydroxy acid loosens the most superficial skin cells, helps to distribute sebum, and helps to keep debris from proliferated cells from blocking pores and skin ducts.

Beta carotene, mixed carotenoids, and the
   related naturally occurring vitamin A
   compounds
Zinc
Essential fatty acids such as linoleic and
   linolenic acid
High fiber diet
Alpha hydroxy acid skin washes

premenstrual accumulation of blood in the endometrium can promote vascular congestion of the extended vasculature and microcirculatory vessels, which can cause a sense of aching, heaviness, and pain. This tea may be used all month long to improve blood flow and to reduce general inflammation in the vasculature.

Salvia miltiorrhiza
Paeonia × suffruticosa
Paeonia lactiflora
Angelica sinensis
Zingiber officinale
Cinnamomum verum

Combine equal parts of the dry herbs and store in an airtight container. Gently simmer 1 teaspoon of the mixture per cup of hot water and then strain. Drink 3 or more cups (720 ml) per day.

Tincture for Premenstrual Acne and Skin Eruptions

Many skin eruptions, including acne of both adolescent females and males, are benefited by alterative herbs such as *Berberis* and *Arctium* in this formula. *Vitex* is mentioned historically for adolescent acne and may benefit the skin via its many known hormonal effects. Other supportive measures would include dietary optimization and treating any constipation, intestinal dysbiosis, or any other gut issues. Treatment of acne is also covered in detail in Volume 1, chapter 5.

Berberis aquifolium	15 ml
Arctium lappa	15 ml
Vitex agnus-castus	30 ml

Take 3 or 4 dropperfuls of the combined tincture per day for a minimum of 3 months.

Formulas for Pregnancy and Related Issues

A handful of natural medicine books and herbals are available that address in detail herbs to use specifically during pregnancy or to support fertility. Due to space limitations, I cannot provide a highly detailed review of herbs for pregnancy and related issues in this chapter, but I can share some of the formulas that are best known as well as my own experience. I do not specialize in obstetrics, but I do have experience with herbal therapies for improving fertility and have seen excellent results.

There has not been extensive study regarding herbs during pregnancy, so most herbalists limit the use of herbs high in alkaloids and, in fact, herbs in general for pregnant women to err on the side of caution. Herbs that have been the most common culinary and tea herbs are presumed safe, but even these have not been rigorously investigated. Therefore, most books that offer lists of herbs safe versus unsafe during pregnancy come from the authors' best guess. Herbs with known mutagenic, abortifacient, and other strong pharmacologic actions would of course be unsafe, whereas most of the long-standing culinary spices such as garlic, mint, pepper, and turmeric are generally presumed safe.

I believe that most, if not all, herbs can be passed in the breast milk, as evidenced by spicy herbs inducing colic in nursing babies. (Garlic, pepper, and cruciferous vegetables are common offenders.) Using this knowledge to advantage, nursing mothers may find that drinking large amount of fennel, mint, and carminative tea helps treat colic in their nursing infant.

Herbs can be extremely helpful for improving fertility in cases of PCOS, repetitive miscarriage, and other hormonal imbalances. Obstructed fallopian tubes are a possible cause of infertility and, as a mechanical problem, will not respond to herbs. Abnormal gonadotropins and feedback loops are a common cause of anovulatory cycles, such as with Cushing's syndrome, hypothyroidism, and PCOS, and will respond to herbal hormone-balancing therapies. Lab testing may reveal elevated prolactin, insufficient progesterone, elevated androgens, imbalances of hormone cycles, or other signs of adrenal or thyroid disorders. Botanical considerations for such conditions include *Vitex*, *Angelica*, *Panax*, *Serenoa*, *Salvia*, and others.

Tincture for Infertility

The formula exemplified here takes a "kitchen sink" approach: It combines leading herbs for addressing common infertility issues, such as thyroid imbalance, chronic miscarriage, and PCOS. All such conditions should be addressed directly as well. *Angelica* in this formula may improve circulation in the reproductive organs, and *Panax* is a classic chi and fertility tonic for both men and women as traditionally used in China. *Lepidium* has a similar reputation in South America, and *Viburnum* is especially indicated for chronic miscarriage and uterine issues. See also "Formulas for Polycystic Ovarian Syndrome (PCOS)" on page 144, and chapter 2 provides in-depth information on supporting adrenal and thyroid dysfunction.

Angelica sinensis	15 ml
Panax ginseng	15 ml
Lepidium meyenii	15 ml
Viburnum opulus	15 ml

Take 1 dropperful of the combined tincture, 3 or 4 times daily, continuing for 3 to 6 months. Reevaluate results after 6 months of regular use.

Fertili-Tea

This is a general all-purpose formula for infertility using herbs chosen for their palatability in teas as well as for their mode of action.

Glycyrrhiza glabra
Pueraria candollei var. *mirifica*
Angelica sinensis
Astragalus membranaceus
Reynoutria multiflora
Berberis aquifolium

Combine equal parts of the dry herbs, blend, and store in an airtight container in a dark cool cupboard. Simmer 1 heaping tablespoon of the mixture per cup of water for 10 minutes, cover and let stand for 15 minutes, and then strain. Drink a minimum of 3 cups (720 ml) per day, continuing for 3 to 6 months.

Inositol-Containing Herbs for Fertility

Many inositol compounds are found naturally in the human body, often in the form of phosphoinositides, and they play important roles in fertility. Mammalian ova bear inositol receptors involved with maturation.[91] Higher levels of naturally occurring myoinositol in human ova and follicular fluid are associated with higher quality oocytes, and supplementation with inositol appears to reduce hyperinsulinism and help restore fertility in women with PCOS.[92] Inositol compounds including myoinositol, pinitol, and d-chiro-inositol are demonstrated to improve insulin response as well as to improve ovarian function and metabolic and hormonal parameters that support fertility.[93] Legumes contain pinitol at approximately 200 to 600 milligrams/100 grams of fresh weight.[94] Chickweed (*Stellaria media*) and *Ginkgo biloba* also contain pinitol, while green coffee extracts (*Coffea arabica*) contain myoinositol,[95] and dandelion (*Taraxacum*) contains d-chiro-inositol.[96] Carob syrup may contain almost 95 grams of pinitol per kilogram,[97] is readily available in South American and Mediterranean markets, and has been shown to improve insulin sensitivity and glycemic control with regular consumption.[98]

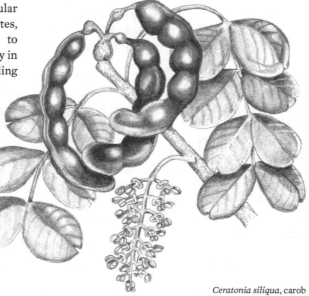

Ceratonia siliqua, carob

Infertility Tincture for PCOS Patients

Polycystic ovarian syndrome (PCOS) is commonly associated with insulin resistance, even in those with a normal body mass index. Insulin-sensitizing compounds not only improve insulin resistance, but they may also improve ovarian response to endogenous gonadotropins and reduce hyperandrogenemia, the hallmark of PCOS. Therefore insulin-sensitizing compounds may establish normal menstrual cycles and ovulation and improve fertility.[99] This tincture is especially indicated for PCOS due to elevated androgens and/or prolactin frequently suppressing ovulation or menses altogether.

Vitex agnus-castus
Glycyrrhiza glabra
Serenoa repens

Combine equal parts of the tinctures in a 2- or 4-ounce bottle. Take ½ teaspoon, 3 times daily for at least 6 months. Regular menstruation is often restored in several months.

Legume Tea to Support Fertility

Plants containing large amounts of the fertility-supporting inositol compounds include many legume family plants: red clover (*Trifolium pratense*), kudzu (*Pueraria candollei* var. *mirifica*),[100] carob (*Ceratonia* spp.),[101] milk vetch (*Astragalus membranaceus*),[102] licorice (*Glycyrrhiza glabra*),[103] and alfalfa (*Medicago sativa*).[104] This formula may help treat insulin resistance, PCOS, and other hormonal imbalances causing infertility.

Trifolium pratense
Pueraria candollei var. *mirifica*
Astragalus membranaceus
Glycyrrhiza glabra
Medicago sativa

Combine equal parts of each herb, and store in an airtight container. Use 1 heaping tablespoon of the mixture per cup of hot water, steep for 10 to 15 minutes, and then strain. Drink 3 or more cups (720 ml) per day.

Tea for Nausea of Pregnancy

There is no one-size-fits-all therapy when it comes to morning sickness. Many women figure out what works for them based on trial and error. Some women find that eating upon waking and avoiding an empty stomach is helpful, some women respond well to B vitamins, and others find that specific diets and avoidance of problem foods works the best. Most women find ginger, mint, and chamomile helpful, although for some women, they are

Herbs for Infertility Formulas

These herbs can be included in formulas for infertility, whether the cause is PCOS, hormonal insufficiency, or poor circulation to the pelvis.

HERB	ACTION
Angelica sinensis	Improves circulation and nutrition to reproductive symptoms
Glycyrrhiza glabra	May affect amenorrhea, elevated testosterone, PCOS
Lepidium meyenii	May affect low libido and infertility with fatigue and weakness
Mentha spicata	May reduce elevated testosterone
Panax ginseng	A chi tonic to support adrenal hormone output
Paeonia lactiflora, P. × suffruticosa	A blood and yin tonic
Reynoutria multiflora	A traditional fertility tonic in TCM noted to be high in resveratrol offering antioxidant protective effects
Salvia officinalis	May help when elevated prolactin suppresses ovulation
Serenoa repens	May reduce elevated testosterone
Urtica urens, U. dioica	May help reduce elevated testosterone
Viburnum opulus, V. prunifolium	Tones an atonic or hyperexcitable uterine muscle that can induce miscarriage
Vitex agnus-castus	Helps raise low progesterone via dopaminergic effects

not enough to really stay comfortable. When the nausea is constant and severe or when hyperemesis occurs, this tea may be part of a broader protocol that includes homeopathic remedies, carbonated liquids such as ginger ale, dietary approaches, or even pharmaceutical measures. Some women find that eating frequently throughout the day and never allowing the stomach to be fully empty is helpful, explaining the wise woman tradition of nibbling on soda crackers to soak up stomach acid.

Mentha piperita leaves	4 ounces (120 g)
Matricaria chamomilla flowers	4 ounces (120 g)

Combine the dry herbs. Steep 1 tablespoon of the mixture per cup of hot water and then strain. Drink as much as possible, taking frequent small sips. When even drinking a tea such as this is nauseating, prepare the tea, freeze it in ice cube trays, then sliver the ice cubes by placing a few inside a clean towel, crush with a hammer, then consume a little at a time by the spoonful—this approach is sometimes more tolerable and effective.

Ginger Tea for Nausea of Pregnancy

Ginger is another classic remedy for morning sickness, and indeed for nausea of all types. It doesn't work for everyone, but it is cheap, easy to prepare, readily available, and worth a try. Ginger syrups are also available and may be placed in carbonated water or ginger ale and consumed in small sips as a quick substitute for this recipe.

Zingiber officinale, freshly grated or sliced root	1 heaping tablespoon
Lemon or orange juice, fresh squeezed	1 to 2 tablespoons, to taste
Honey or maple syrup	1 to 3 teaspoons, to taste

Gently simmer the ginger root in 2 to 3 cups (480 to 720 ml) of water for 10 minutes. Strain and add the juice, if

---

Supplements for a PCOS Fertility Protocol

Women with PCOS struggling to conceive may have good results when diet, exercise, and herbs, as discussed in this section, are complemented with the following nutritional supplements and foods.

B vitamins
Vitamin D
Chromium
Magnesium
N-acetylcysteine
Myoinositol or d-chiro-inositol powder
Maca powder (in smoothies)
Brewer's yeast (in food or smoothies)
Prickly pear juice (in beverages)
Carob syrup (in beverages, smoothies, or oatmeal)

---

tolerated, and honey or maple syrup. Drink in frequent small sips throughout the day. Refrigerate and chill if desired. Some women tolerate carbonated drinks better than plain water when they feel nauseated and mixing this tea with carbonated water creates a homemade, sugar-free "ginger ale" that may be helpful to some.

Tincture for Mother's Cordial

A *partus preparator* is an historical term referring to an herbal formula that helps prepare the body for parturition (labor and delivery). The Eclectic physicians of the 1800s and early 1900s used this combination for a partus preparator and referred to it as "Mother's Cordial." Herbalists consider all the herbs in this formula to be uterine tonics.

Viburnum opulus
Actaea racemosa (also known as *Cimicifuga racemosa*)
Mitchella repens
Caulophyllum thalictroides

Combine in equal parts. Take ½ to 1 teaspoon, 3 times daily in the last trimester of pregnancy, for those with a history of difficult labors or subinvolution postpartum or for anyone who has had multiple childbirths. This formula may also be used acutely for threatened miscarriage, dosing ½ teaspoon every 10 to 15 minutes for several hours.

Tincture for Miscarriage

Uterine tonic and antispasmodic herbs have also played important roles in miscarriage prevention, presumably by reducing miscarriage due to hyperspasticity or endometrial insufficiency. This tincture can be used by those with a prior history of miscarriage and those facing an acute threat of spontaneous abortion. *Lobelia* may seem like a curious choice, being known mostly as a lung antispasmodic, but it will also relax the uterine muscle due to antagonistic effects at uterine muscle β receptors. *Dioscorea* is also an effective uterine antispasmodic; *Viburnum* may improve uterine tone and hormonal balance.

Dioscorea villosa	25 ml
Viburnum spp.	25 ml
Lobelia inflata	10 ml

Take a dropperful of the combined tincture every 15 to 30 minutes, decreasing gradually as danger subsides. For those with a history of miscarriage, this formula may also be taken for many months to prepare the body prior to a future pregnancy.

Viburnum to Support the Uterus

Also known as the guelder rose, cramp bark is one of the most reliable uterine tonic herbs. Poor uterine tone predisposes to postpartum hemorrhage, thus any woman with a previous incidence of uterine hemorrhage might use the herb *Viburnum opulus* or *V. prunifolium* in the last trimester of pregnancy. The risk of hemorrhage also increases with multiple pregnancies and deliveries, and women who have delivered more than five children might also use *Viburnum* in the last trimester or last month of pregnancy. Women with uterine fibroids are more likely to experience subinvolution of the uterus postpartum and associated hemorrhage and thus they, too, might use *Viburnum* prophylactically. The bark of *Viburnum* contains iridoid glycosides, viopudial, and scopoletin, all contributing to the antispasmodic effects, but the plant has not yet been rigorously studied in terms of chemistry and mechanisms of action. *Viburnum* has also been traditionally used as a hypotensive agent suggesting relaxing effects on multiple muscle types.

Viburnum opulus,
cramp bark

Uterine Hemostatics

Hemostatics are agents capable of controlling bleeding wounds when topically applied or for bleeding from the GI or uterus when orally ingested. These herbs (and one homeopathic remedy) are emphasized in the folkloric literature for having astringent action, and thus the ability to control excessive menstrual flow and even hemorrhage.

Achillea millefolium	*Geranium maculatum*
Alchemilla vulgaris	*Secale cornutum*
Capsella bursa-pastoris	(homeopathic
Cinnamomum spp.	remedy)

Miscarriage Tea for Acute Use

Similar to the Tincture for Miscarriage, this tea features uterine antispasmodic herbs. *Cinnamomum* may help control premature bleeding or spotting and, along with *Glycyrrhiza*, imparts a pleasing flavor to the tea.

Glycyrrhiza glabra	2 ounces (60 g)
Cinnamomum spp.	1 ounce (30 g)
Dioscorea villosa	1 ounce (30 g)
Viburnum opulus	1 ounce (30 g)
Valeriana officinalis	1 ounce (30 g)

Combine the dry herbs. Simmer 2 teaspoons of the mixture per cup of water and then strain. Drink a full cup (240 ml) every hour until the symptoms of miscarriage abate.

Tincture for Postpartum Hemorrhage

This formula can help in situations of postpartum hemorrhage. Purified ergotamine alkaloid is commonly used in hospital obstetrical wards and by midwives and naturopathic physicians treating actue hemorrhage, and although its source, the ergot fungus, is a natural agent, it is not commonly available as a tincture or dry herb and is used in pill form. *Achillea*, *Cinnamomum*, and *Erigeron* can also be very effective in controlling heavy postpartum bleeding, used individually or in combination, and are often employed by naturopathic obstetricians and midwives for less severe situations. For other formulas for heavy menses/menorrhagia, see "Formulas for Treating Menstrual Cycle Disorders" on page 115.

Achillea millefolium	15 ml
Cinnamomum spp.	15 ml
Erigeron canadensis	15 ml
Viburnum opulus	15 ml

Take 1 teaspoon of the combined tincture every 10 to 30 minutes, reducing as bleeding subsides.

Tea for Nursing Mothers

I believe galactagogues are best consumed as teas, both because drinking tea will help encourage optimal water intake and because the simple act of brewing and drinking a cup of tea has a relaxing effect on a busy new mom. In fact, relaxation may be part of the "medicine" of this formula, even though fennel, fenugreek, and *Medicago* are reported galactogogues.

Urtica spp.	2 ounces (60 g)
Medicago sativa	2 ounces (60 g)
Foeniculum vulgare	2 ounces (60 g)
Trigonella foenum-graecum	2 ounces (60 g)
Glycyrrhiza glabra	2 ounces (60 g)
Althaea officinalis	2 ounces (60 g)
Rosa damascena petals	2 ounces (60 g)
Mentha spicata	2 ounces (60 g)

Combine the dry herbs to make 1 pound (480 g) of dry bulk tea and store in a jar or resealable plastic bag. Steep 1 tablespoon of the mixture per cup of hot water and then strain. Drink freely, at least 3 cups (720 ml) per day, ideally at part of a self-care ritual.

Salvia Simple for Excessive Lactation

Excessive lactation is an uncommon problem, but if you do encounter it, remember that *Salvia officinalis*, common culinary sage, is an antigalactagogue. Such a remedy may be needed when a woman gives up a newborn for

Folkloric Galactagogues

Galactogogue herbs have no pharmaceutical equivalent and are essential herbal tools of both ancient and modern midwives and herbalists. These herbs are all reported to have significant galactogogue effects in folkloric herbals and can be prepared into teas to drink 3 or more cups (720 ml) per day. The exception in this list is black cumin (*Nigella sativa*), which is consumed in foods or pressed into oil.

Marshmallow (*Althaea officinalis*)
Blessed thistle (*Cnicus benedictus*)
Fennel (*Foeniculum vulgare*)
Goat's rue (*Galega officinalis*)
Black cumin (*Nigella sativa*)
Milk thistle (*Silybum marianum*)
Fenugreek (*Trigonella foenum-graecum*)
Nettle (*Urtica dioica*)

adoption, and I have seen several cases of excessive lactation that caused constant breast engorgement and painful mastitis for which *Salvia* was effective.

Salvia officinalis

Prepare a 2-ounce tincture and dose ½ to 1 teaspoon, 3 or 4 times daily when aiming to halt lactation altogether. When aiming to reduce lactation, take ½ teaspoon each day, building up gradually only if needed.

Formulas for Menopausal Complaints

Ovarian activity declines when viable ova are depleted over the menstrual years, and this can cause hormonal levels to fluctuate rapidly, and sometimes dramatically, leading to a variety of acute or early symptoms. As the reproductive hormones settle to lower levels, there may also be late symptoms or long-term consequences including osteoporosis, acceleration of cognitive impairment, skin changes, and genitourinary atrophy. While vaginal dryness is a well-known phenomenon, changes in the urinary mucosa may also occur. Many women never have a UTI their entire life until postmenopausally when estrogen decline weakens the mucosal cells and barrier function of the bladder.

Menopause is a natural transition and not a disease, and yet many women experience some uncomfortable symptoms for which herbal or other medicines may be helpful. Roughly half of perimenopausal women suffer from one or more of the most common symptoms: hot flashes, insomnia, sweating, anxiety, palpitations, headaches, poor concentration, and loss of libido. Bleeding irregularities may occur in the perimenopausal phase.

Hormone replacement therapy (HRT) has been a standard therapy for decades, however health professionals have come to realize that HRT increases the risk of stroke, coronary artery disease, as well as breast cancer and possibly other hormonal consequences.

I find herbal therapies highly effective in treating the acute symptoms of menopause without the vascular risks. Herbal therapies may also help prevent the late symptoms and be effective in maintaining bone density, vaginal lubrication, neuronal function, libido, and mood. Herbs may help allay menopausal symptoms via estrogen receptor agonism or effects at the level of the neuroendocrine system. Herbs may be chosen specifically for adaptogenic and nervine effects to treat insomnia and mental-emotional disturbances. Because many menopausal women present distinct combinations of symptoms, such as vasomotor symptoms and insomnia, or insomnia and heart palpitations, or heart palpitations and anxiety, herbal formulas should be tailored to the individual. In addition to the formulas here, the "Formulas for Other Female Reproductive Conditions" on page 151 may be helpful for menopausal women experiencing a loss of libido. When significant heart palpitations, blood pressure changes, or other cardiac symptoms are present, refer also to Volume 2, chapter 2 for detailed information on these potentially serious conditions.

Tincture for Menopausal Hot Flashes

Salvia officinalis is the most effective herb I have tried for hot flashes; it would work well as a simple in some cases. A majority of women who experience hot flashes also have menstrual irregularities, insomnia, mood swings, or other perimenopausal complaints, so I more often combine *Salvia* with hormonal-balancing herbs in a formula such as this one. Here, *Salvia* is combined with the most popular and well-studied herbs to support adrenal function, dopaminergic systems, and pituitary gonadotropin output.

Salvia officinalis
Vitex agnus-castus
Actaea racemosa
Glycyrrhiza glabra

Combine in equal parts. Take 1 dropperful, 2 to 8 times daily, depending on the severity of the symptoms. The dose may be reduced as hot flashes improve.

Tea for Hot Flashes

The mechanism behind hot flashes is not completely understood. Although vasomotor symptoms are obviously associated with estrogen decline, the onset and severity of hot flashes do not directly correlate to estrogen or other hormone levels. Hot flashes may be provoked by hormone fluctuations and especially by sudden hormone decline. Because perspiration and vasodilation are controlled by the thermoregulatory nucleus of the hypothalamus, a leading hypothesis is that women become overly sensitive to small changes in core temperature, where the cooling mechanism of sweating is easily triggered. The decline in estrogens at menopause affects feedback loops to the brain in a manner that increases norepinephrine and serotonin leading to thermoregulatory changes in the hypothalamus. The changes in thermoregulation increase peripheral circulation to assist the body in losing heat via hot flashes and night sweats. Nervines, such as *Hypericum* or *Actaea*, may stabilize serotonin and norepinephrine levels. The most effective herb in my experience is *Salvia officinalis*; however, it has not been rigorously investigated. This tea features *Salvia* combined with alteratives, nervines, and adrenal- and hormone-supportive herbs. *Humulus*, *Glycyrrhiza*, *Pueraria*, and *Foeniculum* may be hormonally active, while *Salvia officinalis* has cooling and drying effects. The flavors may be amended as desired, taking care to get enough *Salvia* in the blend to do the job. Dehydroepiandrosterone (DHEA) is a proandrogen produced by the adrenal glands, and supplementation with DHEA may improve hot flashes and other menopausal symptoms and may complement this formula.

Salvia officinalis	4 ounces (120 g)
Glycyrrhiza glabra	2 ounces (60 g)
Pueraria spp.	2 ounces (60 g)
Foeniculum vulgare	2 ounces (60 g)
Mentha spicata	2 ounces (60 g)
Citrus aurantium peel	2 ounces (60 g)
Hypericum perforatum	2 ounces (60 g)

Lemon juice and/or honey, as desired

This recipe will yield 1 pound (480 grams) of tea. It is best stored as two separate mixtures. In one bag combine the *Salvia*, *Glycyrrhiza*, and *Pueraria*. This mixture is prepared as a decoction. The other bag would hold the rest of the ingredients, which can be infused as a tea. Decoct 1 teaspoon of the first 3 ingredients in 2 cups (480 ml) of water for 10 minutes, then add 1 tablespoon of the remaining herbs to infuse. Strain and add fresh lemon juice and/or honey to the brewed tea as desired. Drink 3 or more cups (720 ml) per day. Increase during times of stress.

Tea for Postmenopausal Chronic UTIs

Vaginal estrogen creams are often effective for urinary urgency, frequency, and nocturia in postmenopausal women and may decrease the tendency to postcoital and general UTIs.[105] Women who begin to have urinary tract infections postmenopausally may benefit from phytosterols and mucous membrane supportive herbs such as *Glycyrrhiza*, *Asparagus*, *Althaea*, and *Centella*. *Pueraria* is a high-isoflavone plant included here for its hormonal-supportive effects on the bladder lining. *Hypericum* may reduce urinary hypersensitivity, and *Calendula* may improve microcirculation to the epithelium. *Serenoa* is a classic genitourinary tonic and may help chronic infections but is not good tasting, so it may be used as a complementary tincture or capsule. This formula may reduce the chronicity of urinary infections but is not highly antimicrobial, and separate approaches may be necessary for acute infections. (See Volume 1, chapter 4, for guidance in treating UTIs.) Teas make the most effective medicines for urinary complaints. This tea is not antimicrobial itself and is aimed at preventing repeat infections by supporting general health and cell integrity and turnover of the uroepithelia.

Glycyrrhiza glabra
Asparagus racemosus
Althaea officinalis
Pueraria spp.
Centella asiatica
Hypericum perforatum
Calendula officinalis

Combine equal parts of the dry herbs; add more *Glycyrrhiza* to taste, if desired. Steep 1 tablespoon of the mixture per cup of hot water and then strain. Drink freely.

Tincture for Menopausal Insomnia

Insomnia is an exceedingly common menopausal symptom. Insomnia due to hormonal swings might respond better to adaptogens and hormonal-balancing agents than to sedative nervines such as *Valeriana*. While *Valeriana* and sedatives may be helpful when taken just before bed, most women will get a deeper response from effective hormonal support. *Actaea* is used for many menopausal symptoms and is noted to contain isoflavones as well as to affect dopamine and thereby the HPA axis. *Withania* is a good adaptogen choice for insomnia, because in addition to supporting adrenal function and reducing elevated nighttime cortisol, it binds GABA receptors and offers a calming and relaxing effect.

Actaea racemosa
Withania somnifera

Combine in equal parts. Take 1 dropperful, 3 to 6 times daily to improve hormonal contribution to insomnia. In addition to this tincture taken through the day, valerian capsules and a calcium/magnesium supplement may be taken before bed; additional adrenal support and/or tryptophan supplements may also be added for stubborn cases.

Tincture for Menopausal Before-Bed Insomnia

This "before bed" formula has greater immediate relaxing effects than the Tincture for Menopausal Insomnia. A formula such as this should not be expected to correct the underlying cause of insomnia, and although it might improve sleep following each use, it would need

Herbs for Menopausal Insomnia

These herbs can be considered in teas, tinctures, and encapsulations to improve sleep quality of those suffering from menopausal insomnia. The condition is so severe and recalcitrant in some women that it may take a combination of several such medicines to normalize sleep.

CONDITION	HERBS
Insomnia with restlessness	*Humulus lupulus* *Passiflora incarnata* *Valeriana sitchensis,* *V. officinalis*
Insomnia with nervous exhaustion	*Actaea racemosa* *Avena sativa* *Hypericum perforatum* *Rhodiola rosea*
Insomnia with adrenal exhaustion	*Eleutherococcus senticosus* *Glycyrrhiza glabra* *Panax ginseng* *Withania somnifera*
Insomnia with anxiety	*Hypericum perforatum* *Lycopus virginicus* *Scutellaria lateriflora* *Valeriana sitchensis,* *V. officinalis*
Insomnia with muscular stiffness	*Actaea racemosa* *Piper methysticum* *Valeriana sitchensis,* *V. officinalis* *Withania somnifera*

Pueraria for Hormonal Support

Several species and varieties of *Pueraria*, commonly called kudzu by Western herbalists, have been used in Asia for centuries for vascular and hormonal support, among other metabolic and hepatoprotective actions. *Pueraria* is used in classic antiaging and rejuvenation formulas in Thailand under the name kwao krua and also used for menopausal symptoms, facial rejuvenation, and support of the bone, hair, and fingernails. It is sometimes claimed to support sexual longevity, including vaginal lubrication and breast size. *Pueraria montana* var. *lobata* is combined with *Salvia miltiorrhiza* and *Angelica* species and used as a cardiotonic and may be included in postmenopausal formulas as a nourishing tonic base herb to support the heart and vasculature[106] as well as to support the skin, vaginal mucosa, bones, brain, and cardiovascular system in postmenopausal women. Clinical studies suggest that dosages of 50 or 100 milligrams of *Pueraria candollei* var. *mirifica* can improve menopausal symptoms and slightly increase serum estradiol.[107]

Kudzu preparations may also improve bone density[108] without evidence of proliferative effects on the endometrium[109] or breasts,[110] with some effects credited to the isoflavone formononetin.[111] Kudzu isoflavones may also improve blood lipids[112] and reduce the accumulation of abdominal fat that may follow rapid declines in estrogen levels.[113] One small clinical study compared *Pueraria* to placebo on serum lipids of postmenopausal women. After 2 months, HDL was 34 percent higher and LDL 17 percent lower in the group receiving *Pueraria*.[114] A phase III clinical trial compared kudzu to conjugated equine estrogen on perimenopausal symptoms and hormone levels in the blood. All groups showed comparable relief of vasomotor, urogenital, and psychological symptoms. No significant differences in FSH, LH, or serum estradiol levels were seen between the groups, all indicating that kudzu was as effective as conventional pharmaceutical therapy.[115]

Kudzu phytosterols,[116] miroestrol and coumestrol, act on both α- and β-estrogen receptors, while daidzein and genistein are more active on β-estrogen receptors.[117] In general, α receptors direct cellular proliferation and are most active in early childhood, while the β subtype directs differentiation and apoptosis and dominates in puberty and adulthood. When the balance between α and β receptors is disrupted, reproductive health is impaired. Miroestrol blocks excessive stimulation of estrogen receptors in cases of breast or endometrial cancer yet acts as an agonist to support cardiovascular health and alleviate menopausal symptoms. Formononetin upregulates the expression of β-estrogen receptors without stimulating α-estrogen receptors in endometrial and vaginal cells of ovariectomized rats.[118] Formononetin also displays bone-building effects and helps to reduce bone inflammation[119] without exerting proliferative effects on the endometrium[120] and displaying only weak proliferative effects on the breast epithelia.[121] Puerarin, another isoflavone in *Pueraria*, may also support bone density via estrogenic effects,[122] without having proliferative effects on the endometrium.[123]

Pueraria montana var. *lobata*, kudzu

to be used repetitively. To improve sleep quality significantly over time, use the Tincture for Menopausal Insomnia instead.

Valeriana officinalis	30 ml
Actaea racemosa	15 ml
Humulus lupulus	15 ml

Take 1 dropperful of the combined tincture, 1 hour before bedtime, another 30 minutes before bedtime, and another dropperful right before retiring. GABA (γ-aminobutyric acid) is a nutraceutical supplement that could complement the tincture.

Fine-Tuning Menopausal Transition Formulas

As with PMS, the hormonal fluctuations of the menopausal transition can promote mental emotional disturbances. The choice of herbs—anxiolytic, antidepressant, adaptogenic, and trophorestorative for the nervous system—to use in teas and tinctures for treating perimenopausal mood disorders will depend on whether the presentation takes the form of anxiety or depression.

SPECIFIC INDICATIONS	HERBS TO EMPHASIZE
Nervous exhaustion and debillity	Avena sativa Matricaria chamomilla Verbena hastata
Anxiety, panic, tension	Actaea racemosa Hypericum perforatum Melissa officinalis Leonurus cardiaca Passiflora incarnata
Depression, gloom, lethargy	Actaea racemosa Albizia lebbeck, A. julibrissin Melissa officinalis Panax ginseng
Insomnia	Humulus lupulus Passiflora incarnata Valeriana sitchensis, V. officinalis Withania somnifera
Exhaustion, poor stamina, poor concentration	Eleutherococcus senticosus Panax ginseng Rhodiola rosea

Tincture for Menopausal Anxiety with Palpitations

Compare and contrast this formula with the Tincture for Menopausal Palpitations without Anxiety (page 139). Many women experience functional heart palpitations due to hormonal and emotional triggers at menopause, while many others find that blood pressure and cholesterol become elevated following menopause. This formula is an example of a nervine and hormonal-support formula when heart palpitations are believed to be functional, while the following formula better addresses possible heart disease or risk factors for emerging heart disease. Choose a formula such as this one when cholesterol and blood pressure are normal and palpitations are accompanied by panic or anxiety states. *Leonurus* and *Actaea* are both hormonal-balancing nervines, making them appropriate for this use, and are complemented by *Piper methysticum* (kava), which can be removed from the formula once the other herbs are exerting their effects.

Leonurus cardiaca	30 ml
Actaea racemosa	15 ml
Piper methysticum	15 ml

Take 1 or 2 dropperfuls of the combined tincture, 3 to 6 times daily depending on the severity of the symptoms.

Tincture for Menopausal Anxiety Disorders

Hormonal swings are notorious for contributing to emotional instability, and for some women rapid changes in hormone levels, with significant highs and lows, may cause anxiety, worry, irritability, feelings of urgency, or an inability to cope with stress. Formulas for such presentations might include adaptogenic herbs for adrenal support, general nervines for nervous system trophorestoration, and in some cases, hormonal-balancing agents for reducing wide hormonal swings. This particular formula has no solid adaptogenic herb, which would be essential if fatigue, poor concentration, and muscle weakness accompanied anxiety. Instead, this formula features simple hormonal-balancing nervines emphasized in traditional literature for menopause and proven effective in my own experience. Complement with an adrenal support product or encapsulation.

Hypericum perforatum	15 ml
Leonurus cardiaca	15 ml
Melissa officinalis	15 ml
Scutellaria lateriflora	15 ml

Take 1 to 2 dropperfuls of the combined tincture at a time, as often as 6 times a day for acute anxiety, or just once or twice a day as a maintenance therapy. Add an adrenal support tea or encapsulation where indicated.

Tincture for Menopausal Palpitations without Anxiety

In cases where women present with increasing blood pressure, cholesterol, triglycerides, or other markers of cardiovascular disease, heart-supportive herbs and diet and exercise therapies may be more important than nervines and hormonal support. *Leonurus* has both hormonal and vascular effects and is specific for heart palpitations, so is used as a foundational herb in this formula. *Crataegus* and *Lepidium* are nutritive trophorestorative agents used as basic cardiovascular support. *Angelica* is a blood mover and *Passiflora* is a hypotensive nervine. Because these herbs all work by difference mechanisms, they are complementary and make a good general starting formula. Depending on the situation, more aggressive hypotensives (such as *Rauvolfia serpentina* or *R. vomitoria*), hypolipidemics (such as *Allium* and *Commiphora mukul*), or even cardiac glycosides (such as *Convallaria majalis*) may be added to the formula or used in complementary treatments. For detailed information on herbal formulas for those with heart disease, see Volume 2, chapter 2.

Leonurus cardiaca	20 ml
Crataegus spp.	10 ml
Lepidium meyenii	10 ml
Angelica sinensis	10 ml
Passiflora incarnata	10 ml

Take ½ to 1 teaspoon of the combined tincture, 3 times daily, long term.

Leonurus cardiaca for Heart Palpitations

Legend has it that *leo*, meaning lion, and *cardiaca*, referring to the heart, together suggests that this plant gives one the heart of a lion, as ancient folklore emphasized it not only for irregularities in heart rhythm, but also to restore the spirit when weakened or enfeebled by stress, overwork, and exhaustion. *Leonurus* can be included for hypertension and is specifically indicated for heart palpitations and arrhythmias that are not due to heart disease, but rather due to disruptions in cardiac innervation and hormonal control. *Leonurus* contains rosmarinic acid, which may reduce hyperthyroid-driven tachycardia and episodes of irregular heart action. *Leonurus* alkaloids leonurine and stachydrine may prevent fibrotic infiltration of the uterine muscle and improve dysmenorrhea pain.[124] Leonurine has also demonstrated considerable neuroprotective effects in animal models of ischemic stroke, Parkinson's disease, and Alzheimer's disease, and early evidence suggests that anti-inflammatory effects on neurons and monoamine neurotransmitter effects contribute to the antidepressant effects of *Leonurus*.[125] *Leonurus* can be included in formulas for anxiety and depression, especially when occurring in association with the menstrual cycle or as part of the menopausal transition.

Research also suggests that *Leonurus* may reduce benign hyperplasia (BPH) by reducing testosterone-driven proliferative effects,[126] while increasing estradiol.[127] *Leonurus* has a folkloric reputation as a diuretic, but I suspect that it primarily improves urine flow when urine is retained due to an enlarged prostate gland. Modern molecular research has shown *Leonurus* to have antifibrotic, antiatherosclerotic, and cardioprotective effects. *Leonurus* also has numerous effects on platelets and vascular inflammation.

Leonurus cardiaca, motherwort

Menopausal Complaints

Tincture for Severe Cardiac Symptoms in Menopause

In some cases, heart irregularity and palpitations during menopause can be so severe as to warrant an EKG and thorough lab workups to rule out organic disease. This formula contains cardiac herbs that work via various mechanisms to regulate functional heart pathologies. This formula is more appropriate for severe and acute heart palpitations, tachycardia and irregularity, and a sense of pressure or constriction in the chest than the Tincture for Menopausal Hypertension. Formulas such as this one have been effective for many women I have

Black Cohosh for Mood and Menopause

Actaea racemosa, which has also been called *Cimicifuga racemosa*, is a North American perennial herb used by Native Americans for the treatment of dysmenorrhea, rheumatism, and other complaints and is presently widely used for relief of menopausal symptoms. Black cohosh was popularized by Eclectic medical practitioners in the late nineteenth and early twentieth centuries. They referred to the plant as macrotys, recommending it for ovaritis, endometriosis, amenorrhea, and dysmenorrhea. Black cohosh root was included in the *United States Pharmacopeia* (USP) from 1820 to 1926.

Pharmacologically active compounds in *Actaea* include cycloartane-type triterpene glycosides (actein, deoxyactein, and cimicifugoside), aromatic acids (caffeic and ferulic acids), organic acid esters (including the isoferulic acids fukinolic and cimicifugic acids), and alkaloids (cytisine, cimicifugine). Even though isoflavones have been identified that may act as estrogen-receptor agonists, *Actaea* does not appear to act solely via estrogen agonism. Rather, multiple mechanisms of action appear to have hormone-regulating effects. Multiple effects on hormone-regulating genes may have anticancer effects, protecting against excessive proliferations and encouraging apoptosis in tumor cells.[128] Black cohosh fractions enriched for triterpene glycosides such as actein are shown to inhibit the development of breast cancer and induce apoptosis in human breast cancer cell lines.[129]

Research has shown *Actaea* to bind mu opiate receptors,[130] helping to provide anodyne and anxiolytic effects. Some of *Actaea*'s positive effects on PMS and menopause may occur via the opiate system, which is involved with regulation of mood, temperature, and sex hormone levels. *Actaea* may also have serotonergic and dopaminergic effects. Dopamine is known to regulate the output of pituitary hormones FSH, LH,

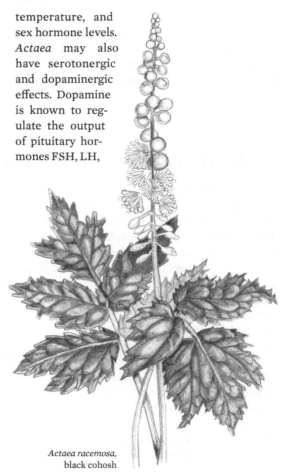

Actaea racemosa,
black cohosh

and prolactin, and *Actaea* suppresses LH. Clinical trials have suggested that *Actaea* can help treat menopausal hot flashes (vasomotor symptoms)[131] and mood disturbances.[132]

seen who report a sensation of the heart "flip-flopping" in the chest. This formula is intended for use acutely at the onset of such symptoms, continuing with daily dosing for several months, at which point the situation is often improved and the frequency of the dose can be reduced. Hormonal support and other thoughtful supportive measures will typically help correct the problem such that use of tinctures like this one are not needed long term. For actual heart failure, herbs such as these might be combined with *Selenicereus* and herbs that contain cardiac glycosides, such as *Convallaria*. For detailed information on herbal formulas for those with cardiomyopathy, see Volume 2, chapter 2.

Lobelia inflata	10 ml
Rauvolfia serpentina	10 ml
Viscum album	10 ml
Leonurus cardiaca	10 ml
Allium sativum	10 ml
Mentha piperita	10 ml

For acute use, take ½ to 1 teaspoon of the combined tincture every 10 to 30 minutes as needed. For general maintenance, take ½ to 1 teaspoon, 3 times daily for several months, as part of a comprehensive treatment plan that might include calcium/magnesium, B vitamins, nervines, or other approaches as specifically indicated.

Tincture for Menopausal Hypertension

Hypertension in general may be treated slightly differently than hypertension that correlates strongly with the menopausal transition. Because there are estrogen receptors on blood vessels, shifting reproductive hormones may play a role in hypertension initiated at the climacteric. This formula is aimed at using general blood-moving anti-inflammatory agents as a first-tier therapy. If not sufficiently effective on its own, consider nervines, peripheral vasodilators, and stronger beta blockers next.

Crataegus spp.
Ginkgo biloba
Angelica sinensis
Leonurus cardiaca

Take ½ to 1 teaspoon of the combined tincture, 3 to 5 times daily. A follow up visit in 2 weeks should reveal initial improvements and if not seen, other approaches can be chosen.

High Mineral Tea to Prevent Osteoporosis

This formula is aimed at providing organic mineral complexes (through inclusion of *Equisetum*, *Foeniculum*, *Centella*, *Taraxacum*, *Urtica*, and *Mentha*) as well as high-isoflavone plants (*Pueraria*, *Glycyrrhiza*, *Medicago*) as a means of supporting bone density. *Piper nigrum* (black pepper) is noted to enhance the absorption of many compounds, including minerals, and may be included in this tea formula to taste.

Equisetum arvense, E. hyemale
Foeniculum vulgare seed powder
Centella asiatica
Taraxacum officinale leaf
Urtica urens, U. dioica
Mentha piperita
Pueraria montana var. *lobata, P. candollei* var. *mirifica*
Glycyrrhiza glabra
Medicago sativa
Piper nigrum powder

Combine in equal parts or to taste. Steep 1 tablespoon of the mixture per cup of hot water and then strain. Drink freely, 3 or more cups (720 ml) a day, continuing long term.

Tincture for Menopausal Muscle Pain and Stiffness

Many menopausal women experience chronic myalgia, especially of a tight and stiff character that is worse upon waking in the morning and gradually improves with the day's activity. Research is emerging regarding little-studied hormones such as relaxin that suggests levels may decline with age and menopausal transitions.[133] *Actaea* (also known as *Cimicifuga*) is emphasized in historical texts for chronic muscle stiffness and is often effective for fibromyalgia and other types of nonarthritic muscle pain. This specific herb is complemented by other muscle relaxers (*Piper methysticum* and *Viburnum* in this formula) and should begin to take effect in less than a week. If no improvements are noted in this time frame, other therapies should be initiated.

Actaea racemosa	15 ml
Boswellia serrata	15 ml
Piper methysticum	15 ml
Viburnum opulus	15 ml

Take 1 teaspoon of the combined tincture, 3 to 6 times daily, reducing dose and frequency as pain improves.

Formulas for Breast Conditions

The common complaint of premenstrual breast tenderness, also known as cyclic mastalgia, responds well to natural and herbal medicine. The concepts of hyperestrogenism or unopposed estrogen, discussed throughout this chapter, also impact cyclic mastalgia. Diets rich in fresh fruits and vegetables are noted to offer some protection against fibrocystic breast disease and breast cancer, one reason being that these foods support glutathione detoxification systems in the liver that help metabolize estrogen.[134] Discrete individual cysts in the breast and fibrocystic breast disease also respond well to hormonal balancing and natural therapies, but must be differentiated from breast cancer. Breast cancer is outside the scope of this book overall, but I cover some general treatment possibilities at the end of this chapter.

Recommending BCPs for PMS and Other Disorders

Allopathic physicians sometimes offer birth control pills (BCPs) to treat PMS, fibrocystic breast disease, mastodynia, and occasionally other female complaints. The rationale is that if one reduces the up-and-down, cyclical nature of the normal menstrual cycle, symptoms will improve. This is rarely effective in any long-term or deep-acting way, because BCPs do not correct any underlying hormonal imbalance. In fact, taking BCPs often creates more work for the liver and can contribute to hormonal difficulties over time. Birth control pills can decrease insulin sensitivity and aggravate glucose and lipid metabolism,[135] which can worsen ovarian cysts, hormone-related weight gain, and the severity of metabolic insufficiency and can promote vascular inflammation. For these reasons, alternative practitioners often feel it is unwise to treat such complaints with BCPs; instead, they aim to create a more deep-acting therapy.

Cyclic mastalgia is most common in women in their twenties and thirties and is obviously associated with the menstrual cycle. Allopathic medicines include hormone-blocking pharmaceuticals such as danazol or tamoxifen but are not preferred by many practitioners due to concern that they are superficial remedies that do little to correct the underlying hormone imbalance and may even worsen hormone imbalance. Women with elevated prolactin, such as those with PCOS, may have persistent mastalgia, possibly associated with amenorrhea. Bromocriptine is often prescribed to treat elevated prolactin, but *Vitex agnus-castus* may be just as effective. Essential fatty acids, vitamin E, and vitamin B_6 supplementation may also be helpful.

Stress may induce elevated prolactin, and dopamine is a natural regulator of prolactin release from the pituitary. Therefore, nervines, adaptogens, and herbs known to promote dopamine may be helpful in formulas for mastalgia. *Vitex* promotes dopamine, which in turn inhibits prolactin release.[136] This is one of several mechanisms whereby *Vitex* is effective for mastalgia. Other dopaminergic herbs include *Hypericum*, *Melissa*, *Verbena*, and *Actaea* (*Cimicifuga*). There has been little research on *Ceanothus* and *Phytolacca* for the treatment of breast cysts and disorders, but both are emphasized in folkloric traditions for lymphatic circulation, breast pain, and cysts, and I have found them to be effective in formulas for breast pain and complaints.

Topical Breast Oil

For breast cysts, fibrocystic breast disease, and mastitis, a topical oil such as this one is useful as part of a hydrotherapy treatment or hot pack application. *Phytolacca* may improve lymphatic circulation and is classically indicated for breast pain, breast masses, and lymphatic stagnation. Iodine supports thyroid function, and both iodine deficiency and hypothyroidism are risk factors for breast cancer.[137] The mammary gland possesses the same sodium-iodine symporter as the thyroid gland in order to concentrate iodine as an important breast milk nutrient. Just as iodine may protect against immune-driven inflammation and proliferation in the thyroid gland, iodine may also protect against proliferation and inflammation in breast tissue. Animal investigations have shown iodine to deter breast cancer. The administration of Lugol's iodine (10 percent potassium iodide) or iodine-rich wakame seaweed to rats treated with the carcinogen dimethylbenzanthracene suppressed the development of mammary tumors. Lugol's iodine and wakame seaweed may induce apoptosis in human breast cancer cells with greater potency than that of fluorouracil, a chemotherapeutic agent used to treat breast cancer.[138] Iodine is shown to affect at least a dozen genes involved in hormone metabolism in breast tissue.[139] Iodine has excellent transdermal absorption, and other than staining clothing, topical iodine can do no harm; it is included in this formula as an anticancer, anti-inflammatory, and anti-proliferative agent for the breasts. Women with a history of breast cancer may use this oil to help prevent a recurrence, or women with hypothyroidism may use it as a preventive.

Phytolacca americana oil	2 ounces (60 ml)
Castor oil	2 ounces (60 ml)
Potassium iodide	100 drops

Combine in a 4-ounce (120 ml) bottle and shake well prior to each use. Apply ½ to 1 teaspoon to the breasts and cover with heat. Repeat as often as hourly for acute pain, inflammation, or infections. If nursing, wash off with soap and water prior to allowing the baby to nurse. If used for cysts and fibrocystic disease, the oil does not need to be washed from the skin.

Tincture for Mastalgia and Fibrocystic Breast Disease

Fibrocystic breast disease may result from hyperestrogenism and excessive stimulation of hormonally sensitive mammary tissue. Since exogenous estrogens and a high toxin load in the body may also initiate or exacerbate breast cysts and mastalgia, alteratives such as *Taraxacum* may be supportive herbs. *Vitex* is often helpful for mastalgia[140] and can reduce hyperprolactinemia often associated with mastalgia[141] by promoting progesterone production to oppose estrogen. However, *Vitex* is not folklorically emphasized for actual cystic lesions as are *Ceanothus* and *Phytolacca*. Both of the latter are traditionally said to be lymphatic herbs and highly specific for breast swelling, pain, cysts, and lymphatic congestion.

Taraxacum officinale root	15 ml
Vitex agnus-castus	15 ml
Ceanothus americanus	15 ml
Phytolacca americana	15 ml

Take ½ to 1 teaspoon of the combined tincture, 3 to 5 times daily, reducing over several months' time as symptoms improve. This formula will often help within a month's time but it is optimal to use it for 3 or even 6 months or more. If liver and bowel health and diet, especially reducing

exposure to exogenous estrogens, are addressed simultaneously, most cases will see significant improvement.

Tincture for Mastalgia

Mastalgia, also referred to as mastodynia, is commonly cyclical, in relation to hormonal rhythms. *Vitex* can be very effective for chronic mastodynia when due to hyperestrogenism. *Ceanothus* and *Phytolacca* are excellent adjuvant herbs due to effects on lymphatic circulation and specificity for breast cysts and congestion. *Hypericum* is included here for its general anti-inflammatory and anodyne effects on irritated nerves. *Hypericum* is most indicated for acute pain and could be removed from the formula for a lower-grade condition.

Vitex agnus-castus
Ceanothus americanus
Phytolacca americana
Hypericum perforatum

Combine in equal parts. Take ½ to 1 teaspoon, 3 times daily, reducing over several months as pain improves.

Tincture for Acute Mastitis

Acute mastitis occurs almost exclusively in nursing women due to engorgement. This formula in conjunction with simple heat packs will often resolve acute mastitis in nursing women quickly.

Echinacea purpurea	25 ml
Ceanothus americanus	25 ml
Phytolacca americana	10 ml

Take a dropperful of the combined tincture every ½ to 1 hour, reducing over the day as symptoms improve. Mastitis typically responds rapidly, so if there has been no improvement in 48 hours, the case should be reconsidered.

Herbs for Breast Cysts from Folkloric Literature

The following herbs may help resolve breast cysts when taken acutely and also may reduce the tendency for cysts to form when taken long term. These herbs can be included in formulas for fibrocystic breast disease, as well as in formulas for rapidly developing and painful breast cysts.

Ceanothus americanus	*Phytolacca americana*
Curcuma longa	*Taraxacum officinale*
Galium aparine	*Vitex agnus-castus*
Iris versicolor	

Formulas for Ovarian Cysts

Multiple ovarian cysts associated with polycystic ovarian syndrome (PCOS) are the most common types of ovarian cysts, but single or recurrent follicular cysts also occur. In the case of PCOS, hyperandrogenism also contributes to cysts,[142] as do complex hormonal imbalances; the term "polycystic" may be a misnomer in that not all cases will display ovarian cysts. As with most of the conditions discussed in this chapter, hyperestrogenism contributes to ovarian cysts, and supporting liver clearance is one prong of herbal therapies. Cinnamon can reduce insulin resistance and is included in some formulas for ovarian cysts related to PCOS.[143] (See also "Formulas for Polycystic Ovarian Syndrome [PCOS].") *Vitex* may promote progesterone and oppose hyperestrogenism. *Serenoa*, *Mentha spicata*, and *Urtica* may help reduce elevated testosterone. *Phytolacca*, *Iris*, and *Bryonia dioica* are specific for the cysts themselves and may be included in formulas for pain.

Formula for Anovulation in PCOS Based on TCM

Tanshinones from *Salvia miltiorrhiza* may reduce hyperandrogenism in women with PCOS and may improve fertility and increase the conception rate.[144] This formula, based on a formula from Traditional Chinese Medicine, may be prepared as a tincture or a tea and is reported to reduce hyperandrogenism and hyperinsulinism, helping to restore cyclical ovulation. [145]

Glycyrrhiza glabra
Paeonia lactiflora
Angelica sinensis
Salvia miltiorrhiza
Panax notoginseng
Lycopus virginicus

Combine tinctures in equal parts, or mix dried herbs in equal parts to make a tea blend. Take 2 or 3 dropperfuls of the combined tincture per day. For a tea, gently simmer 1 teaspoon of the dried herb mixture per cup of hot water for 10 minutes, let stand in a covered pan 10 minutes more, and then strain; drink 2 or 3 cups (480 or 720 ml) per day. Continue for several months and then evaluate results.

Turska's Formula for Acute Ovarian Cysts

This classic naturopathic formula is named after Dr. William Turska, who practiced medicine from the 1930s on, passing away in 1995. Dr. Turska was one of my earliest teachers, and he himself studied with some of the Eclectic era physicians. This formula is in the style of the Eclectics: Potentially toxic herbs are heavily diluted with water. The resulting diluted tincture is taken by the teaspoon only, to avoid the possibility of a toxic reaction.

Phytolacca americana	4 ml
Gelsemium sempervirens	2 ml
Bryonia alba	2 ml
Aconitum napellus	2 ml

Place the tinctures in the bottom of a 4-ounce (120 ml) bottle, and then fill with water. Take 1 teaspoon, 4 times a day, continuing for 3 months or more. If treating an acute ovarian cyst, the dose may be reduced to just once or twice a day after the pain resolves.

Formulas for Polycystic Ovarian Syndrome (PCOS)

Polycystic ovaries are estimated to occur in 5 to 15 percent of women, and the condition is associated with complex underlying endocrine biochemical imbalances. In addition to "cysts" in the ovaries—which can be evidenced with pelvic ultrasound—the hallmarks of the condition are mild obesity, hirsutism, and menstrual irregularities, typically amenorrhea or oligomenorrhea. The cysts are actually multiple mature follicles that persist without fully developing and being released from the ovaries, but these are not present in all cases.

The underlying endocrine imbalances include elevated androgens, which contribute to the hirsutism, and can also promote acne, altered feedback loops with pituitary gonadotrophins, and frequently some degree of insulin resistance, which contributes to obesity. The insulin resistance is often associated with mild hyperglycemia and lipid elevation, and therefore PCOS is often thought linked to metabolic syndrome. Many women with PCOS have anovulatory cycles with difficulty conceiving. Estrogens are often elevated in tandem with the androgens, leaving women at risk of excessive estrogenic stimulation of the breasts or endometrium. Progesterone is sometimes prescribed to reduce the risk of endometrial cancer. Also associated with the complex endocrine imbalances may be altered levels of FSH, prolactin, and sometimes TSH and thyroid hormones.

Mild cases of PCOS may go undiagnosed, or the symptoms may be minimal, and thus women do not seek treatment. In other cases, the desire for pregnancy may promote an infertile woman to seek treatment for PCOS, or irregular menses and amenorrhea may cause women to seek therapy. And in other cases, desire for solutions for weight loss, elevated glucose and lipids, or hirsutism or acne may cause women to seek therapy.

Herbal therapy is often effective for all the symptoms of PCOS, as well as for improving the underlying endocrine imbalances. *Vitex* is an important foundational herb in formulas for PCOS because it can help restore regular menses via effects on the neuroendocrine system. *Vitex* is also appropriate for hormone-related acne. *Salvia officinalis* may help reduce prolactin when elevated and a cause of infertility and amenorrhea. *Serenoa* and *Urtica* can help with elevated androgens, and *Glycyrrhiza* may be included for hormone-balancing effects. Clinical trials have shown restoration of menses in amenorrhic women using *Glycyrrhiza*, as well as a reduction in testosterone in women with PCOS. When lipids and glucose are elevated, *Allium*, *Berberis*, and *Commiphora mukul* might be added. Berberine, a component of *Berberis* and *Coptis*, can improve insulin resistance and improve metabolic function in women with PCOS with overlapping obesity or impaired metabolic functions.[146] *Foeniculum*, like many Apiaceae family members, promotes estrogenic effects, and some research reports the herb can reduce hirsutism when applied topically. Drinking *Foeniculum* tea or including *Foeniculum* in tinctures for PCOS may help combat high testosterone levels and reduce hair growth. *Mentha*

Testosterone-Reducing Herbs for Women

Recent molecular research shows that these herbs may reduce elevated testosterone in women. For more details, see "Specific Indications" on page 165.

Glycyrrhiza glabra
Mentha spicata
Serenoa repens
Urtica spp. root

spicata (spearmint) is also reported to have antiandrogenic effects and may help reduce testosterone.

Tincture for PCOS

This all-purpose formula supports regular menses, fertility, and metabolism and lowers elevated androgens.

Vitex agnus-castus
Serenoa repens
Salvia officinalis
Glycyrrhiza glabra

Combine in equal parts. Take 1 teaspoon, 3 to 4 times daily, long term.

Tincture for PCOS with Acne

Berberis supports liver function and conjugation and elimination of excess hormones to help improve acne. For more formulas for acne, refer to Volume 1, chapter 5.

Vitex agnus-castus	30 ml
Glycyrrhiza glabra	15 ml
Berberis aquifolium	15 ml

Take 1 teaspoon of the combined tincture, 3 to 4 times daily.

Tincture for PCOS with Metabolic Syndrome

This formula would be effective when PCOS overlaps with hypothyroidism and metabolic syndrome resulting in elevated lipids or glucose due to the additions of *Allium* and *Commiphora mukul* to support basal metabolism and lipid processing.

Vitex agnus-castus
Serenoa repens
Allium sativum
Commiphora mukul

Combine in equal parts. Take 1 teaspoon, 3 to 4 times daily.

Tincture for Hirsutism of PCOS

All the herbs in this formula are noted to either decrease the synthesis of dihydrotestosterone or enhance its breakdown and metabolism. Lowering testosterone will help prevent hair follicles from producing coarse whiskerlike hairs. For additional formulas for hirsutism, refer to Volume 1, chapter 5.

Glycyrrhiza glabra
Melissa officinalis
Foeniculum vulgare
Vitex agnus-castus

Combine in equal parts. Take 1 teaspoon, 3 to 6 times daily.

Glycyrrhiza for Ovarian Cysts and Amenorrhea

Glycyrrhiza has been called "the great harmonizer" due to the belief that it can travel everywhere in the body and pull together the actions of other herbs in botanical formulas. Licorice is used for a wide variety of conditions including adrenal disorders, infertility, and hormonal problems. Glycyrrhizic acid, a steroidal saponin in licorice, improves insulin resistance[147] and fat metabolism in animal models of metabolic syndrome.[148] Both animal and human investigations have shown *Glycyrrhiza* to reduce fat deposition in tissues[149] and the accumulation in abdominal fat.[150]

Glycyrrhiza glabra,
licorice

Licorice tea can be a simple and inexpensive prong of a broader protocol for treating polycystic ovarian syndrome (PCOS), improving insulin resistance, and restoring regular menstrual cycles. *Glycyrrhiza* is also indicated for women with infertility due to amenorrhea, reducing elevated androgens and prolactin, and clinical studies have shown licorice to restore regular menses in amenorrheic women[151] as well as in women with PCOS.[152] Both human and animal studies have suggested licorice to reduce androgens when elevated.[153] Licorice is often used in Japan combined with *Paeonia* to treat PCOS in a traditional herbal medicine called shakuyaku-kanzo-to, which is reported to reduce elevated testosterone[154] and improve diffuse cysts in the ovaries.[155] Licorice may normalize elevated testosterone via a number of mechanisms including reducing the synthesis of testosterone,[156] inhibiting dehydrogenase enzymes[157] involved with testosterone synthesis,[158] inhibiting 5α-reductase enzymes, reducing the conversion of testosterone to the more active dihydrotestosterone,[159] enhancing the metabolism of testosterone, and reducing activity at testosterone receptors.[160] Several clinical studies have evaluated the effects of licorice in women with PCOS taking spironolactone to reduce elevated androgens. Spironolactone therapy frequently impairs the regulation of body minerals via mineralocorticoid receptor agonism, and licorice is found to reduce such side effects[161] and, of note, did not elevate blood pressure. Licorice may even be an effective alternative to spironolactone therapy, as licorice alone may reduce elevated testosterone and restore regular menstruation.

Licorice may also reduce elevated prolactin, which is common in women with PCOS and is also an undesirable side effect of some drugs, particularly antipsychotics. The combination of *Paeonia* and *Glycyrrhiza* has been shown to reduce elevated prolactin in women with hyperprolactinemia induced by the antipsychotic medication risperidone.[162]

Because many women with PCOS seek treatment for infertility, a common therapeutic dilemma is whether to continue the use of herbal medicines should pregnancy occur. Licorice appears fairly innocuous during pregnancy and may be valuable to correct elevated androgens and prolactin to restore ovulatory cycles, yet small studies have suggested that licorice should not be used throughout pregnancy or consumed in large dosages by pregnant women. Licorice does not appear to be teratogenic or increase the risk of inducing miscarriage but may increase the risk of preterm delivery. Heavy licorice exposure—more than 500 milligrams of glycyrrhizin per week—is associated with a greater risk of preterm delivery in one outcome study,[163] without evidence of effects on birth weight or maternal blood pressure,[164] and the rate of stillbirths may be marginally higher among women who use licorice compared to those who do not.[165] Maternal consumption of glycyrrhizin may increase salivary cortisol levels in newborns just as the consumption of synthetic corticoids do.[166] Due to this emerging research, it appears best to limit the use of licorice during pregnancy.

There have been widely publicized reports that licorice may promote or exacerbate hypertension, although the actual occurrence of this side effect is quite rare, and recent studies elucidate that licorice may aggravate high blood pressure more often in those with a preexisting complex adrenal disorder involving altered potassium levels and dehydrogenase enzyme activity.[167] Nonetheless, high blood pressure is a concern for some women with PCOS occurring concomitantly with metabolic syndrome. Licorice actually reduced blood pressure in an animal model of metabolic syndrome,[168] but those with abnormally low potassium levels should avoid licorice due to the possibility of serious muscle weakness and high blood pressure with the consumption of licorice. To be on the safe side, it is a simple matter to monitor blood pressure when taking licorice as a daily medication.

Tea to Decrease Testosterone

These herbs may help to reduce testosterone and are chosen due to their palatability in teas.

Mentha spicata	4 ounces (120 g)
Glycyrrhiza glabra	4 ounces (120 g)
Foeniuculm vulgare	4 ounces (120 g)
Urtica spp.	4 ounces (120 g)

This recipe makes 1 pound (480 g) of tea. Store in an airtight resealable plastic bag or glass jar for use long term. Steep 1 tablespoon per cup of hot water and then strain. Drink freely.

Tincture for PCOS with Hypothyroidism

Commiphora mukul and *Melissa* are noted to support metabolism and reduce thyroiditis and/or impaired thyroid function.

Commiphora mukul
Melissa officinalis
Coleus forskohlii
Vitex agnus-castus

Combine in equal parts. Take ½ to 1 teaspoon, 3 to 6 times daily for many months.

Tincture for PCOS with Anxiety and Adrenal Imbalance

This formula combines the adaptogen *Withania* with the nervines *Melissa* and *Hypericum*, and the hormone-regulating agents *Vitex* and *Actaea* to create a broad formula for PCOS.

Withania somnifera
Melissa officinalis
Hypericum perforatum
Vitex agnus-castus
Actaea racemosa

Combine in equal parts. Take ½ to 1 teaspoon, 3 to 6 times daily for many months.

Polycystic Ovarian Syndrome (PCOS)

Vitex agnus-castus for Reproductive Complaints

Vitex agnus-castus goes by the common names monk's pepper and chaste tree berry because the small pungent peppercorn-like fruits were used by men in religious orders to reduce libido and maintain chastity. *Vitex* also has a long history of use in treating PMS, mastalgia,[169] gynecomastia in men, adolescent acne, amenorrhea, ovarian cysts, hot flashes and menopausal complaints, and menstrual irregularities. I use *Vitex* extensively in my practice and find it highly effective and well tolerated. *Vitex* is also indicated in formulas for infertility and hormonal imbalance in women with PCOS. One double-blind placebo-controlled trial showed *Vitex* to be effective in treating mastalgia, both in terms of intensity and duration compared to placebo.[170]

A growing number of molecular and mechanistic studies demonstrate *Vitex* to affect reproductive hormones via neuroendocrine effects.[171] *Vitex* has opiate and dopaminergic effects, and dopamine regulates hypothalamic and pituitary hormones including prolactin, FSH, and LH. Via effects on these pituitrophins, *Vitex* may impact downstream production of reproductive hormones. *Vitex* promotes opiate pathways in the brain,[172] which in turn promotes dopamine activity, but *Vitex* is not an opiate itself nor is it sedating or addictive.

Vitex fruits contain flavonoid glycosides including casticin, vitexin, penduletin, orientin, and apigenin, as well as iridoid compounds aucubin,[173] agnuside,[174] and labdane diterpene alkaloids.[175] Apigenin, vitexin, and penduletin are shown to be selective estrogen receptor agonists[176] able to competitively bind α and β estradiol receptors.[177] *Vitex* may also affect estrogen-induced genes.[178] Cell culture studies suggest *Vitex* to stimulate progesterone expression in endometrial cultures.[179] *Vitex* may also exert antiandrogenic effects, explaining its fairly reliable ability to treat adolescent acne, restore regular menses in hyperandrogenic women, and reduce excessive libido. The antiandrogenic effects may reduce fertility in men, reducing sperm count and motility.[180] A related species, *Vitex negundo*, has been shown to have antiandrogenic effects at a dose of 10 milligrams/kilogram parenterally, blocking the effects of testosterone propionate supplementation in dogs. The flavonoid fraction has been credited with these actions and was significant enough to interfere with metabolic activities in the testes and spermatogenesis.[181]

Due to its hormonal and neuroendocrine effects, *Vitex* should be not consumed in pregnancy and likely not during lactation either, although *Vitex* is sometimes listed as a possible galactogogue by some sources. There are conflicting reports as to whether *Vitex* increases or decreases lactation.[182]

Vitex agnus-castus, chaste tree berry

Formulas for Endometriosis

The presence of endometrial tissue outside the uterus is a frequent cause of both infertility and severe dysmenorrhea. Commonly called endometriosis, this condition is an estrogen-dependent inflammatory disease that occurs in women of reproductive age and generally resolves with menopause, unless a woman uses postmenopausal hormone therapy. Extensive endometriosis can be difficult to control with botanical medicine alone, but therapies noted to improve estrogen clearance, as discussed throughout this chapter, are appropriate. The use of liver and alterative herbs, along with botanicals, nutraceuticals, and dietary approaches that reduce estrogen and promote progesterone are good starting points and should be considered as long-term therapies.

Endometriosis is associated with pelvic pain, and the endometrial lesions are noted to develop their own nerve supply and are thereby influenced by neuronal input, not only hormone signals.[183] About one-third of women undergoing medical work-ups for chronic pelvic pain are found to have endometriosis with other causes of pelvic pain being interstitial cystitis, abdominal migraines, or irritable bowel syndrome. Hormonal therapies and surgeries are the main treatments offered by general medicine, but pelvic pain often returns, confounding patients and clinicians alike. Reducing hyperestrogenism is a primary goal of alternative practitioners when treating endometriosis. The formulary here is sparse, and because there is such overlap in the treatment of the underlying hyperestrogenism, the formulas detailed here are similar to those for ovarian cysts and uterine fibroids. While uterine fibroid formulas may require more hemostatic herbs, endometriosis may require more anodyne herbs due to the painful nature of the condition. Agents that support estrogen metabolism such as liver-supportive alterative herbs and B vitamins would be valuable complements. DIM (diindolylmethane) and the glucosinolates in crucifer family foods would also support estrogen sulfation and clearance. Bromelain and *Curcuma* encapsulations may be both anti-inflammatory and useful for acute pain.

Endometriosis Pain Management Tincture

This formula is less useful for treating the underlying hormonal imbalance and more aimed at helping to mitigate pain and dysmenorrhea. It can be used all month long but may be especially useful during the menses.

Also use a heat pack and topical applications where needed. Complement this formula with other therapies aimed at mitigating hyperestrogenism and inflammatory disease, which should be taken all month long for many months.

Dioscorea spp.
Angelica sinensis
Viburnum opulus
Piscidia piscipula
Zingiber officinale

Combine in equal parts. Take 1 dropperful every half hour for acute pain, reducing as symptoms improve. Complement with simple castor oil packs to the abdomen, or the Topical Anodyne for Endometrial Pain formula that follows. The use of bromelain and/or *Curcuma* capsules may also be complementary to allay pain.

Topical Anodyne for Endometrial Pain

There is no research on the effectiveness of herbs or isolated constituents for endometrial pain. This formula is based on logic: *Hypericum* and *Aesculus* oils for vascular pain, *Foeniculum* and *Mentha* essential oils as antispasmodic anodynes, and castor oil to pull the herbs into the tissues. *Hypericum* oil is commercially available, but the *Aesculus* oil may need to be prepared for one's self. The topical use of cannabidiol (CBD) products prepared from the nonintoxicating cannabidiol constituents of marijuana may also complement this formula. They are gaining popularity and research is showing efficacy for a wide variety of painful conditions.

Hypericum perforatum oil	30 ml
Aesculus hippocastanum oil	30 ml
Castor oil	30 ml
Foeniculum vulgare essential oil	15 ml
Mentha piperita essential oil	15 ml

Combine all oils and shake well before each use. Apply 1 teaspoon to the skin of the lower abdomen and cover with heat such as a heating pad or a hot moist pack. Reapply and refresh the heat every 30 to 90 minutes as desired.

Tea for Endometriosis Based on TCM

Based on a formula from Traditional Chinese Medicine, this anti-inflammatory tea features many herbs with an affinity for the vasculature. These herbs offer anti-in-flammatory, anodyne, and hormone-balancing effects.

Rehmannia glutinosa prepared root (steamed and dried)
Berberis aquifolium
Paeonia lactiflora (white peony root)

Scutellaria baicalensis
Poria cocos

Combine equal parts of the dry herbs, and decoct 1 tea-spoon of the mixture per cup of hot water. Aim to drink 3 to 6 cups (720 to 1,440 ml) per day.

Formulas for Uterine Fibroids

Uterine fibroids, also called leiomyomas, are common reproductive-age benign tumors that contribute to infertility. Large fibroids may induce prolonged or heavy uterine bleeding that can be hard to control and can lead to anemia. In some cases, fibroids may induce uterine hemorrhage. Uterine fibroids occur at a much higher rate in African-American women compared to Caucasians and are a leading cause of hysterectomy in all races. Small uterine fibroids may be found in 40 percent of women over the age of 40 and may be asymptomatic. Uterine fibroids may grow rapidly in perimenopause and may shrink just as rapidly once menses and hormonal stimulation cease postmenopausally, making it desirable to stall on surgical therapies as long as possible. However, controlling the bleeding until this point in time can be challenging.

Despite how common uterine fibroids are, there has been little rigorous research. Consuming meat and alcohol, especially beer, can cause an increase in fibroids. Fibroids decrease the more vegetables and fruits are consumed, especially citrus.[184] Vitamin D deficiency is associated with uterine fibroids, and the greater the deficiency, the greater the size of the fibroids, particularly in African American women.[185] Obesity doubles to triples the risk of developing uterine fibroids. Fibroid cells secrete high levels of collagen and resist apoptosis, and there are no effective pharmaceutical options. Surgical techniques such as uterine artery embolization may be attempted for women of childbearing age who wish to have children; the technique may shrink fibroids by inducing ischemic necrosis. High-intensity ultrasonography through the abdominal wall may also be attempted to cause coagulative necrosis in specific fibroids. Fibroids may regrow after embolization and high-intensity ultrasound treatments, however, because the underlying hormonal imbalances remain uncorrected.[186]

Isoflavones from legumes may reduce the occurrence of uterine fibroids because the phytosterols exert an amphoteric effect. Legumes are the most common dietary source of isoflavone phytosterols, but the medicinal herb *Paeonia lactiflora* (red peony root) is a nonlegume noted to be 30 times higher in isoflavones than legumes. *Paeonia* is a long-standing lead herb in women's formulas and often used to build and nourish the blood, often an important action for women with menorrhagia and anemia due to uterine fibroids. Supporting the liver with alteratives and lipotropic agents and using *Vitex* to reduce hyperestrogenism could be the foundation of formulas for uterine fibroids. Hemostatic herbs, such as *Achillea*, are also usually needed in an effort to control the bleeding that large fibroids often promote. (See "Uterine Hemostatics" on page 133.) Although *Viburnum* is not a true hemostatic, it may help control bleeding in cases of uterine fibroids by improving uterine tone and providing muscular tension around the fibroids and affected blood vessels. In some cases of uterine fibroids, a separate formula of pure hemostatics to dose every 15 minutes may be needed to control bleeding. *Leonurus* has been noted to inhibit uterine fibroids in mice. Researchers attempted to isolate the constituent responsible for the antineoplastic and antifibroid effects in mice but found the synergistic action of all the components to be most effective.[187] One small clinical trial has shown *Camellia sinensis* to reduce the size of uterine fibroids and improve anemia and excessive blood loss.

Uterine Fibroid Decoction Based on TCM

Traditional herbal approaches in China include recipes very similar to the formulas I offer in this chapter for ovarian cysts. Such recipes include *Pueraria, Poria, Paeonia, Cinnamomum,* and *Salvia miltiorrhiza.* Legume family herbs *Astragalus* and *Glycyrrhiza* are also included here for their hormone balancing and flavor-enhancing effects.

Pueraria montana var. *lobata*
Paeonia lactiflora (red peony root)
Cinnamomum cassia or *C. verum*
Salvia miltiorrhiza
Astragalus membranaceus
Glycyrrhiza glabra
Poria cocos

Combine equal parts of the dry herbs and decoct 1 tablespoon of the mixture for every 4 cups (960 ml) of water. Simmer gently for 15 minutes, let stand in a covered pan until cool enough to drink, and then strain. Drink 3 or more cups (720 ml) per day.

Tincture for Uterine Fibroids

Based on folklore and my own experience, this formula can help shrink fibroids and help control heavy menses or frequent spotting. Additional hemostatic formulas as exemplified in this chapter may also be needed for menorrhagia due to fibroids. *Leonurus* has shown an anticancer and antifibroid effect in mice.[188]

Vitex agnus-castus	12 ml
Leonurus cardiaca	8 ml
Viburnum opulus	4 ml
Achillea millefolium	4 ml
Phytolacca americana	4 ml

Take ½ to 1 teaspoon of the combined tincture, 3 or 4 times daily.

Traditional Tea for Blood Stasis

A TCM formula called gui zhi fu ling tang has been used for centuries to help alleviate discomfort from uterine or menstruation problems as well as to treat uterine fibroids, all conditions considered to be symptoms of blood stasis. Cinnamon has warming, blood-moving properties that in TCM is thought to unblock channels and dissolve blood stasis. The fu ling mushroom, *Poria*, resolves stasis by treating underlying "dampness." Peony roots from various species and processing techniques of *Paeonia* are broadly used for many types of vascular stasis including pelvic stagnation.

Cinnamomum cassia
Poria cocos
Paeonia lactiflora
Paeonia × suffruticosa (also called *P. moutan*)

Combine equal parts of the dry herbs and blend. Gently simmer 1 teaspoon of the mixture per cup of water for 15 minutes and then strain. Drink 3 cups (720 ml) per day for at least 3 months, preferably for 6 months.

Formulas for Other Female Reproductive Conditions

Painful intercourse (technically known as dyspareunia), uterine prolapse, and low libido occur often enough to deserve coverage in this chapter.

I have not found low libido in women to respond well to the same therapies that menopausal or menstrual issues respond to readily. While libido certainly has physiologic and hormonal regulators, psychospiritual influences are also crucial for women. Herbal formulas may offer adrenal support to improve progesterone and support healthy DHEA and cortisol ratios. Offer nervines where needed to treat stress and tension in the body. And consider adaptogens as chi tonics to build the life force to which the libido is closely linked. It is more difficult to choose what emotional support to offer each individual, but careful listening will often reveal relationship issues that block sexual desire, any number of physical ailments that cause fatigue, emotional issues from anxiety to depression that interfere with normal human connections, and simple exhaustion in working moms. For these reasons, I know of no powerful or broadly effective herbal formula for low libido. Try adrenal support and adaptogens as suggested throughout this volume. Consider nervines, B vitamins,

or DHEA, depending on the person, and treat any concomitant thyroid, metabolic, or digestive disorder. See also "Sexy Herbs" on page 152 for libido formula options to consider combining with nervines and adaptogens on a case-by-case basis, an example being the Aphrodisiac Tincture on page 154.

Herbal Lubricant for Painful Intercourse

Dysparuenia is sometimes due to psychological issues best addressed gradually over time with counseling, couple's therapy, building trust, and sexual liberation. In other cases, however, thinning or drying of the vaginal mucosa causes dyspareunia, as may infections or vaginal lesions. A pelvic exam can help determine if vaginal lesions are present and if antiherpetic or antimicrobial medications are warranted. This formula for vaginal lubrication is appropriate for postmenopausal vaginal atrophy and as a sexual lubricant where indicated. When dyspareunia is atrophic and occurs postmenopausally, the local use of hormones can do wonders. Estriol creams or a combination of estrogens are used vaginally on a daily basis for several weeks, and I find that the effects can often be maintained with

The "Sexy" Herbs

Although some vendors and websites exaggerate and sensationalize the abilities of herbs to boost sexual desire and performance, a few herbs may indeed support sexual drive and function into old age. While most of the products and research are targeted toward men, many such herbs may benefit the female libido and reproductive health as well. Most "sexy" products appear to be aimed at increasing the frequency or duration of the sex act itself, but for many people sexual satisfaction also involves feelings of intimacy and connection, eroticism, playful and creative variety, and caring and nurturing expressions of love that no herbal research has explored. Therefore, while research on herbs that enhance erectile function is certainly appreciated by those in need, for those who find the emotional aspects of lovemaking crucially important, the use of the following herbs, popularly regarded as libido, erectile, and virility enhancing,[189] may still fail to provide the spark needed to kindle the romance. Many psychological factors are known to contribute to failing libidos in postmenopausal women, including retirement from a career and the social status or sense of purpose it provided and thereby self-esteem that extends to body image. A history of sexual assault and abuse or issues with one's partner can interfere with sexual desire at any age, but may come to the fore in middle age when hardworking mothers may find themselves with more time on their hands or with more emotional maturity and fortitude to address long-repressed issues. Physical factors can interfere with sexual desire, too, such as vaginal dryness causing dyspareunia in women, as well as general decreases in vitality and failing health in both genders. The same factors apply to same-sex couples and nonbinary individuals as well. Androgen-promoting substances are the core pharmaceutical option for enhancing libido, and many traditional herbs purported for their libido-enhancing properties are now being shown to promote androgens.

Epimedium grandiflorum, E. koreanum, E. brevicornu. Horny goatweed contains the flavonol icariin shown to promote nitric oxide synthesis to enhance penile circulation and erectile function.

Eurycoma longifolia. Tongkat ali may increase sexual desire.

Lepidium meyenii. Maca may promote nitric oxide and other circulatory effects and increase sperm count and motility. It is not shown to activate androgen receptors or promote androgen levels.

Mucuna pruriens. Velvet bean or cow-itch seeds contain alkaloids, including L-dopa, that increase spermatogenesis, libido, and hormone levels. *Mucuna* may increase gonadotropins if deficient and support healthy levels of testosterone.

Panax ginseng. Ginseng is one of the best-known sexual and reproductive herbs, used for centuries as an antiaging and longevity tonic and purported to maintain sexual function into old age. Animal and clinical investigations suggest regular use to enhance spermatogenesis, support fertility, increase libido, and increase erectile function. The ginsenosides may promote endothelial nitric oxide production and luteinizing hormone release from the pituitary, which in turn may boost serum testosterone.[190]

Pausinystalia yohimbe. Yohimbe is used to treat impotence with activity credited to yohimbine and related alkaloids shown to promote penile vasodilation via α-2-adrenergic blockade and monoamine oxidase inhibition, increasing serotonin and epinephrine in the brain.

Tribulus terrestris. Puncture vine contains steroidal saponins, including protodioscin, which may be converted in the body into DHEA[191] in both men and women. These steroidal saponins may lead to an increase in testosterone in men, however, testosterone has not been shown to increase in women, and animal studies suggest that *Tribulus* may support the maturation of ovarian follicles.[192] One study found *Tribulus* to support DHEA levels in women, which was associated with increased sexual satisfaction, and to promote penile circulation and libido in male animals.[193]

***Turnera diffusa*, *T. aphrodisiaca*.** Damiana is a shrub native to the United States and has a folkloric reputation for enhancing the libido, as well as for treating menstrual and pregnancy disorders and as having a nervine effect. Although the aphrodisiac claims may be exaggerated and with sparse corroborating scientific research, the plant has been shown to increase sexual activity in aging rats. *Turnera* is in the Passifloraceae family, a family noted for its calming, relaxing, and GABAergic activity.

***Withania somnifera*.** Ashwagandha is highly regarded as an adaptogen; animal studies suggest an inhibitory effect on the libido, perhaps due to the calming and GABAergic effects. Testosterone, gonadotropins, and nitric oxide are all increased by *Withania*, however, along with a spermatogenic effect.

weekly or twice weekly application thereafter. This formula may help other cases or be a simple nonirritating and nonpetroleum-based sexual lubricant.

Hypericum perforatum oil	⅓ ounce (10 ml)
Calendula officinalis oil	⅓ ounce (10 ml)
Vitamin E oil	⅓ ounce (10 ml)
Essential oils of rose and/or sandalwood	2 or 3 drops each (optional)

Combine the first three oils in a 1-ounce glass bottle and add the essential oils if desired. Use as a vaginal lubricant.

Tincture for Uterine Prolapse

Uterine prolapse will not be resolved by herbal formulas alone but may sometimes be satisfactorily improved with weight loss and strengthening of pelvic and pubococcygeal muscles. The herbs in this formula are emphasized in folkloric literature to help improve the tone of the uterine muscle itself and tighten the uterine vasculature. Do not expect dramatic results from the formula on its own.

Viburnum opulus or *V. prunifolium*
Caulophyllum thalictroides
Collinsonia canadensis
Hamamelis virginiana

Combine in equal parts and take ½ teaspoon, 4 times daily.

Aphrodisiac Tea and Bath and Ritual

This formula is aimed at promoting relaxation and a sensual feeling—mother nature can more easily take her course from here. There is nothing particularly magical about this bath; this is just an example of finding self-care techniques that are stress relieving and soul nourishing and help women to feel sensuous. Therapies will be most effective when individualized and may include anything from massage to creating a romantic and sacred space, to the more long-term work of making peace with one's body via diet and exercise or gratitude, to couples therapy and deep relationship work.

TEA INGREDIENTS

Rosa canina petals
Rosa canina hips, chopped
Stevia rebaudiana
Lavandula angustifolia flowers
Jasminum sambac

Combine the dry herbs in equal parts or to taste. Steep 1 tablespoon of the mixture per cup of hot water and then strain. Drink the tea in the bath or afterward in your silky lingerie.

BATH INGREDIENTS

4 cups Epsom salts
1 cup dry lavender flowers
1 cup dry patchouli leaves
2 oranges, sliced into thin rounds
Fresh flower petals such as rose, whole passionflower, or clematis
Rose or other scented skin oil or lotion

Fill the bath with hot water and dissolve the Epsom salts in the water. Enclose the lavender and patchouli in a muslin bag and use as a large tea bag to float in the water. Float the orange slices and flower petals on the top of the bath water. Light some scented candles, turn on soft romantic music, and soak for 20 minutes or more. After bathing, rub the oil or lotion into the skin before toweling off to help seal the moisture in the skin. Dress in silk and move the candles into the bedroom.

Aphrodisiac Tincture

Folkloric herbals often list damiana (*Turnera diffusa*, also known as *T. aphrodisiaca*) as an aphrodisiac, but I don't find anything immediate or pronounced in its action. Horny goatweed (*Epimedium* species) is reported to improve the male libido, but may also support the sex drive in women. Shatavari (*Asparagus racemosus*) is used as a sexual rejuvenator for women in Ayurvedic traditions, and kudzu (*Pueraria*) is claimed to be a female sexual tonic in Thai folklore.

Turnera diffusa
Epimedium brevicornu
Asparagus racemosus
Pueraria montana var. *lobata*
Panax ginseng

Combine in equal parts. Take 1 teaspoon, 3 times a day for 3 to 6 months. Evaluate whether the results warrant continuing the medicine or perhaps reducing the dose.

Herbal Therapies for Reproductive Cancers

Many of the factors that influence reproductive cancers may be mitigated through diet, herbal therapies, and avoidance of environmental toxins. Phytoestrogens and xenoestrogens—environmental compounds capable of binding to estrogen receptors—are important influences on reproductive cancers. The mechanisms influencing hormone receptor expression and receptor signaling are complex and still being unraveled. I give examples of the research on single phytosterols including coumarins and triterpenes throughout this chapter, but quality human clinical trials on herbal therapies for cancer are sparse. A great deal of the research has examined the ability of an herbal extract or an isolated compound to induce apoptosis, prevent cancer initiation, or up- or downregulate enzyme systems or hormone signaling in cancer cell cultures. Other studies have examined the ability of specific diets, single nutrients, or various herbal therapies to prevent or treat reproductive cancers in animal models of endometrial, ovarian, prostate, or breast cancer. None of these studies offers a deep or broad understanding, but the sheer number of studies stacking up are providing some guidance in crafting diets or hormonal support to prevent reproductive cancers. To a lesser degree, the research is also exploring more aggressive interventions to treat specific types of reproductive cancers. Although plant phytoestrogens exert effects on the reproductive organs, the research suggests they are protective and of positive benefit when consumed in normal dietary levels, rather than disrupting to endocrine balance as are so many synthetic chemicals and pharmaceutical hormones.[194]

Through voluminous research, it is also becoming increasingly clear that industrial pollutants and pesticides act as endocrine disruptors that may increase the risk of cancer. Many such compounds act as xenoestrogens that influence cell proliferation and gene transcription. Steroidal metabolism and signaling may be disrupted early in life through exposure to such hormonally active synthetic chemicals. Phytoestrogens may mitigate the severity of endocrine disruptors and offer positive hormonal influences on the breast and prostate. Embryologic development of the breasts, prostate, and reproductive organs involves both estrogens and androgens. Exposure to some exogenous estrogens (such as hormonally active synthetic chemicals) in the neonatal period may increase the risk of hormone-related cancers later in life. Furthermore, genes that control cellular and hormonal receptor quantities and response are also shown to be unfavorably influenced by exogenous chemicals[195] and favorably influenced by phytosterols.[196]

Natural plant-based compounds such as the phytoestrogens may reduce the chances of hormone-related reproductive cancers later in life. Phytosterols have been shown mostly to mitigate the action of synthetic endocrine disruptors, competing for the same binding sites and decreasing cell proliferation and offering protective effects on genes, enzyme systems, and hormone signaling.

Androgen receptors can also have proliferative effects on the prostate gland and are implicated in prostate cancer. Androgen-blocking drugs are a leading therapy for prostate cancer but can induce hot flashes, gynecomastia, metabolic syndrome and insulin resistance, cardiovascular disease, reduced lean body mass and muscle strength, osteoporosis, depression, and sexual dysfunction. Herbal medicine such as phytoestrogen-containing plants and nutritional supplements such as selenium, fatty acids, calcium, and vitamins D and E may help mitigate these side effects[197] and may serve as possible alternatives to pharmaceutical therapies. Pharmaceutical agents that interfere with LH release from the pituitary also reduce androgen levels, but when powerfully inhibiting, their use can result in "medical castration" due to severe androgen suppression. Plants that inhibit the conversion of testosterone into the more active dihydrotestosterone, such as those containing 5α-reductase inhibitors, can be employed as a less drastic method of reducing testosterone.

Phytosterols are found in the oils of nuts and seeds, mushrooms, cereal grains, legumes, and some medicinal plants such as *Hippophae rhamnoides* fruits (sea buckthorn), which are extremely high in phytosterols, *Serenoa* fruits (saw palmetto), and *Urtica* (nettle) roots. Nettle roots also contain phytosterols including stigmasterol and campesterol shown to have positive effects on hormone-binding globulin, aromatase enzymes, epidermal growth factor, and sex steroid receptors on prostate membrane, all serving to protect hormone-sensitive prostate cells against hyperplasia. Whole fruits and vegetables including carrots and broccoli have lesser amounts of phytosterols but can still contribute to the

anticancer effects of a whole foods diet. Even cocoa beans contain phytosterols and contribute to some of the purported anticancer effects of eating dark chocolate, including antimetastatic effects on prostate cancer.

Phytosterols are analogous to cholesterol in that phytosterols are the main lipid in plant cell walls and cholesterol is a component of animal cell membranes. Phytosterols have beneficial effects on cholesterol levels and metabolism and affect cholesterol's conversion to sex hormones. Phytosterols have also been shown to improve availability of lipid-based compounds, such as carotenoids, in the body, a mechanism that may also contribute to anticancer effects. β-sitosterol and ergosterol are two phytosterols that can inhibit growth in some hormone-sensitive cancer cell lines and promote apoptosis. Phytosterols also include isoflavones, coumestans, lignans, and stilbenes shown to act as partial estrogen agonists and provide other benefits.

Isoflavones and coumestans are particularly high in the legume family, which includes soy and many edible beans including lima, kidney, garbanzo, pinto, navy, and green beans, as well as lentils and peas. Isoflavones also contribute to the numerous medicinal virtues of

Sulfotransferase Enzymes for Hyperestrogenism

Sulfotransferases are detoxifying enzymes found in the tissues. The sulfotransferases add atoms of sulfur to estrogens, which renders them less active[198] because sulfated estrogens do not bind estrogen receptors.[199] Estrogen sulfotransferases are more active in healthy breast cells than in cancer cells, and the ingestion of phytoestrogens such as genistein,[200] quercitin, resveratrol, and other flavonoids have been shown to promote the sulfotransferases.[201] This is yet another reason to include high flavonoid berries and genistein-containing legumes in the diet to prevent hormonal cancers. Consuming these medicinal foods can also be part of an overall program to treat hormonal cancers. Curcuminoids from *Curcuma longa* are also shown to support sulfotransferases.[202]

Dehydrogenase Enzymes for Hormonal Cancers

The 17-β hydroxysteroid dehydrogenase family of enzymes are responsible for oxidizing steroids, making them less active, and controlling the amounts of active hormones circulating in the bloodstream. There are at least 15 subtypes of dehydrogenase enzymes, and breast cancer, prostate cancer, and endometriosis involve imbalances of these enzymes and the hormones that they regulate.[203] Many phytoestrogens including flavonoids, coumarins, and coumestans are noted to inhibit 17-β dehydrogenase.[204] One example of an herb that is shown to inhibit dehydrogenases is *Glycyrrhiza glabra*, which contains genistein and other phytoestrogens.[205]

Trifolium, *Astragalus*, *Pueraria*, *Medicago*, and *Glycyrrhiza*. Individual isoflavones include genistein and daidzein, the subjects of hundreds of studies showing hormonal and metabolic benefits. In any given legume food or herb, the isoflavones genistein and daidzein predominate, and lesser quantities of other isoflavones such as biochanin A and formononetin also are present. Some research claims formononetin is a more powerful estrogenic than daidzein and genistein. Biochanin A and formononetin are metabolized in the colon into genistein and daidzein, which are then metabolized into equol. Biochanin A has shown chemopreventive effects against various types of cancer (e.g., prostate, breast, colon). Most of the research in humans has suggested that the formation of equol in the intestinal lumen is desirable because as equol is absorbed, it acts as an estrogen antagonist, mitigating the increased risk of hormonal cancers by blocking endogenous and exogenous hyperestrogenism. However, in livestock who ingest relatively much greater quantities of legumes than humans, the amount of equol produced may be so great as to interfere with normal estrogenic actions.

Lignans are fiberlike compounds common in numerous whole grains and seeds, fruits, and vegetables. Lignans also exert a hormonal influence in the body after being altered by intestinal bacteria. Lignans are altered into enterodiol and enterolactone, which may be absorbed from the intestines and into general circulation, exerting weak estrogenic effects. All such phytosterols are believed to offer chemopreventive effects against hormonal cancers. Compounds found in cruciferous vegetables (discussed throughout this volume) are also found to deter many types of cancer, especially hormone-related cancers.

Research on Estrogenic Herbs and Cancer

Only a decade ago the use of phytoestrogens at all for cancer patients was controversial and of concern to practitioners due to valid concerns over possible proliferative effects on hormone-sensitive tissues. As research mounts, not only do phytosterols appear safe for consumption by patients with previous or existing hormonal cancers, but they also appear to have preventive and clinical benefits for breast and prostate cancer patients. The research is broad and it is still too early to specify dosage regimens of phytoestrogens for reproductive cancers, but the data on preventing hormonal proliferation with phytoestrogens is now strong. It appears that legume and phytosterol consumption

by pregnant women may be of benefit to the hormonal balance of their offspring, and that regular consumption of phytosterol-containing whole foods over a lifetime is to be encouraged. Although phytoestrogens can exert weak estrogenic effects, the research is mounting that such hormonal actions do not overstimulate hormone dependent cancers, but rather serve to block or mitigate the body's endogenous hormones via effects on hormone receptors, enzyme systems that process hormones, and hormone-regulating genes within the cells. At the present time, phytoestrogens appear to be valuable tools to include in the overall treatment protocols for breast, prostate, and other hormonal cancers.

Breast, endometrial, and prostate cancers are often hormone-dependent, where hormone receptors can be found on the cancer cells and are involved with cell proliferation. Histologic analysis of biopsied tumor tissue will screen for estrogen-, progesterone-, and androgen-dependent cell types, and classify cancers as estrogen receptor (ER) positive or negative. Estrogen binds to receptors and affects transcription factors, which leads to altered gene expression. In mutated cancer cells, this leads to excessive cell proliferation, oxidative stress, and further damage to the DNA. Natural products may protect against initial oncogenesis, reduce oxidative stress, and affect estrogen signals. Such products include sulforaphane from crucifers, isoflavones from legumes, epigallocatechins from green tea, resveratrol from berries, flavonoids from fresh fruits and vegetables, shogaol from ginger, and curcumin from turmeric. Many medicinal herbs may also exert hormonal activity and appear to exert anticancer effects via effects on estrogen receptors and/or estrogen signaling. Phytoestrogens bind α and β estrogen receptors but are weaker agonists than the endogenous estrogens, creating complex interactions.[206] This action is of great benefit to alleviate menopausal symptoms and mitigate the risk of developing hormonal cancers, but a long-standing question has been whether such herbs are safe in those with existing hormonal cancers. Although the question has not been answered definitively, herbs with estrogenic activity do not necessarily stimulate ER-sensitive cancers, and in fact, they may balance hormones in a manner having an anticancer effect.

For example, *Actaea*, one of the leading herbs used to manage menopausal complaints, has been investigated for having a possible stimulating effect on breast cancer cell lines in vitro. In these investigations, *Actaea* has *not* promoted estrogen-like effects in ER positive breast cancer cells[207] and was noted to inhibit the cell-proliferating effects of actual estrogens.[208] This suggests that the use of *Actaea* for a relief of menopause symptoms is not contraindicated in breast cancer patients. Another example is *Leonurus*, a folkloric remedy for fertility, uterine tone, lactation, and menopause. *Leonurus* is shown to have antineoplastic effects on mammary tumors.[209]

Actaea and *Leonurus* exert hormonal-balancing effects mostly via neuroendocrine mechanisms—affecting dopamine in the brain and regulating the pituitary's release of LH and FSH, affecting estrogen downstream.

The leguminous herbs high in isoflavones, however, have direct effects on estrogen receptors, but these medicinal plants are not shown to be a significant

Naturally Occuring Molecules with Chemopreventive Effects

Current chemopreventive strategies for ER-positive breast cancer have targeted the ER, estrogen, or estrogen metabolites. It is well established that many bioactive natural products have beneficial effects on estrogen-driven breast cancer. These natural products have shown promise in preventive and therapeutic settings and point toward tumor initiation and progression as points of intervention. However, use of natural products and derivatives has many hurdles to overcome as chemopreventive agents because of low bioavailability, poor absorption, and lack of specific cellular targets. Due to the role of breast cancer stem cells in tumor initiation, maintenance, and progression, recent studies have identified the following natural compounds that may inhibit estrogen-induced cancer stem cells.[210]

Allyl isothiocyanate reduces oxidative stress and promotes antioxidant and detoxification enzyme systems such as superoxide dismutase and glutathione peroxidase.

Apigenin is a natural flavonoid shown to induce apoptosis and inhibit cell proliferation in breast cancer.

Anthocyanidins bind to estrogen receptors and significantly reduce estrogen-induced signaling.

Isoflavones such as genistein inhibit estrogen-induced DNA synthesis and proliferation of estrogen receptor positive breast cancer cells.

Coumarins may bind estrogen receptors and protect cellular DNA from damage and prevent cancer initiation.

Caffeine may improve the metabolism of carcinogens.

Curcumin affects estrogen metabolism and induces apoptosis in various cell types.

Diallyl sulfides protect cellular DNA and promote cytochrome, superoxide-dismutase, and glutathione-detoxification pathways. They also improve estrogen and xenoestrogen metabolism.

Diosgenin has antioxidant and possibly hormone-modulating effects.

Organic acids such as ellagic and ferulic acids may mitigate estrogen metabolism and promote detoxification enzyme pathways.

Epigallcatechin has antiproliferative, antimutagenic, and chemopreventive effects.

Hesperetin and tangeretin are *Citrus* flavonoids that may regulate aromatase transcription, reducing serum levels of estrogen and oxidative stress.

Indole-3-carbinol enhances the metabolism of estrogen, xenoestrogens, and environmental toxins.

Carotenoids such as lycopene and β-carotene inhibit estrogen-signaling pathways by decreasing estrogen-induced activation of both α and β estrogen receptors.

Isothiocyanate offers chemopreventive effects against hormonal cancer, and sulforaphane is particularly protective against breast and other hormonal cancers, promoting detoxification pathways.

Quercetin has antioxidant, anti-inflammatory, antiallergy, and anticancer effects and is most active when consumed with other substances.

Resveratrol may induce apoptosis, protect against estrogen-driven inflammatory cascades, and modify genes that regulate estrogen receptors.

Thymoquinone is a potent antioxidant that promotes cytochrome and other detoxification enzyme systems.

concern to use in those with hormonal cancers either. Studies investigating blood concentration of legume isoflavones genistein and daidzein have been conducted on women with benign proliferative breast conditions, active breast cancer, and age-matched controls. The study demonstrated an inverse relationship between blood isoflavone levels and cancer risk, meaning the higher the blood isoflavone levels, the better; less breast cancer is seen in those with higher serum genistein and daidzein. One meta-analysis looked at soy consumption and reported that two or three servings per day, containing 25 to 50 milligrams of isoflavones, offered protection against breast cancer and its recurrence. Other studies corroborate that soy does not increase circulating estradiol or proliferate estrogen-sensitive tissues,[211] nor does the use of soy by women taking tamoxifen increase the incidence of breast cancer recurrence. Another human clinical study on women with a family history of breast cancer reported that *Trifolium* isoflavones exerted no proliferative effects on breast or endometrial tissue.[212] Human investigations of the legume *Pueraria* showed that supplementation for the purpose of reducing bone resorption and turnover showed no proliferative effects on the breasts or the endometrium.[213] This data suggests that legumes rich in isoflavones are not contraindicated in those with a history of breast or endometrial cancers.

Patients with hormone-sensitive reproductive cancers should consult with a knowledgeable physician. It is not a good idea to consume unusually large amounts of phytosterols, but the consumption of beans is not contraindicated nor is the use of herbal medicines to manage menopausal symptoms in women with a history of cancer or those experiencing surgical menopause. There is no evidence that the consumption of leguminous herbs or dietary soy and other beans causes harm, but the research is ongoing to confirm safety in all circumstances.

Radish Sprouts as a Source of DIM

Glucosinolates are believed to support detoxification pathways, as well as inhibit cancer cell proliferation, including breast cancer.[214] Radishes, especially radish sprouts and black radishes (*Raphanus sativus* var. *niger*), contain more glucosinolates—commonly referred to as diindolymethane (DIM)—than other cruciferous vegetables[215] and cost pennies compared to a pill supplement. I especially encourage radishes and DIM for men and women with hormonal reproductive cancers and for women with any manner of hyperestrogenism,

such as fibrocystic breast disease, uterine fibroids, endometriosis, and severe PMS. It is recommended to eat a cup a day, which is easily accomplished by mixing into salads, topping beans or cooked vegetables with the raw sprouts, or even liquefying in various smoothies and green drinks. Other cruciferous sprouts and microgreens are also beneficial, especially over a lifetime, but it's never too late to start. In some regions, you may need to grow them for yourself; in others, you may find them at the local grocery.

1 tablespoon radish seeds
2 cups (480 ml) pure water
Sprouting jar or tray

Place the seeds and water in the jar or tray and allow to soak overnight in a dark cupboard. Drain and rinse in the morning, and then rinse 2 or 3 times a day faithfully. Continue to store the jar or tray in the dark. When the first tiny leaves emerge, place the jar or tray in a sunny window and continue to rinse 2 times a day. Once the leaves are fully green, remove from the sprouter, and place in the refrigerator. Consume within 2 days, while also starting a new batch.

Phytoestrogen Tea for Hormonal Cancers and Chemoprevention

Because many types of breast[216] and prostate[217] cancer involve altered and excessive hormonal stimulation of these organs, phytoestrogens may be an important tool for improving hormonal balance in these cancers. Genes that control cellular and hormonal receptor quantities and response are also shown to be unfavorably influenced by exogenous chemicals[218] and favorably influenced by phytoestrogens.[219] These herbs all have additional medicinal virtues and are tasty enough to enjoy on a regular basis.

Glycyrrhiza glabra	3 ounces (90 g)
Trifolium pratense	1 ounce (30 g)
Medicago sativa	1 ounce (30 g)
Urtica urens	1 ounce (30 g)
Camellia sinensis	1 ounce (30 g)
Humulus lupulus	1 ounce (30 g)

Mix the dry herbs and store in an airtight container. Steep 1 tablespoon of the mixture per cup of hot water for 10 minutes, strain, and drink freely. This tea may simply be enjoyed as a preventive or used at a dose of 3 or more cups (720 ml) per day to attempt to reduce estrogen dominance.

Legume Truffle Superfood for Chemotherapy Patients

Patients undergoing chemotherapy are often nauseated and exhausted, and they have a hard time getting the supernutrition that would be best. These truffles can be a meal for those unable to eat regular food and feature *Astragalus* noted for its numerous anticancer effects, *Medicago* as a superfood, and licorice for further adrenal and hormonal support. Due to the numerous metabolism-supportive ingredients, these truffles would also be suitable for improving fertility, treating PCOS, and providing a more innocent sweet for those with diabetes and insulin resistance.

2 cups (470 ml) cashews (or cashew butter)
½ cup (120 ml) carob syrup
1 tablespoon inositol powder
2 teaspoons *Astragalus membranaceus* powder
2 teaspoons *Medicago sativa* powder
2 teaspoons *Glycyrrhiza glabra* powder
1 teaspoon brewer's yeast
10 drops vitamin D3 liquid

Phytosterols as Selective Estrogen Response Modifiers

There are multiple types of estrogen receptors, including α and β estrogen receptors and subtypes of each. The various types of estrogen counteract and balance one another. In general, the α subtype directs cellular proliferation and the β subtype directs differentiation and apoptosis.[220] These subtypes occur on reproductive organs, as well as on bone, vasculature, and many other organs and tissues. Estrogen receptors help to direct growth, tone, and, of course, reproductive function. The different types of estrogen (estriol, estradiol, and estrone) have differing ligand affinities for the receptor subtypes, and the distribution of the different subtypes of estrogen receptors on different tissues helps give the body the ability to control growth selectively. For example, the body can stimulate bone growth via estrogen receptors without overstimulating the endometrial lining. The different sensitivities of various receptor subtypes may help support vascular tone without the overproliferation of hormone-sensitive breast tissue. Pharmaceutical agents for building bone that lack such specificity of action are undesirable because they may promote breast cancer or coronary artery disease. Agents that bind to estrogen receptors on the bone to build bone density yet will not bind to estrogen receptors in the breast or endometrial lining would allow the body to gain the benefits without an increased risk of hormonal cancer. Such hormonal agents are commonly referred to as selective estrogen response modifiers (SERMs). These pharmaceuticals have been developed to be a safer alternative to blanket hormonal therapies. The pharmaceuticals tamoxifen and raloxifene are commonly prescribed for breast cancer patients and as chemopreventives for women at high risk for breast cancer. They are considered selective in that they may act as antagonists to estrogen receptors in the endometrium and breasts yet offer agonist activity for the bones.

Herbal phytosterols are natural SERMs and may be even safer than pharmaceutical SERMs. Intensive research is underway to develop therapies that target specific subtypes of estrogen receptors as a means of treating estrogen-dependent cancers and disease. Among the most studied natural SERMs are the isoflavone phytosterols found mostly in legume family plants—both edible beans and medicinal herbs. Although plant phytoestrogens exert effects on the reproductive organs, research suggests they are protective and of positive benefit when consumed in normal dietary levels, rather than disruptive to endocrine balance as are so many synthetic chemicals.[221] Phytoestrogens may offer a positive influence on hormonal balance when included in the diet on a regular basis, as early as in utero.

When phytoestrogens bind estrogen receptors, they act variably, as either agonists or antagonists. For more information on this action, see "Phytoestrogens and Amphoterism" on page 127.

Reproductive Cancers

Pour enough boiling water over the cashews to cover them, and let sit overnight. In the morning, drain the water and grind the nuts in a blender or nut grinder. Transfer the ground nuts or cashew butter (if using) to a small mixing bowl and blend with the carob syrup using a fork. Work in the inositol powder, herbal powders, brewer's yeast, and vitamin D liquid, adding a bit more carob syrup if too dry and a bit more herbal powder if too moist to achieve a cookie dough–like consistency. Shape the dough into small spheres, from ½ to 1 inch in diameter. Consume 2 or 3 per day as part of a broader protocol for patients going through chemotherapy. This recipe should yield around 20 truffles.

Dairy-Free Isoflavone Bean Chowder

I have a collection of bean recipes and I encourage people to eat at least one legume every day due to their ability to support hormonal balance, reduce cholesterol, improve insulin resistance and blood sugar regulation, and support a healthy intestinal ecosystem. In this recipe, pureed navy beans create a white and creamy texture, yet the delicious "chowder" is dairy free.

3 cups (600 g) dry white navy beans
1 to 2 cartons (950 to 1,900 ml) of soy milk
2 ears fresh sweet corn, kernels cut from cob
  (save the cobs)
1 large onion, diced
4 large cloves garlic, minced
5 slices turkey bacon, cut into small pieces (optional)
1 12-ounce can clams with juice (optional)
1 cup (75 g) chopped shiitake mushrooms
2 tablespoons miso paste
2 tablespoons olive oil
1 can (14 ounces or 450 g) pinto beans,
  rinsed and drained
1 can (14 ounces or 450 g) red kidney beans,
  rinsed and drained (or fresh peas when available)
1 large potato or yam, chopped into small pieces
2 carrots, chopped
2 stalks celery, chopped
1 teaspoon dry thyme
Salt
Pepper
Radish sprouts, for serving
Fresh parsley, chopped, for serving
  (can be added in last 10 minutes of cooking)

Soak the white navy beans overnight in water. Strain and rinse in the morning, place in a large saucepan, add fresh water to cover, and simmer until soft. Rinse again, drain, place in a blender or food processor with 3 or more cups (720 ml) of soy milk, and blend until it reaches a grainy puree. Meanwhile, place corn cobs in 4 cups (960 ml) of water and gently simmer for 30 minutes. Remove the cobs and scrape any remaining corn debris into the water, using the edge of a spoon. Place the onion, garlic, bacon, clams, and mushrooms in the saucepan with the miso paste and olive oil and heat until onions are translucent, stirring occasionally. Add the white bean puree and the corn water. Add the rinsed pinto beans, kidney beans, corn, potatoes, carrots, and celery. Add the thyme plus salt and pepper to taste. Add additional soy milk as needed to maintain the liquid. Gently simmer, covered, approximately 1 hour, stirring occasionally. When the carrots are tender, reduce the heat until ready to serve. Top each bowl with fresh radish sprouts and chopped parsley and serve with a fresh salad. This recipe may be cooked in the slow cooker after pureeing the navy beans and makes a slow-cookerful of chowder. It may be halved to make a smaller quantity. The turkey bacon and clams may be omitted for a vegetarian version. The recipe should yield at least 10 bowls of soup.

Formulas for Male Reproductive Disorders

Erectile dysfunction, benign prostatic hyperplasia (BPH), and low sperm count are among the most common male reproductive complaints. (Infectious disorders of the prostate are covered in Volume 1, chapter 4.)

The inability to achieve or maintain an erection can have both psychological and physical causes. Psychological causes such as depression, fear of intimacy, guilt, or other emotions, of course, need to be addressed directly with counseling and cognitive approaches. The diagnosis of a psychological cause of ED is implied if a man has morning erections or can achieve an erection with masturbation. The physical causes of erectile dysfunction include the multitude of disorders that affect circulation (atherosclerosis, diabetes, alcoholism, etc.) or hormonal status (hypothyroidism, hypopituitarism). Drugs such as antihypertensives and sedatives commonly interfere with erectile ability. Less common causes include neurological pathologies such as multiple sclerosis, spinal cord lesions, or lesions within the genitalia. Hence, a complete physical exam including blood work-ups with glucose, thyroid, and reproductive hormone assessments are a necessary part of the evaluation for men presenting with erectile dysfunction.

Both circulatory-enhancing and hormone-supporting botanicals may be helpful to men with erectile dysfunction. *Panax ginseng* has been shown to improve erectile ability in clinical trials.[222] If androgens are found to be low, *Panax ginseng* may be an alternative to supplementation with actual testosterone, which can exacerbate prostate enlargement. For patients with diabetes or atherosclerosis or for older men with poor circulation, *Ginkgo biloba* may improve circulatory status and thereby erectile ability. If patients are on beta blockers to manage hypertension, attempting to wean with the use of dietary, exercise, and herbal and nutritional therapies may restore erectile function. (See "Formulas for All-Purpose Vascular Support" in Volume 2, chapter 2.)

The pathophysiology of erectile dysfunction involves vascular inflammation over a lifetime and results in blood vessel and microcirculatory damage. Therefore, agents that manage hyperlidemia and hyperglycemia, hypertension, and excessive platelet activation will be appropriate throughout a lifetime, and less can be expected of herbs, or any medicine, *after* the vessels have already sclerosed. High flavonoid herbs and consistent flavonoid intake over a lifetime[223] are important preventive measures and helpful therapeutic agents. Fresh fruits, especially berries, are one of the best sources of flavonoids, such as anthocyanins. Smooth muscle fibers in the vasculature of the penile corpus collosum may be gradually replaced by collagen with aging, impairing erectile function. Flavonoids and nitric oxide have a protective effect on the muscle cells, and a long list of herbs have been found to act as nitric oxide promotors,[224] including herbs traditionally thought to be "blood movers"—*Angelica* species, *Ginkgo biloba*, *Salvia miltiorrhiza*, *Paullinia cupana*, *Ptychopetalum olacoides*, *Zingiber officinale*,[225] and many more. *Mucuna pruriens*, *Tribulus terrestris*, and ashwagandha (*Withania somnifera*) are among the herbs purported to support erectile function. Many of the molecular and mechanistic investigations have been in animals, and human clinical trials are lacking. However, the majority of these herbs do indeed have a folkloric reputation for supporting male libido, muscle mass and stamina, and erectile function.

Benign prostatic hyperplasia (BPH) is a common chronic disease in aging men, affecting more than 50 percent of those 60 years of age and older. BPH is due in part to proliferation of smooth muscle and epithelial cells in the prostate, leading to urinary frequency, straining, urgency, weak stream, incomplete emptying of the bladder, and nocturia. Because the urethra passes through the prostate gland, BPH can cause urinary symptoms and retention; herbal formulas for BPH are also covered in Volume 1, chapter 4. (Volume 1 also offers therapies for the treatment of prostatic cancer and bladder cancer.) Because hormone-driven proliferation is involved in prostatic enlargement, I also review herbal remedies that affect estrogen and testosterone clearance, stimulation, and interactions with the male genitourinary tissues in this volume.

BPH pathophysiology involves proliferative effects of hormones on prostate cells, especially increased dihydrotestosterone (DHT), increased 5α-reductase activity, and elevated prostatic specific antigen levels. Blockade of α1-adrenergic receptors is helpful due to resultant relaxing effects on prostate smooth muscle, improving urine flow. Blockade of the 5α-reductase enzyme is also helpful because this enzyme converts testosterone into

Coumarins for Erectile Dysfunction

Angelica species and other umbel family plants such as *Ligusticum striatum* and *Cnidium monnieri* have been particularly well studied for their effects on the vasculature. Formulas for vascular reactivity and inflammation, including impotence, could often include at least one Apiaceae (umbel) family ingredient. These plants contain coumarin compounds, the most basic molecular example being osthole. Extensive molecular research shows that osthole and other Apiaceae coumarins have vasodilating effects in the penis[226] via promotion of nitric oxide pathways and other mechanisms ranging from hypotensive effects to anti-inflammatory and hormonal actions.[227] Osthol may also possess a gonadotropin-like effect, promoting FSH and LH and downstream testosterone levels.[228]

DHT, which has a five-times more powerful proliferative effect on the hormone-sensitive prostate gland than testosterone. Numerous botanical agents have been noted to block 5α-reductase including *Serenoa repens*, berry liposterols, *Prunus africana* (also known as *Pygeum africanum*), and *Urtica* roots. The most commonly used pharmaceuticals for BPH are α-receptor blockers and 5α-reductase inhibitors such as finasteride. However, strong suppression of 5α-reductase has undesirable side

---

Botanical 5α-Reductase Inhibitors

The enzyme 5α-reductase converts testosterone into the more powerful, active, and potentially proliferative dihydrotestosterone (DHT). Because DHT is more stimulating to the prostate gland compared to testosterone, inhibition of this enzyme is a therapy in the management of prostate enlargement and cancer. Finasteride (Prosgar) is a synthetic pharmaceutical used for this purpose. Genes that control 5α-reductase in the prostate are induced by high-fat diets and inhibited by genistein.[229] *Serenoa repens*, *Prunus africana*, *Urtica* species, and *Camellia sinensis* catechins are botanical agents noted to inhibit 5α-reductase activity.[230] *Serenoa repens* is noted to inhibit tumorigenesis and induce apoptosis in animal models of prostate cancer.[231] Human investigations have confirmed the ability of *Serenoa* to inhibit 5α-reductase and reduce circulating levels of active testosterone.[232] The following plants are also shown to reduce the activity of this enzyme or downregulate its production in hormone-sensitive tissues such as the prostate.

Avicennia marina	*Magnolia grandiflora*
Benincasa hispida	*Mangifera indica*
Camellia sinensis	*Mentha spicata*
Carthamus tinctorius	*Panax ginseng*
Chrysanthemum morifolium	*Phyllanthus emblica, P. niruri*
Citrullus colocynthis	*Pueraria* spp.
Curcuma spp.	*Reynoutria multiflora*
Cuscuta reflexa	*Serenoa repens*
Ganoderma lucidum	*Sphaeranthus indicus*
Glycyrrhiza glabra	*Thuja occidentalis*
Lepidium meyenii	*Urtica dioica*

---

effects on libido, erectile, and ejaculatory function, so the pharmaceuticals have limitations. Herbal medicines that have the same mechanisms of action tend to have fewer side effects than finasteride, but need to be used long term and as part of a comprehensive protocol, to be a satisfactory therapy for BPH.

Maca Milkshake

Maca (*Lepidium meyenii*) is a folkloric herb used in the Andes for libido, stamina, and circulatory support, and this formula uses the good-tasting powder as a medicinal food. Hemp milk is used in this formula, but other milk substitutes such as almond or soy milk may also be used.

2 cups (480 ml) hemp milk, cold
¼ cup (30 g) fresh berries, chilled
1 to 2 tablespoons *Lepidium meyenii* powder
1 tablespoon cashew butter
1 teaspoon pumpkin seed oil

Place all ingredients in a blender and puree. Transfer to a tall glass and drink promptly.

Serenoa Simple for BPH

Serenoa is one of the most common herbs used for BPH due to its ability to reduce excessive stimulation of the prostate gland. *Serenoa repens*, *Prunus africana*, *Urtica* species, and *Camellia sinensis* can all inhibit 5α-reductase activity,[233] thereby reducing the conversion of testosterone into the more active dihydroestosterone.[234] *Serenoa repens* is noted to inhibit tumorigenesis and induce apoptosis in animal models of prostate cancer.[235] Many commercial products containing a combination of herbs such as these are available. *Prunus africana* is also known as *Pygeum africanum*, and most commercial herbal prostate formulas may use the latter term.

Serenoa repens (saw palmetto) capsules

Take 2 pills, 3 times a day for several years' time. Commercial products may contain other ingredients such as zinc, liposterols, or other supportive herbs.

Pumpkin Seed Oil for BPH

Pumpkin (*Cucurbita pepo*) seeds are a traditional remedy for prostatic disease, and modern research shows them to be exceptionally high in zinc, a crucial nutrient for the prostate gland to protect the prostate from testosterone-driven hyperplasia. Animal and human

Maca for Fertility and Libido

The mighty maca (*Lepidium meyenii*) is actually a small turnip that grows in the high Andes. Maca has a reputation for supporting libido and erectile function but may simply improve oxygen utilization and circulatory function in low oxygen situations, as well as enhance mood. The plant is a staple food to the high-Andean people who dry the tubers and use them in soups, grind them into flour, and prepare them in porridges. Over a lifetime, maca may support sexual functioning simply by enhancing mood and keeping people's energy high and cardiovascular health strong.

Maca contains alkaloids, steroids, tannins, saponins, anthocyanins, and cardiotonic glycosides, in addition to glucosinolates, isothicyanates, macamides, and alkamides. Maca may exert hormonal effects via activity at hormone receptors. GABAergic effects are reported in human subjects using red maca, and enhanced libido was reported in 50 percent of study participants.[236] The macamides, a group of fatty acids in maca, are reported to increase endocannabinoid signaling by inhibiting the breakdown of anandamide, also contributing to mood-enhancing effects.

Animal studies show maca to enhance male fertility by increasing spermatogenesis and to enhance female fertility as demonstrated by increased litter size and by increased uterine weight in ovariectomized animals.[237] Human studies show maca to significantly enhance sperm motility and morphology in young healthy subjects, after 12 weeks of consuming 3 grams per day,[238] an effect not seen with the consumption of 1.5 grams per day. Maca may correct impaired spermatogenesis in animals when due to lead toxicity.

As with other cruciferous vegetables, regular consumption of maca may reduce the risk of prostate cancer, due to the antiproliferative and proapoptotic actions of glucosinolates.[239] Maca contains the specific glucosinolate, glucotropaeolin, which upon ingestion is converted to isothiocyanates by gut

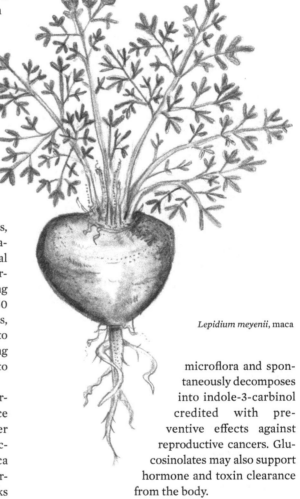

Lepidium meyenii, maca

microflora and spontaneously decomposes into indole-3-carbinol credited with preventive effects against reproductive cancers. Glucosinolates may also support hormone and toxin clearance from the body.

Maca may reduce androgen-driven proliferation of the prostate gland without negatively impacting serum hormone levels. While most studies have reported no effects on serum testosterone, some researchers have reported increased levels of progesterone in female mice and testosterone in male mice with large doses of maca, a quantity that is unrealistic to consume in the diet. When hormones are low, however, maca may help to increase estrogen levels.[240]

clinical studies have suggested pumpkin seed oil to improve nocturia and other urinary symptoms in cases of BPH and overactive bladder.[241] Look for a quality pumpkin seed oil, organically grown, and cold pressed in small batches.

1 to 2 teaspoons pumpkin seed oil

Take the oil right off the spoon or use on salad and to make salad dressings. Work the pumpkin seed oil into the daily diet or at least 3 or 4 times a week.

Tea for BPH

Serenoa does not make a tasty tea, so it is typically used in capsules as with the *Seronoa* Simple for BPH. These more palatable herbs offer additional 5α-reductase activity[242] and would be complementary to tinctures or capsules of *Serenoa*.

Urtica dioica root, finely shredded
Glycyrrhiza glabra root
Solidago chilensis leaves and flowers
Agrimonia eupatoria leaves
Camellia sinensis leaves

Combine in equal parts or to taste. Gently simmer 1 tablespoon per cup of hot water, strain, and drink freely, long term.

Tincture for Male Infertility

Low sperm counts can be due to hypothyroidism, genetic disorders, diabetes, and hormonal imbalances. Vitamin E is important to steroidogenesis and might be supplemented in cases of low sperm counts. *Panax ginseng* is a notable chi tonic in China. *Panax* is able to improve sperm counts[243] and improve low testosterone levels in men. *Achillea millefolium* is shown to protect testicular cells from damaging effects of nicotine and support spermatogenesis.[244]

Panax ginseng	30 ml
Achillea millefolium	10 ml
Serenoa repens	10 ml
Glycyrrhiza glabra	10 ml

Take 1 dropperful of combined tincture, 3 or 4 times per day for at least 3 months.

Tea for Prostate Cancer

Hormones are known to stimulate and proliferate cells of the prostate gland in men. As discussed throughout this chapter, herbal therapies that help the body process hormones can both help prevent and treat prostate enlargement. For example, prolonged exposure to genistein in early life is reported to reduce prostate size and may reduce the risk of BPH and prostate cancer later in life.[245] Legume isoflavones may inhibit prostate cancer,[246] enhance the efficacy of radiation therapy,[247] and slow cancer progression without noticeable side effects or toxicity.[248] Umbelliferone, a coumarin in the seeds of the umbel family such as anise and fennel, may inhibit dehydrogenase enzymes in human testes, an activity that may decrease prostatic hyperplasia and have therapeutic potential for prostate and testicular cancer. Teas may be particularly beneficial due to an ability to pass directly through the genitourinary system.

Astragalus membranaceus	8 ounces (240 g)
Glycyrrhiza glabra	4 ounces (120 g)
Foeniculum vulgare seeds	4 ounces (120 g)
Ganoderma lucidum slices	4 ounces (120 g)

Combine the dried herbs and store in an airtight container. Gently simmer 1 teaspoon per cup of water for 20 minutes and then strain. Drink 3 or more cups (720 ml) per day.

Soy Milk for Prostate Cancer

Soy milk can be consumed as is or used as a base to make smoothies or golden milk or to use on oatmeal. One study treated men with recurrent prostate cancer with soy milk quantified to deliver a standardized amount of genistein, dosed 3 times a day for 1 year. For nearly all men, serum equol was increased, and for many men the upward trend of PSA levels was either stabilized or in some cases reversed.[249] Be sure to choose a non-GMO, unsweetened brand of soy milk, or prepare for one's self by gently simmering soybeans until soft and pureeing in a blender to use in beverages or cooking. See also the Dairy-Free Isoflavone Chowder on page 160 as another example of including soy milk in the daily diet.

Soy milk

Consume a cup 3 times a day, or use to prepare smoothies, pour over oatmeal, or create medicinal soups.

Specific Indications:
Herbs for Reproductive Endocrine Conditions

Many types of herbs are appropriate for use in formulas for reproductive conditions including blood movers, adaptogens, alteratives, and those herbs that have direct effects on the neuroendocrine and reproductive pathways. Modern research has elucidated numerous mechanisms involving direct effects on estrogen receptors, as well as effects on enzyme systems involved in metabolizing hormones. Folklore has much to offer regarding specific indications for fertility, lactation, bleeding disorders, genitourinary complaints, and other common reproductive issues. The following are among the most emphasized plants of traditional herbalism for treating male and female reproductive complaints.

Achillea millefolium • Yarrow

The young flowers are dried and tinctured and used as a liver tonic, alterative, vasodilator, and antimicrobial and for many other functions. In TCM, *Achillea* is thought to tonify deficiency and clear heart phlegm. All traditions recognize the significant diaphoretic effects of yarrow flowers taken in a hot tea. *Achillea*'s alterative effects make it useful to improve the liver's processing of hormones in cases of estrogen dominance; yarrow teas were used folklorically as an emmenogogue and yet also as a hemostatic in cases of heavy menses. Apigenin in *Achillea* has been shown to stimulate estrogen receptors, both the α and β subtypes, and luteolin is a very weak β estrogen receptor agonist.[250]

Actaea racemosa • Black Cohosh

The roots of *Actaea* are specifically indicated for anxiety and muscular tension, for tension headaches that move from the occiput forward, and for premenstrual breast pain and backache, uterine pain, and neuralgic ovarian pains. *Actaea* is specifically indicated for menstrual cramps that are tight and of muscular squeezing quality, worse jarring or walking. Historically, *Actaea* has also been used for difficult labor due to weak contractions and for postpartum hemorrhage. Other reproductive system indications for *Actaea* include late puberty and dysmenorrhea with scanty flow, acne at puberty, testicular pain and genital pain associated with prostatic enlargement, excessive nocturnal emissions, and menopausal muscle tightness and muscle tension headache.

Actaea is specific for women with fibromyalgia, anxiety and depression, and hormonal imbalance. Tinctures and encapsulations are useful for all manner of heavy, tight, aching muscular pains. The Eclectic physicians often combined *Actaea* with licorice and salicylates. *Actaea* is particularly helpful in cases in which hormonal issues are accompanied by muscle stiffness, anxiety, and depression. Research has shown *Actaea* to bind mu opiate receptors helping to provide anodyne and anxiolytic effects. *Actaea* is often very helpful in fibromyalgia formulas or in formulas for headaches, muscle stiffness, and rheumatic-like muscle pain associated with menopause. Some recent research has suggested that *Actaea* may cause liver irritation and inflammation and should be used with discretion. When long-term use proves helpful, check liver enzymes at least annually, and take weekends or weeks off now and then. *Actaea racemosa* is also known as *Cimicifuga racemosa*.

Alpinia galanga • Galangal

The roots of galangal are used medicinally, and the plant is a relative of turmeric and ginger. *Alpinia* is used traditionally for antiulcer, anticancer, and hypolipidemic effects, and the aromatic constituents have carminative and antimicrobial effects. *Alpinia* may improve sperm counts and motility as well as general reproductive function.[251] *Alpinia officinarum* is used in a similar manner. Galangal has also demonstrated effects against breast cancer cell lines,[252] and β sitosterol has been identified in the roots.[253]

Angelica sinensis • Dong Quai

Angelica has become established in Western herbalism after being introduced from TCM where the roots have has especially been used as a blood mover for vascular congestion, heart disease, inflammation in the vasculature of the brain, organs, and tissues, in general, and as a women's medicine to treat pelvic stagnation, menstrual cramps, and menopausal issues such as osteoarthritis and dementia. *Angelica* has attained a reputation as a female tonic or even "the female ginseng," even though its main actions appear to be circulatory rather than hormonal. Animal models of osteoporosis show *Angelica* to benefit bone density without affecting estrogen levels,[254] probably via enhancing microcirculation or reducing inflammation in trabecular bone. *Angelica* polysaccharides may protect a variety of tissues such as the heart, neurons,[255] and reproductive organs from oxidative stress via blood- and endothelial-protective effects.[256] *Angelica* and *Astragalus* are a common duo for neuro- and circulatory-enhancing effects. *Angelica* and *Epimedium* are a common duo for bone health in many traditional formulas. *Angelica* and *Paeonia* are another duo to support fertility in many TCM formulations.

Arctium lappa • Burdock

Arctium roots are a time-honored alterative agent, used for liver and digestive processing and to improve hepatic conjugation and clearance of hormones, which is useful for treating adolescent acne and skin disorders. *Arctium* roots contain inulin, which may support a healthy microbiome and thereby impact hormone metabolism and excretion. *Arctium* is specifically indicated for hyperestrogenism, intestinal dysbiosis, and hyperlipidemia. *Arctium* may optimize hepatic metabolic and detoxification pathways, reducing oxidative stress in the body. Decreasing oxidation stress can improve spermatogenesis and reproductive hormones in diabetic patients or others with toxicity, elevated lipid peroxidation, or other stressors.[257] Arctigenin is a lignan in burdock seeds that has anti-inflammatory effects and is shown to enhance chemosensitivity of several types of cancer cells to cancer drugs and also to exert activity against breast cancer[258] and prostate cancer.[259]

Asparagus racemosus • Shatavari

The roots of asparagus have long been used in Ayurvedic medicine as a genitourinary tonic, recommended for urinary atony and irritations, as well as for postmenopausal vaginal dryness. The common name shatavari translates as "possesses a hundred husbands" due to the folkloric reputation for increasing libido and improving vaginal dryness. Shatavari is considered a female fertility and vitality tonic—it is sometimes referred to as a "queen" herb and touted to promote love and devotion. It is used in many restorative formulas, especially for the genitourinary and other mucous membranes. The plant is also traditionally used as an antiaging and longevity tonic and has a history of being used to prevent miscarriage and promote lactation. Many of the anti-inflammatory and tissue-restorative effects are credited to the shatavarins, a group of steroidal saponins. Shatavarins are shown to have immunomodulating activity.[260]

Astragalus membranaceus • Milk Vetch

Astragalus roots have been used in Traditional Chinese Medicine for thousands of years to support vital energy and body immunity and *Astragalus* is very often a supportive and nonspecific herb in formulas for circulatory weakness, inflammatory disease, cancer, and diabetes. As such, *Astragalus*, also called huang qi, is a primary chi tonic used to support longevity. *Astragalus* polysaccharides, flavonoids, and triterpenoid saponins (known as astragalosides) are credited with antiaging effects by numerous mechanisms including telomerase-protective, antioxidant, anti-inflammatory, immunoregulatory, anticancer, hypolipidemic, antihyperglycemic, and hepatoprotective activities.[261] *Astragalus* teas, tinctures, and medicinal foods are specific for frequent infections that linger, or that are associated with fatigue and exhaustion. *Astragalus* is commonly combined with *Angelica* species in China to improve circulation in the pelvis, treat menopausal symptoms, and improve fertility. The fertility-enhancing effects of *Astragalus* are not due to direct hormonal effects, but rather to an ability to protect sensitive gonadal cells and reproductive tissues from oxidative and circulatory stress.[262] Research has shown *Astragalus* to have very broad anti-inflammatory, circulatory-enhancing, and immune modulating effects. *Astragalus membranaceus* may enhance sperm motility and semen quality via reducing toxicity and reducing oxidation in the testes.[263] Isoflavones and triterpene saponins in *Astragalus* are credited with hormone-modulating effects.

Avena sativa • Oats

Oat groats are nourishing and can be used as a gentle gut and nervous system restorative food for those with nervous debility. Mature oat seeds are high in amino acids and "milky oat" medications are prepared from the ripe seed stalks before the juicy quality has been transformed into starch. Milky oats may be used as a complementary ingredient in formulas for hormonal issues accompanied by exhaustion, irritability, and symptoms related to long-term stress. Folkloric uses for milky oats include amenorrhea and dysmenorrhea associated with weakness, debility, or poor circulation.

Berberis aquifolium • Oregon Grape

The outer root bark of *Berberis* is used in herbal medicine, having a long history as an alterative agent. Modern research has focused on berberine in the root bark for its diverse effects—on metabolism, enhancing insulin, and its anti-inflammatory action. Traditionally, *Berberis* has been specifically recommended for vaginal pain and dyspareunia, especially when associated with mucosal infections or with pelvic stagnation. Intestinal dysbiosis, vaginitis, cystitis, and fungal infections are other specific indications. As an alterative agent, *Berberis* may improve the liver's processing of hormones and help treat symptoms and diseases associated with estrogen dominance. *Berberis* powder can be included in vaginal douche formulas, especially for thick mucus-like vaginal secretions. Oregon grape is also known as *Mahonia aquifolium*.

Ceanothus americanus • Red Root

Also known as New Jersey tea, the roots of *Ceanothus* are specifically indicated for breast cysts and mastodynia, heavy menses, and anemia. Additional traditional indications include chronic vaginal infections with discharges occurring in the obese, and for those with poor circulation, and the doughy skin and chronic infections often seen in long-term diabetics. Many of these symptoms are thought to be related to lymphatic stasis, or a "damp" constitution in folkloric herbalism.

Chamaelirium luteum • False Unicorn

Chamaelirium is not widely distributed—loss of its habitat is threatening the plant, and herbalists no longer can recommend it. The plant is listed here for historical interest and with the hope that the plant can be cultivated and made available again in the future. *Chamaelirium* roots were specifically indicated for a sense of weight, for congestion in the pelvis, and for laborlike expulsive menstrual cramps and recurrent miscarriage. *Chamaelirium* is considered a uterine and reproductive tonic and has been included in traditional partus preparators such as the Tincture for Mother's Cordial on page 132. *Chamaelirium* is also indicated for mental apathy and lack of libido as well as for mental and physical weakness.

Cimicifuga racemosa • See Actaea

Cinnamomum species • Cinnamon

The bark of *Cinnamomum verum*, *C. cassia*, and other species is used medicinally and is a common ingredient in TCM for warming, blood-moving, and anti-inflammatory actions. Cinnamon can have moderate to substantial hemostatic effects and is specific for postpartum hemorrhage, menorrhagia due to fibroids and polyps, and heavy menses made worse with motion and jarring. Cinnamon combines well with *Achillea* for menorrhagia or to complement ergotamine for postpartum hemorrhage. Cinnamon may also reduce bleeding in the digestive, urinary, and respiratory tracts. Cinnamon essential oil can be a useful ingredient in douche formulas when highly diluted and used to treat vaginitis. Cinnamon is also included in numerous traditional formulas to improve circulatory and metabolic distress due to its hypoglycemic and hypolipemic affects. Cinnamon may irritate the intestines in those with IBS, so should be initiated in low doses. Use cinnamon in herbal formulas to treat dysmenorrhea and to reduce pain, heavy bleeding, and nausea and vomiting. Cinnamaldehyde, one of the aromatic compounds in the bark, has demonstrated antispasmodic action, and another aromatic compound, eugenol, has anti-inflammatory activity due to effects on prostaglandins.

Cnicus benedictus • Blessed Thistle

The leaves of blessed thistle may improve sluggish liver function and be used as a supportive herb in formulas to improve hormone metabolism or to treat the symptoms of hyperestrogenism. Improving the liver's processing of hormones, carbohydrates, and lipids can reduce oxidative stress in the tissues. *Cnicus* has been used as a galactagogue and may be included in nursing teas and formulas.

Coleus forskohlii • Coleus

Coleus has been used for allergies, pain, inflammation, obesity, and itching in traditional medicine. The roots of the colorful *Coleus* plant are specifically indicated in formulas where atony, sluggishness of general metabolic functions, or excessive inflammatory response accompany hormonal dysfunction. Numerous anti-inflammatory mechanisms have been demonstrated, including modulating antioxidant enzymes in the liver, decreasing tumor necrosis factor α, and reducing elevated cyclooxygenase in tissues.[264] *Coleus* roots contain the diterpene forskolin shown to raise intracellular cAMP levels via direct effects on adenylyl cyclase.[265] Because cAMP acts as a "second messenger" in cells, helping to transmit the signals from cellular receptors inward, *Coleus* may boost the metabolic activity of many types of cells, explaining the plant's traditional use for asthma, diabetes, obesity, and cardiovascular disorders. Forskolin may also inhibit the glucose transport into some cell types, such as the retina, and help protect against diabetic retinopathy. The plant is also known as *Plectranthus forskohlii*. Though it is not botanically synonymous, *Plectranthus barbatus* is often listed as an alternate name for *Coleus forskohlii*.

Commiphora mukul • Guggul

The resin from the tree bark is used to manage elevated lipids due to diabetes, hypothyroidism, or slow metabolism and may be used as a supportive herb in formulas to reduce oxidative stress in tissues. Specific indications for guggul include a sense of weight or dragging in the pelvis, amenorrhea associated with uterine congestion and constipation, amenorrhea associated with anemia, and leukorrhea. Guggul can be used as a mucous membrane stimulant and antiseptic, specific for excessive discharges.

Coptis trifolia • Goldthread

Coptis is used in TCM for chronic infections of the mucous membranes. *Coptis* may be used in formulas to treat fungal infections of the genitals or prepared as compresses, sitz baths, or a vaginal douche. *Coptis* can also be included in formulas for internal use to treat intestinal dysbiosis and sluggish digestion as well as chronic infections in diabetics. *Coptis* roots contain coptine and berberine, which have been widely studied for antimicrobial and metabolic-supportive effects.

Corydalis yanhusuo • Turkey Corn

Also known as Yan Hu Suo in Asia, the roots of *Corydalis* are used to treat pain and tension and are traditionally included in formulas for chest pain, epigastric pain, and pelvic pain including dysmenorrhea. *Corydalis* is considered one of the 50 fundamental medicines in TCM, where it is used traditionally for its sedative, neuroleptic, and analgesic properties. *Corydalis* is also thought to promote blood circulation, alleviate amenorrhea, and resolve pelvic masses considered to be due to vascular stasis. Tetrahydroprotoberberines are a group of alkaloids with extensive pharmacological activities including antitumor, antinociceptive, antihypertensive, and anti-ischemic properties. These protoberberines are shown to suppress dopamine D2 receptors in the central nervous system, contributing to the analgesic effect. One such agent, tetrahydropalmatine, inhibits cytochrome P450.[266]

Crataegus spp. • Hawthorn

Hawthorn berries, leaves, and flowers are traditionally used to improve heart function, support circulation, and protect the endothelium from prooxidants and elevated blood pressure. Hawthorn may be used as a supportive herb in menopausal formulas for women with heart palpitations and hypertension at climacteric. Hawthorn is also used to protect the blood vessels in cases of hypertension, hyperglycemia, and hyperlipidemia. (See Volume 2, chapter 2, for details on *Crataegus*'s many mechanisms of action on the vasculature.)

Cucurbita pepo • Pumpkin

The seeds of ripe pumpkins are recommended for their quality essential fatty acids and may be used to treat inflammatory processes. The seed oil is especially recommended for the prostate where the liposterols may improve hormonal balance and help treat benign prostatic hyperplasia. The seeds are also an excellent source of zinc, an important mineral for prostate health.

Curcuma longa • Turmeric

Curcuma tubers are a well-known and well-researched spice and medicine with broad antimicrobial, antioxidant, and anti-inflammatory actions. *Curcuma* can be valuable in formulas for reproductive issue where it offers hepatic support and assists in hormone clearance via the liver. *Curcuma* can also improve lipid and carbohydrate metabolism in cases of diabetes and hyperlipidemia and can be included in formulas for PCOS, endometriosis, and other diseases where toxicity and oxidative stress contribute to the pathology. *Curcuma* may also act as a chemopreventive against reproductive cancers. *Curcuma* improves intestinal dysbiosis and propensity to fungal and other infections from simple yeast infections to HIV infections. Curcumin and related flavonoids, the curcuminoids, have been the subject of extensive research that shows the compounds to reduce cholesterol level, prevent low density lipoprotein (LDL) oxidation, inhibit platelet aggregation, suppress thrombosis, inhibit HIV replication, and help treat diabetes, rheumatoid arthritis, multiple sclerosis, and Alzheimer's disease. The numerous antioxidant and anti-inflammatory actions enhance wound healing, increase bile secretion, and protect the liver from fibrosis, fat definition, and toxic injury. Even high doses have not been found harmful. *Curcuma* may, therefore, be used in a wide variety of herbal formulas for reproductive diseases, as a chemopreventive agent, or in formulas for active cancers such as prostate cancer. The hepatoprotective and metabolic-enhancing properties of turmeric are also discussed in Volume 1.

Cynara scolymus • Artichoke

Artichoke leaves are used traditionally as a liver tonic and may improve cholesterol and hormonal imbalances by supporting the hepatic processing of these substances. Artichoke preparations can improve glucose and lipid processing, making both artichoke food and leaf medicines useful for hyperglycemia and hyperlipidemia. Because such metabolic distress also occurs with PCOS and may contribute to cancer initiation, *Cynara* may also be included in teas, tinctures, and encapsulations for hormonal disorders associated with diabetes or with a concern of an increased risk for cancer.

Dioscorea villosa • Wild Yam

The tuberous roots of the wild yam have been traditionally used as an antispasmodic and are especially indicated for colicky pains in abdominal organs including menstrual cramps, poor digestion, and flatulence. One old herbal in my collection reports *Dioscorea* to be helpful for women who have "trouble finding the right names for things," which I often see in menopausal women—I suspect hormonal fluctuations affect neurontransmitter regulation in the central nervous system. In TCM, related species of *Dioscorea* have been considered general chi and yin tonics and used in various formulas when lack of appetite, fatigue, loose stools, shortness of breath, thirst, or sweating accompany various complaints.

Eleutherococcus senticosus • Siberian Ginseng

Eleutherococcus is in the same family as *Panax ginseng* and is similarly used as an adrenal tonic for long-term stress resulting in nervous symptoms and fatigue. *Eleutherococcus* roots are considered an adaptogen. Include *Eleutherococcus* in formulas for endocrine imbalances and immune insufficiency, especially when related to stress and overwork and associated with fatigue, weakness, and emotional instability. Modern research has identified many antibacterial, antifatigue, antioxidant, hypoglycemic, and immunomodulating activities, many of which are credited to lignans (sesamin, eleutheroside E), glycans (eleutherins, eleutheroside D), triterpene saponins (eleutheroside I, K, L, and M), and steroid glycosides (eleutheroside A).[267] The plant is also known as *Acanthopanax senticosus* and has gone by the common name eleuthero.

Epimedium grandiflorum • Horny Goatweed

There are hundreds of species of *Epimedium*, and many go by the common name horny goatweed in the West. Both the roots and leaves have been used medicinally. *Epimedium brevicornu* is also commonly used; it is also known as yin yang huo in TCM. *Epimedium* species have been used traditionally for infertility, impotence, and low libido. *Epimedium* species also have a long history of use in the treatment of estrogen deficiency–related diseases, such as osteoporosis. *Epimedium* has long been used in supporting reproductive function in both sexes, improving fertility, and maintaining sexual function. It has been shown to support spermatogenesis when suppressed by hormonal imbalance and via reducing oxidative stress in the testes.[268] *Epimedium* contains isoprenylated flavonoid glycosides, several of which are potent promoters of estrogen biosynthesis in human ovarian cells.[269] Among the most studied of these flavonoids are icariin and epimedin, shown to inhibit bone resorption and improve erectile function in animal models of aging. Icaritin, the aglycone of icariin, has estrogenic properties and stimulates estrogen-driven cells.[270] Icariin and related flavonoids may also have significant anti-inflammatory and immunoprotective effects, and even though some compounds appear to be estrogenic, the flavonoids are shown to induce apoptosis in cancer cells, including drug-resistant cancers.[271] One study reported *Epimedium* to increase the sensitivity of breast cancer cells to radiation therapy,[272] and another study showed purified icaritin to have antiproliferative effects against endometrial cancer due to estrogen receptor modulation.[273]

Eschscholzia californica • California Poppy

The entire plant is traditionally used for symptoms of anxiety, muscular tension, and poor sleep. *Eschscholzia* may be used as a complementary nervine herb in formulas for adrenal exhaustion due to long-term stress. Some of *Eschscholzia*'s anxiolytic effects are due to promoting GABA neurotransmission.

Eurycoma longifolia • Tongkat Ali

Tongkat ali is regarded as an aphrodisiac and energy tonic in Southeast Asia, and research suggests that it may promote testosterone levels and sperm parameters via androgenic effects.[274] Animal studies indicate that *Eurycoma* root extracts promote steroid synthesis, possibly via enhancing ATP production and enhancing cell membrane signaling cascades. Human studies have shown the herb to improve male fertility, improving semen volume, sperm motility and morphology and increasing pregnancy rates in female partners.[275] Androgen promotion may also reduce bone resorption[276] and help to treat osteoporosis.

Foeniculum vulgare • Fennel

Fennel seeds have numerous medicinal uses including carminative, gastroprotective, and bronchodilating actions. They are also traditionally used to regulate the menstrual cycle, treat menopausal symptoms, and support fertility. Fennel tea and the essential oil (applied topically) have antispasmodic effects on digestive and uterine muscle, and *Foeniculum* tincture can be included in formulas to treat dysmenorrhea, decreasing oxytocin- and prostaglandin-driven uterine spasms.[277] Fennel is also a galactagogue and makes a good tea for nursing women, not only to support lactation, but also because the carminative actions of fennel can be passed in the breast milk and benefit colicky babies. Animal studies have demonstrated *Foeniculum* extracts to increase the weight of mammary glands and reproductive tissues and to promote the maturation of ovarian follicles, suggesting hormonal effects.[278] *Foeniculum* contains the aromatic compound anethole, which is credited with some of the medicinal effects, along with diosgenin, a steroidal saponin also found in the galactagogue *Trigonella* (fenugreek) and in *Dioscorea* (wild yam). Anethole is the primary component of the essential oil fraction and may have estrogenic activity. Fennel can be included in teas and medications for osteoporosis because the seeds are an excellent source of minerals and also because *Foeniculum* is shown to have a proliferative effect on bone cells.[279]

Fucus vesiculosus • Bladderwrack

Seaweeds, including *Fucus* species, are natural sources of iodine and are credited with an ability to support thyroid function. *Fucus* is one of the seaweed species referred to as kelp, and it is traditionally used to treat obesity, goiter, exophthalmia, and constipation—all symptoms of thyroid dysfunction. *Fucus* is also traditionally recommended to treat impotence in hypothyroid men. Estrogen-dependent cancers are highest in Western countries and lower in Asian countries, and Japanese women are also noted to have longer menstrual cycles. Both phenomena are thought to be due to healthier hormone balance in the East compared to the West. Regular consumption of seaweed is credited with some of these hormonal benefits. *Fucus* and other brown seaweeds contain sulfated polysaccharides known as fucoidans, shown to have antitumor and antimetastatic activities and to possibly help detoxify xenoestrogens such as dioxins. The consumption of *Fucus* may exert antiestrogenic and progestagenic action in human subjects and optimize the length of the menstrual cycle;[280] eating *Fucus* may also offer chemopreventive effects against reproductive cancers.[281]

Ginkgo biloba • Ginkgo

Ginkgo biloba is an ancient tree species whose leaves are used to enhance peripheral vascular cerebral blood flow. Their use may improve several medical ailments associated with impaired circulation. *Ginkgo* has been widely studied for its circulatory-enhancing mechanisms and has become established in the treatment of coronary artery disease and cerebrovascular insufficiency and as a means of supporting the vasculature in cases of diabetic retinopathy, nephropathy, and neuropathy. *Ginkgo* may be a supportive herb in formulas for vascular stasis and blood congestion in women with pelvic pain, and in formulas for elderly men with impotence when other signs or symptoms of circulatory insufficiency accompany. For an extensive review of *Ginkgo*, refer to Volume 2, chapter 2.

Glycyrrhiza glabra • Licorice

Licorice is included in every chapter of every volume of these formularies due to the diverse effects of the roots, which in TCM are said to enter every meridian and have medicinal virtues for every organ system. Licorice has many applications for the endocrine system. Licorice is widely credited with adaptogenic effects and can be used in a wide array of teas, tinctures, and medicinal formulas where stress contributes to hormonal imbalance. Glycyrrhetinic acid, one of licorice's steroidal compounds, inhibits 11β-hydroxysteroid dehydrogenase enzymes, reducing cortisol and cortisone levels and inhibiting the conjugation of the metabolic precursors, deoxycorticosterone and dehydroepiandrosterone (DHEA), in the adrenal gland.[282] Supporting adrenal function may help to improve inflammatory issues, as well as improve DHEA-to-cortisol ratios. Supporting DHEA levels can support the production of progesterone and thereby help to oppose hyperestrogenism. This makes *Glycyrrhiza* appropriate in formulas for menstrual irregularities, PCOS, endometriosis, amenorrhea, infertility, breast cysts, and fibroids. Licorice is a traditional remedy in Japan for amenorrhea, combining well with *Vitex* for this purpose. In addition to the inositol compounds, the saponin glycyrrhizic acid may also improve insulin resistance and fat metabolism, as demonstrated in animal models of metabolic syndrome.[283] *Glycyrrhiza* has been found to reduce inflammation and support function in animal models of diabetic kidney damage,[284] improve insulin sensitivity and reduce fat deposition in tissues,[285] and suppress the accumulation of abdominal fat.[286] *Glycyrrhiza* also has many anti-inflammatory and antispasmodic actions on the uterus,[287] possibly via inhibition of oxytocin, prostaglandin, and KCL-induced uterine contractions,[288] as well as those induced by calcium channel activation, nitric oxide synthase, or cyclooxygenase.[289] See also the sidebar on *Glycyrrhiza* on page 146 in this chapter.

Hamamelis virginiana • Witch Hazel

The bark of witch hazel is used for its astringency, topically on wounds or over hemorrhoids, and to treat vascular trauma. Some herbalists use the plants' leaves in addition to the bark. *Hamamelis* preparations, including sitz baths, may be helpful in treating vaginal infections and pelvic congestion when associated with poor circulation and profuse discharges. Formulas for men with varicocele might combine *Hamamelis* with *Ginkgo* and *Aesculus*. Formulas for women with perineal and heavy, aching genital pain associated with premenstrual vascular congestion might combine *Hamamelis* with *Caulophyllum* and/or *Angelica*. *Hamamelis* may also be included in formulas for heavy menses due to excessive endometrial buildup or abnormal shedding, as is the case with fibroids and polyps; it can be combined with *Achillea* and *Cinnamomum* for more powerful hemostatic effects.

Humulus lupulus • Hops

Hops strobiles (a flowerlike plant part resembling a pinecone) are traditionally used as a digestive bitter and nervine to relieve stress, anxiety, and insomnia. The strobiles contain the powdery resin lupulin, credited with a calming and sleep-enhancing effect. Due to the anxiolytic effects, hops may be a complementary ingredient in adrenal and stress formulas for hormonal imbalances or menopausal symptoms accompanied by irritability or poor sleep. Hops have also long been noticed to have hormonal effects,[290] and women who hand-harvest hops are reported to experience menstrual cycle changes due to presumed absorption of estrogenic compounds through the skin. Hops contain one of the more potent phytoestrogens, 8-prenylnaringenin, which is shown to bind to α estrogen receptors. Clinical trials have demonstrated efficacy for menopausal hot flashes.[291] Hops were also an Eclectic remedy for genitourinary symptoms in men associated with nocturnal seminal emissions. The phytosterols in hops may offer chemoprotective effects against reproductive cancers.[292]

Hydrastis canadensis • Goldenseal

Goldenseal roots contain strong antimicrobial agents such as hydrastine, and the plant is most known for treating bacterial infections of the mucous membranes.

However, goldenseal has also been recommended as an astringent tonic in formulas for uterine hemorrhage, metrorrhagia, or menopausal bleeding disorders due to underlying uterine fibroids or atony. *Hydrastis* is specific for congested mucous membranes that oversecrete mucus and thereby become a breeding ground for opportunistic infections. *Hydrastis* is useful internally as a douche for vaginitis with profuse thick discharge. *Hydrastis* has been used as an ingredient in "vag pack" medications (meaning an internal vaginal medication) in the treatment of uterine fibroids, polyps, cervical and vaginal ulcerations, lesions such as warts and vegetative growths, and chronic vaginal infections. (For instructions on making a vag pack, see The Vag Pack: A Traditional Eclectic Formula on page 114.) *Hydrastis* is also specific topically and locally for chronic fungal infections seen in diabetics. Orally, *Hydrastis* is used to treat intestinal dysbiosis seen in diabetes and/or chronic urinary, skin, and vaginal infections.

Hypericum perforatum • St. Johnswort

The young buds of *Hypericum* are commonly prepared into medicines to treat anxiety and depression and may be used as a synergist in formulas for menopause and adrenal dysfunction associated with anxiety. *Hypericum* is also antiviral and may help treat the neuralgic pains of genital herpes lesions, due to both antiherpetic and neural anti-inflammatory and anodyne effects. *Hypericum* is specifically indicated for shooting lancinating pains due to herpes infections, nerve trauma, and neuronal inflammation. *Hypericum* may also be a supportive ingredient in herbal formulas for the polyneuropathies of diabetics or for severe premenstrual breast pain. *Hypericum* may also reduce the symptoms of irritable bladder and improve postmenopausal bladder weakness. St. Johnswort is one of the most heavily studied botanicals for treatment of depression, with mood-stabilizing actions credited to multiple neurotransmitter effects. *Hypericum* may be included in formulas for premenstrual dysphoria and for menopausal and perimenopausal mood disorders. *Hypericum* can also benefit men with urinary pain or anxiety. *Hypericum* combines well with *Actaea* for menopausal anxiety, or with *Humulus* and *Withania* for menopausal insomnia. *Hypericum* can speed the liver's clearance of drugs, so individuals tenuously managed on precise dosages of medications should be monitored by a physician if adding *Hypericum* to their daily therapy. There are isolated reports of *Hypericum* leading to birth control failure when used in tandem with oral contraceptives.

Juniperus communis • Juniper

Juniper berries are used as a diaphoretic, circulatory stimulant and as an antimicrobial urinary stimulant. Small amounts of *Juniperis* tincture can be included in formulas for atony of the genitourinary system of elderly men or in formulas for postmenopausal tendency to chronic bladder infections. Avoid juniper in acute inflammatory disorders of the urinary system and use small doses only for those with atony and poor circulation. Include juniper berries in teas or small amounts of tincture in tonic formulas for women who begin to suffer from acute painful UTIs and urinary symptoms postmenopause as declining estrogen levels fail to maintain the urinary mucosa.

Leonurus cardiaca • Motherwort

The leaves of motherwort are historically used to calm anxiety, restlessness, and heart palpitations, especially when associated with hyperthyroidism, menopause, or other endocrine imbalances. The plant and related species are used in TCM to treat blood stasis and menstrual irregularities and are said to clear heat and toxins from the blood. *Leonurus* is specific for thyroid disorders manifesting at the climacteric and for endocrine disorders associated with restless anxiety, nervous debility, fitful sleep, tics, twitches and tremors, and cardiac excitability and irregularity. Motherwort is useful in formulas for amenorrhea and infertility occurring as part of a larger endocrine disorder, such as PCOS. *Leonurus* is also helpful for back and pelvic pain that has a bearing-down, expulsive sensation.

Lepidium meyenii • Maca

Maca is a turniplike tuber in the crucifer family that grows exclusively at high elevations of the Andes. The tuber has been cultivated since Incan times and is dried and eaten as a staple part of the Andean diet; few other vegetables thrive at that extreme altitude. Traditionally, dried maca tubers are ground into flour and used in a variety of baked goods, and it is believed to improve energy and stamina, fertility, heart function, mood, and longevity. A survey of Andean maca users compared to nonusers reported generally better health and better longevity, based on questionnaires, physical performance assessments, and basic lab indices.[293] Early herb texts claimed *Lepidium* to be specific for breast pain when squeezing or lancinating in quality and reported maca as helpful for menorrhagia. *Lepidium* may be included in formulas for male and female infertility and to support general reproductive health.

Lobelia inflata • Pukeweed

The young seedpods and flowering tops of *Lobelia* are most well known to relax bronchial smooth muscle, but *Lobelia* also has a spasmolytic effect on uterine muscle. *Lobelia* has been used orally as well as topically for failure of the cervix to dilate during labor, a rigid os, or perineal and vaginal rigidity that delays the progression of labor. To avoid nausea, *Lobelia* is most often used topically in these conditions, or as an enema to elicit reflexive relaxation of the uterus and vagina. Small, frequently repeated doses of the tincture may be used, reducing at the first sign of any nausea. *Lobelia* oil may be applied topically directly to the cervix, or various preparations may be applied to the chest, back, or pelvis, over the area of the spastic uterus, ureters, or bronchi to relax spasms or to promote cervical dilation during labor. Lobeline, a piperidine alkaloid in *Lobelia*, has been widely studied for effects on nicotinic acetylcholine receptors in the lungs and brain, but not yet studied for effects on the uterine muscle or vasculature, despite the traditional usage.

Lycopus virginicus • Bugleweed

Lycopus was included in early American herbals where the leaf preparations were noted to be specific for vascular excitation and a rapid pulse due to sympathetic (adrenergic) stimulation. Research on related *Lycopus* species has shown calcium channel effects and other mechanisms that may help reduce tachyarrhythmias. *Lycopus* is especially indicated for functional arrhythmias, where tumultuous heart action occurs due to anxiety or thyroid disorders. Folkloric indications for *Lycopus* also include hyperthyroid symptoms such as sweating, hypertension, exophthalmia, or goiter. *Lycopus* is also specifically indicated for testicular pain due to vascular congestion or hormonal imbalance and for dysfunctional uterine bleeding that is frequent and scant in quantity. Gyspywort (*Lycopus europaeus*) is used in a similar manner. Asian species of *Lycopus* are considered blood-moving heart tonics[294] and have been shown to protect sensitive vasculature from inflammatory damage.[295] *Lycopus* species contain rosmarinic acid, which can attenuate excessive thyroid function. (See "Rosmarinic Acid for Hyperthyroidism" on page 43.) *Lycopus europaeus* has been shown to reduce elevated TSH and cardiac symptoms[296] and to increase the urinary excretion of T4[297] in hyperthyroid human subjects. Include *Lycopus* in formulas where thyroid imbalance contributes to reproductive imbalances.

Matricaria chamomilla • Chamomile

Sometimes called German chamomile, *Matricaria* is used for many conditions, including anxiety and digestive upset. The name *Matri-* is said to have derived from "matrix," meaning womb, reflecting the herb's many benefits to the uterus, including as an antispasmodic against menstrual cramps. The dried flowers of *Matricaria* are widely used in teas, but also processed into concentrated tinctures and other medicines. For the reproductive system, *Matricaria* is specifically indicated for premenstrual anxiety and irritability and for diarrhea and intestinal upset with menses and menstrual cramps. *Matricaria* can also be included in formulas for amenorrhea and dysmenorrhea where there is a sense of weight in the pelvis and genitals. *Matricaria* is known to bind to GABA receptors in the brain, contributing to its anxiolytic effects, and *Matricaria* is traditionally said to be specific for those who are nervous, restless, irritable, apprehensive, and hypersensitive to pain. GABAergic signals play an important role in the release of gonadotropins from the pituitary, and research shows that *Matricaria chamomilla* affects LH (luteinizing hormone) surge and improves the hormonal control of ovarian follicle maturation. Changes in serum levels of gonadotropin-releasing hormones are associated with the development of ovarian cysts in polycystic ovarian syndrome, and *Matricaria* may inhibit the inappropriate elevation of gondatopin and LH via GABAergic effects.[298] *Matricaria* is high in the flavonoid apigenin, shown to preferentially bind β estrogen receptors, and is also being explored for anticancer and hormone-regulating mechanisms.[299] *Matricaria recutita* is used in a similar manner.

Medicago sativa • Alfalfa

The leaves of alfalfa are highly nutritive and have been used as livestock fodder and in herbal teas for general tonifying and nourishing support. Being a legume, alfalfa also is high in phytosterols, which may offer many benefits including hormonal regulation and hypolipidemic and bone-building actions. *Medicago* also contains flavonoids and triterpene saponins, both credited with anti-inflammatory and hormone-balancing effects. Include alfalfa in teas and medicinal vinegars to offer minerals and nutrients to those with osteoporosis or use as a general nutritive foundational herb when crafting reproductive formulas.

Melissa officinalis • Lemon Balm

Melissa leaves are a traditional nervine used for stress, anxiety, and depression. Much of the modern research has focused on the plant's neuroprotective actions. *Melissa* contains rosmarinic acid, shown to reduce excessive thyroid activity. *Melissa* can be valuable as a mood-stabilizing ingredient in formulas for endocrine imbalances, including thyroid and female-hormonal disorders. Studies show that *Melissa* is an acetylcholinesterase inhibitor,[300] explaining its traditional use for impaired memory. It is also shown to inhibit GABA-transaminase enzymes, thereby enhancing GABAergic neurotransmission. Even inhaling the essential oil has been shown to have benefits on mood and cognition. One clinical trial showed *Melissa* capsules to improve PMS in high school girls.[301] Include *Melissa* in teas, tinctures, and other medications in formulas for premenstrual or perimenopausal agitation or in formulas for endocrine imbalances associated with hyperthyroidism. *Melissa* is also known to have antiviral effects, especially the essential oil fraction, including activity against the herpes family of viruses.[302] Include *Melissa* in oral medications and the essential oil in topical compresses to treat genital herpes, both for acute outbreaks and to reduce the frequency and duration of chronic episodes.

Mitchella repens • Partridge Berry

Mitchella has been traditionally used as a uterine tonic and included in partus preparator formulas. *Mitchella* is especially useful for women who have borne many children, because it may improve uterine muscular tone. *Mitchella* is an ingredient in the classic Mother's Cordial used by Eclectic physicians. (See the Tincture for Mother's Cordial on page 132.)

Ocimum tenuiflorum • Holy Basil

This plant is also known as *Ocimum sanctum* and goes by the common name of tulsi in India, where it has been used in Ayurvedic medicine for thousands of years for medicinal as well as for spiritual and religious purposes. In India, the fresh or dried leaves of tulsi are considered an adaptogen and longevity tonic, balancing hormonal and neurologic processes in the body and supporting organ function. Much of the current research has focused on anti-inflammatory and tissue-protective effects, helpful for preventing harm to the body and nervous system from stress and protecting against harm from toxins and microbes. *Ocimum* is mentioned as being specific for testicular pain, vulvar and vaginal pain, and breast pain. Several studies have reported temporary impairment of semen quality, sperm counts, and sperm motility[303] associated with changes in the ultrastructure of epididymal epithelial cells[304] occurring with the use of *Ocimum*. Several researchers have reported that antiandrogen effects are responsible for the antifertility effects.[305] Serum testosterone levels are not shown to decline and in fact may increase from *Ocimum* use, perhaps due to impairment of local binding.[306] All of these effects are reversible, with hormones and semen parameters returning to normal within 2 weeks of *Ocimum* cessation. Due to the apparent lack of any other toxicity or side effect, *Ocimum* is being explored as a possible male contraceptive agent.

Panax ginseng • Ginseng

Ginseng roots are one of the most well-known adaptogens in the world, traditionally said to be a chi tonic and used to improve energy, stress tolerance, and immunity. *Panax* has long been used to treat both male and female infertility and to support reproductive function into the elder decades for men and help treat postmenopausal issues in women. *Panax* is specifically indicated for exhaustion and endocrine imbalances that follow long-term overwork and stress. *Panax* may improve low sperm counts and low testosterone levels in men and is widely used by herbalists as an adrenal tonic. *Panax* is a warming herb best for those with cold or neutral constitutions because it can sometime cause heat symptoms in the body. See also "The Ginsengs" on page 47.

Pausinystalia yohimbe • Yohimbe

Yohimbe is an African species whose bark is traditionally used to treat impotence, with activity credited to yohimbine and related alkaloids. These alkaloids promote penile vasodilation via α-2 adrenergic blockade and also a monoamine oxidase inhibitor, increasing serotonin and epinephrine in the brain. Due to neurotransmitter—especially adrenergic—effects, yohimbe can result in many side effects, including nervousness, anxiety, insomnia, headaches, panic attacks, increased frequency of urination, and possibly hypertension. Overall, it may have more druglike and less tonifying effects than other sexual tonics mentioned in this chapter, and I do not use the plant in formulas for my patients. Nonetheless, yohimbe may improve erectile function by enhancing penile circulation, even in diabetics and other cases of impaired circulation. Promotion of adrenergic signaling in pelvic ganglia may enhance sexual sensation and improve stamina, and the herb is also reported to improve orgasmic dysfunction. The Food and Drug Administration approved yohimbine as the first plant-derived drug

for treating impotency in the late 1980s, earning the moniker of "herbal viagra" in the marketplace thereafter. The efficacy of yohimbine on erectile function can be increased by simultaneous administration of phosphodiesterase inhibitors that augment the release and action of nitric oxide in the corpus cavernosum of the penile vasculature. However, phosphodiesterase inhibitors can also promote side effects including rhabdomyolysis, visual perception changes, headache, dyspepsia, and back pain. I recommend the use of more nourishing circulatory and reproductive tonics rather than yohimbe.

Petroselinum crispum • Parsley

Parsley leaves, seeds, and stems have all been used medicinally, and numerous studies have shown strong antioxidant, circulatory-enhancing, antispasmodic, and other physiologic actions. Older herbal texts suggest that *Petroselinum* is specific for urinary symptoms with itching, burning, and urethral sensations and suggest that *Petroselinum* can be used as a diuretic for premenstrual fluid retention. *Petroselinum* is also an emmenagogue that may be useful in cases of amenorrhea and dysmenorrhea with scanty flow due to low hormonal levels. Because of folkloric writing about parsley's emmenagogue action, there have been cases of toxicity when women have attempted to induce an abortion by taking large amounts of parsley essential oil. *Petroselinum* has significant anti-inflammatory action and may be included in formulas as a stimulating nutritive tonic, offering carminative, hypoglycemic, diuretic, hypolipidemic, antimicrobial, anticoagulant, and hepatoprotective actions. Parsley is in the Apiaceae family and contains hormonal-regulating and anti-inflammatory coumarins, among many other active constituents.

Phellodendron amurense • Amur Cork Tree

Also known as huang bai in Asia, the bark of *Phellodendron* has been used medicinally for a variety of issues, including infections and prostate, breast, and other cancers. *Phellodendron* is one of the 50 fundamental medicines in TCM, used as an antiviral and a metabolic tonic and against eye and lung infections. *Phellodendron* contains the isoquinoline alkaloids berberine, palmatine, and phellodendrine credited with antimicrobial and anti-inflammatory actions. *Phellodendron* has been developed into an anti-inflammatory agent marketed under the name Nexrutine; it is a concentrate of the berberine alkaloids shown to deter the development and progression of prostate cancer in animal studies and limited human investigations.[307] The human clinical trials dosed Nexrutine at 500 milligrams three times a day, with no reported toxicities.[308] Relora is another trademarked product; it is a blend of *Phellodendron* and *Magnolia* and is used for elevated cortisol.[309]

Phytolacca americana • Pokeweed

The roots of *Phytolacca* have antimicrobial, anti-inflammatory, and anticancer effects, and when handled raw, they can be irritating and even caustic. *Phytolacca* is gentler once processed into medicines, but still has stimulatory effects and has been traditionally used for glandular swelling and inflammation, syphilitic bone pain, sore throat with enlarged tonsils, quinsy, and diphtheria. The roots of *Phytolacca* are thought to have an affinity for white blood cells and the lymphatic channels of the body, including tissue with a large degree of lymphatic support such as the breasts, and of course, the lymph nodes themselves. *Phytolacca* is specific for acute mastitis, breast pain with hard lumpiness, premenstrual breast pain, cystic masses, lymphatic congestion, ovarian neuralgia, testicular pain, hard swelling of the testes, and goiter. *Phytolacca* has been traditionally included in formulas where fluid stasis, tendency to cysts, or a lymphatic constitution underlie a condition such as tonsillitis, stasis ulcers, or mastalgia. *Phytolacca* is not commonly used as a tea but is available as a tincture and also as an oil extract for use topically on the breasts to treat mastitis. *Phytolacca* oil should be washed off the breasts prior to nursing. *Phytolacca* contains immunostimulatory lectins[310] that have a binding affinity for white blood cells and some other cells. *Phytolacca*'s significant antiviral effects are due, in part, to the lectin's interference with viral entry into cells and interference with viral replication. *Phytolacca americana* is also known as *P. decandra*.

Piper methysticum • Kava Kava

The resinous roots of *Piper methysticum* are primarily used as an antispasmodic anodyne and anxiolytic herb. Kava can be used acutely and short term for urinary or prostate pain, for difficulty with urine flow and painful micturition, especially when due to sexually transmitted disease and associated with purulent urethral discharges. Kava may also be useful in formulas for the elderly with atony of bladder and nocturnal enuresis. As an anodyne, kava may offer symptom relief in patients with fibromyalgia and with stress associated with acute muscle spasms. Kava's effects are rapid in onset and

rapid to abate, thus medicines must be dosed frequently for adequate pain relief. Kavalactones may have chemopreventive effects for bladder, prostate, testicular, and other cancers.[311] More details on *Piper methysticum* can be found in Volume 1, chapter 4.

Podophyllum peltatum • Mayapple

The fruits of the mayapple are used to extract the resin podophyllin, from which podophyllotoxin may be purified. *Podophyllum* fruits and leaves are also prepared into tinctures. *Podophyllum* resin is an extremely caustic, yet painless escharotic agent with antiviral properties, useful topically on warts, both veruccus lesions and genital condyloma. Podophyllotoxin has strong antiviral, antiproliferative, and antineoplastic effects, and research on its antitumor effects led to the development of several important anticancer drugs including etoposide and teniposide. *Podophyllum* medications can ablate condylomata acuminata and warts when topically applied,[312] either as a podophyllotoxin 0.5 percent cream or a podophyllin 20 percent cream.[313] Although expensive, *Podophyllum* medicines can be prepared for oneself; see Topical Application of Podophyllin for Warts on page 113. The creams are generally well tolerated and safe to apply at home, but occasionally induce irritation and must be discontinued.[314] *Podophyllum* of any kind should not be consumed during pregnancy, nor should it be topically applied due to teratogenicity.

Prunus africana • African Plum Tree

Also known as *Pygeum africanum*, the bark from this African evergreen rose family tree has been used for the symptoms of benign prostatic hyperplasia (BPH) for centuries. *Prunus africana* is specific for prostatic enlargement and urinary retention due to atony of the urinary passages. It can modulate bladder contractility along with various anti-inflammatory effects that may limit fibroblast production associated with prostatic enlargement. *Prunus africana* may also modulate androgen-driven proliferation of the prostate, protect and restore secretory activity of the prostate epithelium, and may inhibit prostate cancer initiation and progression.[315]

Ptychopetalum olacoides • Muira Puama

Also known as "potency wood," this Amazonian plant is reported to heighten sensory sexual stimuli and increase the libido. The plant is also reported to increase testosterone and improve erectile function, along with emotional desire and response to stimuli. The plant has not yet been the subject of much modern research but is known to contain ferulic and caffeic acids, magnoflorine, and ginsenosides.

Pueraria montana var. lobata • Kudzu

Pueraria species have many names due to extensive use in many locales and herbal traditions of the world. *Pueraria lobata* goes by the name kudzu in the United States, yege or gegen in China, ohwi in Japan, and kwao krua in Thailand, a term for a rejuvenation tonic. *Pueraria* is sometimes called Japanese arrowroot because it can be processed into a starchy paste similar in texture and culinary application as the *Arum* (arrowroot) roots. *Pueraria* roots have many medicinal virtues, including hepatoprotection and vascular and hormonal support. Kudzu has been traditionally used to improve tinnitus, dizziness, and vertigo; its circulatory-enhancing effects may improve circulation to the head. Kudzu is also credited with optimizing metabolism and having antidiabetic, insulin-enhancing, and antiobesity effects. Other traditional uses include an ability to reduce alcohol consumption and craving. In Thailand, *Pueraria* species, especially *Pueraria candollei* var. *mirifica*, is used in antiaging formulas, particularly as a female reproductive and rejuvenation tonic to treat menopausal symptoms. *Pueraria* has also been historically used for facial rejuvenation and support of the bone, hair, and fingernails. It is sometimes claimed to support sexual longevity, including vaginal lubrication, and enhance breast size. Modern research has shown that the isoflavones in *Pueraria* bind to estrogen receptors, support bone density and breast health, and display anticancer effects in hormone-sensitive tissues.

Raphanus species • Radish

Radish tubers contain glucosinolates, which are transformed into indole-3-carbinol in the body that in turn induces hepatic detoxification pathways such as via phase I cytochrome P450 enzymes. Radish glucosinolates may support tissue detoxification and offer chemopreventive effects against some cancers, including breast and reproductive cancers. Reducing oxidative stress in the body may also improve liver and general metabolic function, and radish consumption may improve diabetic parameters.[316] As there is overlap among diabetes, obesity, and hypothyroid issues with reproductive cancers, radish foods and medicines may be part of broad protocols for reducing endometrial proliferation and cancer risk, as well as for breast cancer. See also the discussion on glucosinolates in "Disrupting the Sodium/Iodide Symporter"

on page 38. Specific glucosinolates in radishes include glucoraphanin and glucoraphasatin, which yield isothiocyanates upon digestion including sulforaphene, sulforaphane, and indole-3-carbinol. Sulforaphene, in particular, has been the subject of several studies showing it to inhibit breast cancer. Several studies have also reported an inverse relationship between dietary glucosinolate consumption and the risk of prostate cancer.

Rhodiola rosea • Arctic Rose

The astringent roots of *Rhodiola* are used for fatigue, muscle weakness, the inability to exert oneself, and mental fatigue. *Rhodiola* is considered an adaptogen and can be used as an adrenal supportive herb in formulas for hormonal imbalances. The plant has slightly stimulating effects so is best reserved for lethargic and deficient individuals and may occasionally promote anxiety or restless sleep if taken by those who are already hyper, restless, or high-strung.

Rubus idaeus • Raspberry

Raspberry leaves are emphasized in folklore for uterine disorders and are thought to be safe to consume throughout pregnancy. Raspberry leaves may support labor by improving uterine strength and act as a hemostatic in cases of menorrhagia. As with berries in general, raspberry leaves act as a vascular and circulatory tonic. There has been very little research on *Rubus* leaves, but compounds named urolithins are produced by gut microbiota from the metabolism ellagitannins found in raspberry leaves. Urolithins are reported to have estrogenic effects[317] and reduce androgen-driven proliferation of the prostate gland.[318] The leaves are nourishing and can be included in mineral tonics, nursing teas, and general tonic teas.

Salvia miltiorrhiza • Red Sage

Known as dan shen in TCM, the roots of *Salvia miltiorrhiza* may enhance estrogen pathways in animal models of estrogen deficiency and stimulate estrogen receptors. *S. miltiorrhiza* is also widely used for its broad circulatory-enhancing effects, with numerous studies showing the roots to have antioxidant and anti-inflammatory effects that in turn may translate to chemoprotective and anticancer effects.

Salvia officinalis • Sage

The leaves of *Salvia officinalis* have a number of medicinal applications, from antimicrobial effects to cholinesterase inhibition, but are not widely researched for hormonal effects. Sage, however, is the most effective herb I have found to quickly improve menopausal hot flashes and night sweats and reduce excessive perspiration in general. The leaves are also very effective in reducing lactation, such as when weaning a child, giving up an infant for adoption, or in the rare case of hyperlactation.

Sanguinaria canadensis • Bloodroot

The orange-red roots of *Sanguinaria* can have caustic effects when topically applied and were used by indigenous peoples of the midwestern region of the United States to treat skin lesions and cancers. Some native peoples referred to the plant as puccoon. In the 1850s, an American surgeon, Jesse Fell, combined *Sanguinaria* with another highly caustic agent, zinc chloride, and prepared a salve known as Fell's paste that was used topically on skin cancer. Over the ensuing 150 years, many variations on the formula have become known as "black salve," and sometimes referred to as "nature's scalpel" due to its ability to ablate and destroy tissue. Black salves can be extremely destructive to tissue and therefore are dangerous to use without expert understanding and guidance regarding their application. See the Escharotic Treatments (Traditional Naturopathic Formula) on page 114 for more information on black salve preparation. *Sanguinaria* rhizomes contain quaternary benzophenanthridine and protopine alkaloids that are shown to intercalate with DNA and RNA, altering gene expression, inhibiting microtubule formation and thereby cell division, and inducing oxidative stress.[319] Two such alkaloids, sanguinarine and chelerythrine, are being investigated for anticancer effects and have also been diluted to use in mouthwashes against leukoplakia.[320]

Scutellaria lateriflora • Skullcap

Scutellaria lateriflora leaves may be included in formulas where nervous irritation, anxiety, and fearfulness accompany PMS and menopausal difficulties. Skullcap is specific for anxiety and fear, inability to concentrate, restless sleep and fearful dreams, and insomnia due to worry, nervousness, seminal emissions, or headaches.

Serenoa repens • Saw Palmetto

The olivelike berries of the small palm of extreme southern regions of the United States are most well known for their benefits to the prostate gland. Research on the liposterols in saw palmetto fruits led to the development of important drugs for benign prostatic hyperplasia, due to their ability to inhibit the formation of dihydrotestosterone. Saw palmetto is a natural 5α-reductase inhibitor and is specifically indicated for prostatic hyperplasia with painful or difficult urination and dribbling of urine and prostatic discharges. Saw palmetto, however, can also be used in formulas for women with PCOS for whom reducing testosterone may be beneficial. Include *Serenoa* in formulas for elevated testosterone in women, underdevelopment or atrophy of the testes and ovaries and mammary glands, loss of libido in both men and women, and gonadal pain and inflammation including ovaritis, ovarian pain, orchitis, and epididymitis. For more information on *Serenoa*, refer to Volume 1, chapter 4.

Smilax ornata • Sarsaparilla

The roots of many *Smilax* species are noted to contain steroidal saponins, and many species are used medicinally, especially for hormonal issues. Several species of *Smilax* may go by the common name sarsaparilla. Many *Smilax* species have been used traditionally to detoxify the blood and help remedy dermatitis, arthritis, and other types of chronic inflammation including liver disease, diabetes, and cancer. Asian species of *Smilax* are said to clear heat and dampness from the tissues. *Smilax* is an aromatic alterative agent with hormonal- and adrenal-supportive properties, and it is specific for fatigue, muscle weakness, and poor digestion. The roots have slightly warming properties. A variety of anticancer mechanisms have been identified in *Smilax*.[321] There are well over 300 species of *Smilax* and many have been used as traditional anti-inflammatory, anticancer, and hormonal-balancing medicines, including *S. glabra*, *S. officinalis*, *S. china*, *S. larvata*, *S. aspera*, and *S. sieboldii*. For example, *Smilax china* is known as ba qia or jin gang teng in TCM and is used for many chronic inflammatory ailments including prostatitis and cancer.[322] All these species are noted for their saponins, such as sarsapogenin, which are widely studied for their cytotoxic, immunomodulating, and anti-inflammatory activities.[323]

Terminalia arjuna • Arjuna

Terminalia species have many medicinal applications, and the leaves and bark have been traditionally used in Ayurvedic medicine for centuries to treat rheumatic joint pain, skin disease, hepatitis, and metabolic and cardiovascular complaints. Many *Terminalia* species have also demonstrated significant antiviral effects.[324]

Terminalia may be used to treat acute herpes simplex viral infections or to support the immune system and reduce chronic outbreaks of herpes. *Terminalia* species have also shown activity against HIV infections.[325] *Terminalia chebula*, *T. paniculata*, and *T. catappa* (tropical almond) are also used medicinally in a similar manner.

Thuja species • Cedar

Thuja occidentalis is known as arborvitae or northern white cedar, and *Thuja plicata* is western red cedar, and both were used by indigenous peoples for bronchial catarrh, enuresis, cystitis, psoriasis, uterine carcinomas, amenorrhea, and rheumatism. Due to the vitamin C content of the fresh leaves, cedar species were also used to relieve symptoms of acute scurvy. *Thuja* extracts exert immunomodulating and antiviral properties, including activity against HIV[326] and HPV infections,[327] and *Thuja* is specific for warts, condyloma, and abnormal Pap smears due to sexually transmitted pathogens. *Thuja* extracts and essential oil concentrates have been used topically for genital warts and as a sitz bath for genital discharges. Thujone in the essential oil fraction is credited with antimicrobial activity, as well as anticancer effects,[328] but it can be a powerful irritant to the digestive, vascular, and especially urinary passages with high doses. *Thuja* essential oil has been employed by naturopathic and other physicians in various topical applications to treat cervical dysplasia and cancer.

Tribulus terrestris • Puncture Vine

Tribulus has been used in many traditional healing systems for low libido, sexual debility, erectile dysfunction, premature ejaculation, low sperm counts, prostatitis, male and female infertility, and other medical complaints including diabetes, kidney stones, and heart disease. Both the fruits and roots are used medicinally, with many effects credited to steroidal saponins including spirostanol and furostanol, as well as anti-inflammatory flavonoids. Some popular claims regarding the herb's power to enhance sexual function may be exaggerated and sensationalized, but *Tribulus* has been shown to improve erectile function via circulatory enhancement and to increase libido in postmenopausal women.[329] The saponin protodioscin in the plant may support testosterone production.[330] Luteinizing hormonelike effects have been reported in animal studies, promoting precocious puberty. *Tribulus* may also help treat PCOS, reducing folliculogenesis and the tendency to form ovarian cysts.

Trifolium pratense • Red Clover

Red clover is a common leguminous meadow and pasture plant, and the young blossoms with their leaves are used broadly in traditional herbal medicine as a general alterative, to prepare cough syrups and teas, to prepare external skin plasters, to improve fertility, and in some cases, to treat cancer. *Trifolium* has traditionally been considered a blood mover for stasis or an increased tendency to clots. Some native peoples used red clover for menopausal symptoms, and modern research supports the theory that use of *Trifolium* may improve postmenopausal bone density.[331] *Trifolium*'s folkloric indications include bone-strengthening, fertility-enhancing, miscarriage-preventing, and menopause-supporting actions, all credited to its content of isoflavone phytoestrogens in contemporary medical thinking. Several human and animal studies have demonstrated hypolipidemic, hypoglycemic, and antiatherosclerotic effects, all conditions known to increase postmenopausally. While pharmaceutical hormone therapy is associated with an increased risk of breast cancer, red clover has shown no such risk and, in fact, protects the breasts from overstimulation of endogenous hormones.[332]

Although red clover is commonly used in agriculture to fix nitrogen in the soil and is cultivated as a livestock feed, many farmers report that grazing livestock on red clover increases the risk of reproductive problems, from infertility to mastitis and antigalactogogue effects.[333] As noted above, *Trifolium* contains isoflavones, as do many legumes. *Trifolium* contains higher levels of isoflavones biochanin A and formononetin than do other species of clover or alfala. Typical content is around 4 milligrams of each of the two isoflavones per gram dried herb, or 7 to 14 milligrams total phytoestrogens.[334] Furthermore, formononetin may be a more powerful estrogen agonist than other phytosterols, so red clover may be a stronger estrogenic herb than other legumes such as licorice or *Astragalus*. The estrogenic action in livestock, whose diets may contain much larger amounts than in humans, is so great as to disrupt normal hormonal balance in sheep and possibly other grazers, as well as poultry.[335] *Trifolium* may not readily cause such problems when used in tea blends or briefly in a cough syrup. Large and chronic doses of concentrated *Trifolium* products, however, should be scrutinized carefully for women with hyperestrogenism, and many other herbs may be more appropriate in such cases. Yet for menopause therapy, *Trifolium* has not been shown to have harmful effects at therapeutic dosages.

Some clinical studies have shown little to no relief of hot flashes from *Trifolium*,[336] but other trials have shown efficacy.[337] One meta-analysis reported *Trifolium* isoflavones at a dose of 80 milligrams/day to alleviate hot flashes and pain, and significantly improve mood and cognitive symptoms,[338] without evidence of side effects[339] or adverse events.[340] No negative effects on the breast and endometrium[341] or serum lipids and thrombi risk[342] have been reported. The use of *Trifolium* to treat menopausal symptoms or surgical menopause appears to be safe for women with current or a history of breast cancer,[343] and some studies on isolated *Trifolium* isoflavones have reported activity against breast cancer.[344]

Trigonella foenum-graecum • Fenugreek

The ripe seeds of *Trigonella* are used as a culinary spice and herbal medicine, especially for metabolic support in cases of diabetes and hyperlipidemia, but also as a traditional galactogogue, and are sometimes reported to support breast development in cases of low hormonal status. Fenugreek is hepatoprotective due to antioxidant and other actions and has been reported to protect against breast[345] and colon cancer. Fenugreek saponins, such as diosgenin and protodioscin, and the alkaloid trigonelline are credited with anticancer effects.[346] Because women with PCOS are at increased risk for reproductive cancers, due in part to metabolic imbalances, fenugreek extracts may be a helpful preventive.[347] A fenugreek product containing concentrated protodioscin is reported to improve libido, testosterone level, and sperm parameters in human males.[348]

Turnera diffusa • Damiana

Damiana leaves are a traditional medicine for sexual debility, impotence, incontinence in the elderly, irregular menses at menarche, menstrual and pregnancy support, anxiety and fatigue, and low libido. Damiana is also traditionally regarded as an aphrodisiac, reducing anxiety while improving sexual function. The plant has not been widely studied, but anticancer effects against breast cancer are reported and credited to its general anti-inflammatory action combined with estrogenic yet aromatase-inhibiting actions.[349] Animal investigations have supported the notion that *Turnera* may stimulate the libido,[350] but human clinical trials are lacking. This species is also known as *Turnera aphrodisiaca*.

Urtica dioica, U. urens • Nettle

The highly nutritive leaves of *Urtica* have been used traditionally for insufficient lactation, malnourishment, uterine hemorrhage, menorrhagia, metrorrhagia, and itching and burning of the vulva. *Urtica* teas and tinctures can be used as general mineral tonics and diuretics in cases of fluid retention and premenstrual edema, as well as a genitourinary tonic for male infertility and for prostatic enlargement. Animal studies have shown nettles to boost glutathione in the testes, liver, and kidneys and to protect the seminiferous tubules against toxins. *Urtica* may enhance semen parameters by protecting tissues from mercury,[351] metabolic distress,[352] and nicotine.[353] Nettle root is also used medicinally, especially for the genitourinary system. Nettle root may protect against prostatic hyperplasia,[354] and several human clinical trials have shown nettles to reduce the clinical symptoms of BPH.[355] For more details on *Urtica*, see Volume 1, chapter 4.

Valeriana officinalis • Valerian

Valerian root is a well-known sedative nervine and may improve sleep for women with hormone-driven sleep disturbances, such as perimenopausal insomnia. Valerian is specifically indicated for nervousness, hypersensitivity, obsessive thinking, worry, stress headaches, and long menstrual cycles with scanty flow.

Verbena species • Verbena

The aerial parts of verbena are a gentle nervine, traditionally used for anxiety and depression, including mood disorders that accompany menopausal or thyroid disorders. Other specific indications for verbena include amenorrhea, menstrual disorders, abdominal problems, and rheumatic and thyroid problems.

Viburnum opulus • Cramp Bark

As the common name cramp bark implies, *Viburnum opulus* is a leading herb for menstrual cramps. (See the Viburnum Simple for Menstrual Cramps formula on page 123.) In fact, cramp bark is so reliable in treating the pain of uterine spasms, when it fails to help, I suspect the pain is not of uterine muscle origin, but rather may be due to vascular congestion, neuralgia, or other cause. The root bark of *Viburnum* is a uterine tonic and may reduce chronic miscarriage when due to hyperexcitability or atony of the uterine muscle. Cramp bark is also specific for heightened awareness of internal sexual organs and for long menstrual cycles with scanty flow and much cramping. The herb is also a traditional partus preparator.

Viburnum prunifolium • Blackhaw

Blackhaw roots are specific for atonic or feeble reproductive organs, spastic uterine pains, and for both amenorrhea and profuse menstruation. Like cramp bark, blackhaw has also been used as a partus preparator and to quiet a hyperexcitable uterine muscle in cases of threatened miscarriage.

Vitex agnus-castus • Chaste Tree Berry

The peppercorn-like fruits have been used for hormonal imbalances and were used by the devout to decrease the libido when living in monasteries, hence they have another common name, monk's pepper. *Vitex* may promote progesterone to improve estrogen dominance and thereby help treat dysfunctional uterine bleeding, uterine fibroids, breast cysts, breast pain, and endometriosis—all associated with high estrogen relative to low progesterone. *Vitex* can often restore normal menses in women with amenorrhea associated with PCOS. I have seen *Vitex* promote menses in amenorrheic women in as little as 2 weeks and have also found *Vitex* to be effective in restoring normal menstrual cycles whether too frequent, too infrequent, or highly irregular and erratic at perimenopause. *Vitex* is also specific for premenstrual headaches and PMS that abates as the menstrual flow begins, and for acne at puberty for both young men and women. *Vitex* is also usually highly effective for mastalgia and may correct gynecomastia occurring in adolescent males at puberty. *Vitex* is generally well tolerated, the main side effects being nausea and gastrointestinal irritation, dizziness, dry mouth, headache, menstrual disorders, acne, pruritus, and erythematous rash, all reversible upon cessation of the medication.[356] See also "*Vitex agnus-castus* for Reproductive Complaints" on page 148.

— ACKNOWLEDGMENTS —

I would like to thank Patryk Madrid, ND, class of 2020, for his admirable skills with word processing and editing support.

Thank you to my editor at Chelsea Green, Fern Marshall Bradley, for her professional prowess and for making this much more polished than it would have been without her guidance. I wish to also thank Margo Baldwin, the publisher at Chelsea Green, for making this text possible at all, as well as Pati Stone and the entire production team for helping to perfect the many small details that producing a reference text of this nature entails.

And I extend warm, fond appreciation to my kind and supportive sweetie, Warren Martin, for his patience, political updates, and willingness to pitch in with the day-to-day chores of running a household and a business.

I would like to thank and honor all my teachers and the herb community for being my beloved tribe, filled with powerful, skillful women and gentle, reverent men, and for all the times that the healing plants have brought us together in beauty, to share our gifts and wisdom with one another. If I have never told you to your face (and to many of you I have), I love you!

And last, but not least, to Pachamama herself for her mystery, abundance, and amazing gifts. May we all Walk in Beauty and return her gifts tenfold. Blessed Be!

— SCIENTIFIC NAMES —
TO COMMON NAMES

The following lists include all of the herbs, medicinal fungi, and homeopathic preparations mentioned in the text of this book.

Achillea millefolium	yarrow
Aconitum napellus	aconite, wolfsbane
Actaea cimicifuga (also known as *Cimicifuga foetida*), *A. dahurica*, *A. heracleifolia*, *Actaea* spp.	shengma
Actaea racemosa (also known as *Cimicifuga racemosa*)	black cohosh, macrotys
Aesculus hippocastanum	horse chestnut
Agaricus spp.	white button mushrooms
Agrimonia eupatoria	agrimony
Albizia julibrissin	silk tree, mimosa
Albizia lebbeck	siris, lebbeck tree
Alchemilla vulgaris	lady's mantle
Allium cepa	onion
Allium sativum	garlic
Aloe vera, *A. barbadensis*	aloe
Alpinia galanga, *A. officinarum*	galangal
Althaea officinalis	marshmallow
Amanita muscaria	fly agaric
Ammi visnaga	khella, bishop's weed
Andrographis paniculata	king of bitters, andrographis
Anemone pulsatilla	pasque flower
Angelica sinensis	dong quai, angelica
Apium graveolens	celery
Arctium lappa	burdock, gobo root
Armoracia rusticana	horseradish
Artemisia abrotanum	southernwood
Artemisia absinthium	qing hao su, sweet wormwood
Artemisia annua	sweet Annie, sweet wormwood, qinghao
Artemisia vulgaris	mugwort
Asparagus racemosus	shatavari
Astragalus membranaceus	milk vetch, huang qi
Atropa belladonna	belladonna, deadly nightshade
Avena sativa	oats
Avicennia marina	white mangrove, gray mangrove
Azadirachta indica	neem, bean tree, China tree
Benincasa hispida	wax gourd
Berberis aquifolium (also known as *Mahonia aquifolium*), *B. nervosa*	Oregon grape, mahonia
Beta vulgaris	beet
Boswellia serrata	frankincense, Indian frankincense
Brassica alba (see also *Sinapis alba*)	
Brassica napus, *B. juncea*, *B. rapa*	rapeseed, canola
Bryonia alba	white bryony
Bryonia dioica	bryony
Bupleurum chinense	Chinese thoroughwax, chai hu
Bupleurum falcatum	Chinese thoroughwax, saiko
Bupleurum flavum	bupleurum, chai hu
Calendula officinalis	pot marigold, calendula
Camellia sinensis	green tea
Capsella bursa-pastoris	shepherd's purse

Capsicum annuum (also known as *C. frutescens*)	cayenne
Carthamus tinctorius	safflower
Caulophyllum thalictroides	blue cohosh
Ceanothus americanus	red root, New Jersey tea
Centella asiatica	gotu kola, pennywort
Ceratonia siliqua, Ceratonia spp.	carob
Chamaelirium luteum (also known as *Helonias*)	false unicorn
Chionanthus virginicus	fringe tree
Chrysanthemum morifolium	mum
Cimicifuga racemosa (also known as *Actaea racemosa*)	black cohosh
Cinnamomum cassia	cassia cinnamon, Chinese cinnamon
Cinnamomum polyadelphum (also known as *C. saigonicum*)	Saigon cinnamon
Cinnamomum verum (also known as *C. zeylanicum*)	Ceylon cinnamon, cinnamon
Citrullus colocynthis	bitter apple, bitter cucumber
Citrus aurantium	bitter orange
Citrus paradisi	grapefruit
Citrus reticulata	mandarin
Citrus sinensis	orange
Cnicus benedictus	blessed thistle
Cnidium monnieri	snow parsley, osthole
Cocos nucifera	coconut
Coffea arabica	coffee
Coix lacryma-jobi	Job's tears
Coleus forskohlii (also known as *Plectranthus forskohlii*)	coleus
Collinsonia canadensis	stoneroot
Commiphora molmol, C. myrrha	myrrh
Commiphora mukul, C. wightii	guggul, guggulu
Conium maculatum	poison hemlock, hemlock
Convallaria majalis	lily of the valley
Coptis chinensis	goldthread, duan e huang lian

Coptis trifolia	goldthread
Cordyceps militaris, C. sinensis	cordyceps, caterpillar fungus
Corydalis yanhusuo	corydalis, turkey corn
Crataegus laevigata, C. monogyna, C. oxyacantha	hawthorn
Cucurbita pepo	pumpkin
Curcuma longa	turmeric
Cuscuta reflexa	dodder, giant dodder
Cymbopogon citratus	lemon grass
Cynara scolymus	artichoke
Digitalis purpurea	foxglove
Dioscorea villosa	wild yam
Echinacea angustifolia	coneflower
Echinacea purpurea	coneflower, purple coneflower
Eleutherococcus senticosus (also known as *Acanthopanax senticosus*)	Siberian ginseng, eleuthero, ciwujia
Epimedium brevicornu (also known as *E. rotundatum*)	horny goatweed, yin yang huo, bishop's hat
Epimedium grandiflorum, Epimedium koreanum, Epimedium pubescens	horny goatweed
Equisetum arvense, E. hyemale	horsetail, scouring rush
Erigeron canadensis	Canada fleabane, Canadian horseweed
Eschscholzia californica	California poppy
Eucalpytus globulus	eucalyptus
Euonymus alatus	burning bush
Euphrasia stricta	eyebright
Eurycoma longifolia	tongkat ali
Filipendula ulmaria	meadowsweet
Foeniculum vulgare	fennel
Forsythia suspensa	forsythia
Fritillaria thunbergii	zhe bei mu
Fucus vesiculosus	bladderwrack, kelp
Galega officinalis	goat's rue
Galium aparine	cleavers
Ganoderma lucidum	reishi, reishi mushroom

Gelidium spp.	red algae
Gelsemium sempervirens	false jasmine, yellow jessamine, yellow jasmine
Gentiana lutea	gentian, yellow gentian, bitter root
Geranium maculatum	wild geranium
Ginkgo biloba	ginkgo, maidenhair tree
Glycine max	soybean, soy
Glycyrrhiza glabra	licorice
Gossypium hirsutum	cotton
Gracilaria spp.	red algae
Gratiola officinalis	hedge hyssop
Grifola frondosa	maitake mushroom, hen of the woods
Gymnema sylvestre	gurmar, sugar destroyer, cow plant
Gynostemma pentaphyllum	jiaogulan
Hamamelis virginiana	witch hazel
Hebanthe eriana (also known as *Pfaffia paniculata*)	suma, Brazilian ginseng
Hedeoma patens	oregano chiquito
Helianthus tuberosus	Jerusalem artichoke
Hibiscus rosa-sinensis	Chinese hibiscus
Hibiscus sabdariffa	roselle, flor de Jamaica, hibiscus
Hippophae rhamnoides	sea buckthorn
Hizikia fusiformis (also known as *Sargassum fusiforme*)	hijiki
Humulus lupulus	hops, common hop
Hydrastis canadensis	goldenseal
Hypericum perforatum	St. Johnswort, St. John's wort
Hyssopus officinalis	hyssop
Inonotus obliquus	chaga mushroom
Inula helenium	elecampane
Iris versicolor, I. tenax	wild iris, blue flag, vegetable mercury
Jasminum sambac	Arabian jasmine
Juniperus communis	juniper
Juniperus sabina	savin juniper
Laminaria japonica	kelp, ma-kombu
Larrea tridentata	chaparral, creosote bush
Lavandula angustifolia (also known as *L. officinalis*)	lavender
Lentinus edodes	shiitake mushroom
Leonurus cardiaca	motherwort
Lepidium meyenii	maca
Ligusticum striatum (also known as *L. chuanxiong, L. wallichii*)	Szechuan lovage, chuanxiong, ligusticum
Linum usitatissimum	flax
Lithospermum officinale	gromwell
Lithospermum erythrorhizon	purple gromwell, red gromwell
Lithospermum ruderale	stoneseed
Lobelia inflata	Indian tobacco, pukeweed
Lomatium dissectum	biscuitroot
Lycium barbarum, L. chinense	goji, goji berry, wolfberry
Lycopus europaeus	gypsywort
Lycopus virginicus	bugleweed
Magnolia grandiflora	magnolia
Mahonia aquifolium (also known as *Berberis aquifolium*)	Oregon grape, mahonia
Mangifera indica	mango
Matricaria chamomilla, M. recutita	chamomile
Medicago sativa	alfalfa, lucerne
Melaleuca alternifolia	tea tree
Melissa officinalis	lemon balm
Mentha piperita	peppermint
Mentha pulegium	pennyroyal
Mentha spicata	spearmint
Mitchella repens	partridge berry
Momordica charantia	bitter melon
Mucuna pruriens	velvet bean, cowage, cow-itch
Nigella sativa	black cumin, black seed
Ocimum tenuiflorum (also known as *O. sanctum*)	holy basil, tulsi

Oplopanax horridum	Devil's club
Opuntia ficus-indica (also known as *O. vulgaris*), *O. streptacantha*	prickly pear
Origanum vulgare	oregano
Paeonia × suffruticosa	moutan, peony, tree peony
Paeonia lactiflora	white peony root, bai shao yao, red peony root
Panax ginseng	ginseng, ren shen
Panax notoginseng	san qi, notoginseng
Panax quinquefolius	North American ginseng
Passiflora incarnata	passionflower
Paullinia cupana	guarana
Pausinystalia yohimbe	yohimbe
Perilla frutescens	shiso, perilla, Korean perilla
Petroselinum crispum	parsley
Pfaffia paniculata (also known as *Hebanthe eriana*)	suma, Brazilian ginseng
Phaseolus spp.	beans
Phellodendron amurense	Amur cork tree
Phyllanthus emblica	amla, amalaki, Indian gooseberry
Phyllanthus niruri	stonebreaker
Phytolacca americana (also known as *P. decandra*)	pokeweed, pokeroot, poke
Pilocarpus microphyllus	jaborandi
Pinellia ternata	crow dipper, ban xia
Piper methysticum	kava kava
Piper nigrum	black pepper
Piscidia piscipula (also known as *P. erythrina*)	Jamaican dogwood
Plantago ovata	plantain, ispaghula
Plantago spp.	plantain, psyllium
Podophyllum peltatum	mayapple
Polygonum multiflorum (also known as *Reynoutria multiflora*)	fo ti, he shou wu, Chinese knotweed
Poria cocos	hoelen mushroom, fu ling
Pouteria lucuma	lucuma
Prunella vulgaris	common self-heal, heal-all
Prunus africana (also known as *Pygeum africanum*)	African plum tree
Prunus spp., *P. serotina*	cherry, black cherry
Pterocarpus marsupium	kino tree
Ptychopetalum olacoides	muira puama
Pueraria montana var. *chinensis* (also known as *P. thomsonii*)	kudzu
Pueraria montana var. *lobata*	kudzu, gegen, Japanese arrowroot
Pueraria candollei var. *mirifica*	kudzu, white kwao krua
Pueraria tuberosa	kudzu, Indian kudzu
Punica granatum	pomegranate
Pygeum africanum (also known as *Prunus africana*)	African plum tree
Quercus alba	white oak
Quercus brantii (also known as *Q. persica*)	Brant's oak, Persian oak
Raphanus sativus var. *niger*	Spanish black radish, black Spanish radish
Rauvolfia serpentina, *R. vomitoria*	Indian snakeroot
Rehmannia glutinosa	di huang, Chinese foxglove
Reynoutria multiflora (also known as *Polygonum multiflorum*)	fo ti, he shou wu, Chinese knotweed
Rhamnus purshiana	sacred bark, cascara
Rhodiola kirilowii, *R. crenulata*	rhodiola
Rhodiola quadrifida	rhodiola
Rhodiola rosea	arctic rose, rhodiola
Ricinus communis	castor, castor bean
Rosa canina	dog rose
Rosa damascena	damask rose
Rosmarinus officinalis	rosemary
Rubus idaeus	raspberry, red raspberry
Rumex crispus	yellowdock, curly dock
Ruta graveolens	rue
Salix alba	white willow
Salvia hispanica	chia, salba

Salvia miltiorrhiza	red sage, dan shen
Salvia officinalis	sage
Salvia rhytidea	Persian sage
Sambucus nigra	elderberry, elder, black elder
Sanguinaria canadensis	bloodroot, puccoon
Sargassum fulvellum	brown algae
Sargassum fusiforme (also known as *Hizikia fusiformis*)	hijiki
Sargassum muticum	Japanese wireweed
Sassafras spp.	sassafras
Schisandra chinensis	magnolia vine, wu wei zi, five flavor fruit
Scutellaria baicalensis	scute, huang qin, baical
Scutellaria barbata	barbed skullcap
Scutellaria lateriflora	skullcap
Secale cornutum	spurred rye, ergot
Selenicereus grandiflora	night blooming cactus
Senecio aureus	golden ragwort
Serenoa repens	saw palmetto
Sesamum indicum	black sesame seed
Silybum marianum	milk thistle
Sinapis alba (also known as *Brassica alba*)	mustard seed, white mustard
Siraitia grosvenorii (also known as *Momordica grosvenorii*)	monk fruit, luo han guo
Smallanthus sonchifolius	yacon
Smilax ornata, S. regelii	sarsaparilla, Honduran sarsaparilla, Jamaican sarsaparilla
Smilax spp.	sarsaparilla
Solidago chilensis (also known as *S. repens*)	goldenrod
Sphaeranthus indicus	East Indian globe thistle
Stellaria media	chickweed
Stevia rebaudiana	stevia, sweet leaf
Symphytum officinale	comfrey
Syzygium aromaticum	cloves
Syzygium cumini (also known as *S. jambolanum*)	Java plum, roseapple, jambul

Tamarindus indica	tamarind
Tanacetum parthenium	feverfew
Taraxacum officinale	dandelion
Taxus brevifolia	Pacific yew
Terminalia arjuna	arjuna
Terminalia chebula	myrobalan, haritaki
Terminalia catappa	tropical almond
Terminalia paniculata	Kindal tree
Teucrium polium	mountain germander
Thuja occidentalis	northern white cedar
Thuja plicata	western red cedar
Thymus vulgaris	thyme
Tribulus terrestris	puncture vine
Trifolium pratense	red clover
Trigonella foenum-graecum	fenugreek
Turnera diffusa (also known as *Turnera aphrodisiaca*)	damiana
Ulmus fulva	slippery elm
Undaria pinnatifida	wakame
Urtica dioica, U. urens	nettle, stinging nettle
Usnea barbata	old man's beard, tree moss
Vaccinium myrtillus	blueberry, bilberry
Valeriana officinalis	valerian
Valeriana sitchensis	Sitka valerian
Veratrum viride, Veratrum alba	false hellebore, corn lily
Verbascum thapsus	mullein
Verbena hastata	vervain
Viburnum opulus	cramp bark, guelder rose, snowball bush,
Viburnum prunifolium	blackhaw
Viscum album	mistletoe
Vitex agnus-castus	chaste tree berry, vitex
Vitex negundo	Chinese chaste tree
Vitis vinifera	common grape vine
Withania somnifera	ashwagandha
Zingiber officinale	ginger
Ziziphora clinopodioides	blue mint bush

— COMMON NAMES —
TO SCIENTIFIC NAMES

The following lists include all of the herbs, medicinal fungi, and homeopathic preparations mentioned in the text of this book.

aconite	*Aconitum napellus*
African plum tree	*Prunus africana* (also known as *Pygeum africanum*)
agrimony	*Agrimonia eupatoria*
alfalfa	*Medicago sativa*
aloe	*Aloe vera, A. barbadensis*
amalaki	*Phyllanthus emblica*
amla	*Phyllanthus emblica*
Amur cork tree	*Phellodendron amurense*
andrographis	*Andrographis paniculata*
angelica	*Angelica sinensis*
Arabian jasmine	*Jasminum sambac*
arctic rose	*Rhodiola rosea*
arjuna	*Terminalia arjuna*
artichoke	*Cynara scolymus*
ashwagandha	*Withania somnifera*
bai shao yao	*Paeonia lactiflora*
baical	*Scutellaria baicalensis*
ban xia	*Pinellia ternata*
barbed skullcap	*Scutellaria barbata*
bean tree	*Azadirachta indica*
beans	*Phaseolus* spp.
beet	*Beta vulgaris*
belladonna	*Atropa belladonna*
bilberry	*Vaccinium myrtillus*
biscuitroot	*Lomatium dissectum*
bishop's hat	*Epimedium brevicornu* (also known as *E. rotundatum*)
bishop's weed	*Ammi visnaga*
bitter apple	*Citrullus colocynthis*
bitter cucumber	*Citrullus colocynthis*
bitter melon	*Momordica charantia*
bitter orange	*Citrus aurantium*
bitter root	*Gentiana lutea*
black cherry	*Prunus* spp., *P. serotina*
black cohosh	*Actaea racemosa* (also known as *Cimicifuga racemosa*)
black cumin	*Nigella sativa*
black elder	*Sambucus nigra*
black pepper	*Piper nigrum*
black seed	*Nigella sativa*
black sesame seed	*Sesamum indicum*
black Spanish radish	*Raphanus sativus* var. *niger*
blackhaw	*Viburnum prunifolium*
bladderwrack	*Fucus vesiculosus*
blessed thistle	*Cnicus benedictus*
bloodroot	*Sanguinaria canadensis*
blue cohosh	*Caulophyllum thalictroides*
blue flag	*Iris versicolor, I. tenax*
blue mint bush	*Ziziphora clinopodioides*
blueberry	*Vaccinium myrtillus*
Brant's oak	*Quercus brantii* (also known as *Q. persica*)
Brazilian ginseng	*Hebanthe eriana* (also known as *Pfaffia paniculata*)
brown algae	*Sargassum fulvellum*
bryony	*Bryonia dioica*
bugleweed	*Lycopus virginicus*
bupleurum	*Bupleurum flavum*

burdock	*Arctium lappa*
burning bush	*Euonymus alatus*
calendula	*Calendula officinalis*
California poppy	*Eschscholzia californica*
Canada fleabane	*Erigeron canadensis*
Canadian horseweed	*Erigeron canadensis*
canola	*Brassica napus, B. juncea, B. rapa*
carob	*Ceratonia siliqua, Ceratonia* spp.
cascara	*Rhamnus purshiana*
cassia cinnamon	*Cinnamomum cassia*
castor	*Ricinus communis*
castor bean	*Ricinus communis*
caterpillar fungus	*Cordyceps militaris, C. sinensis*
cayenne	*Capsicum annuum* (also known as *C. frutescens*)
celery	*Apium graveolens*
Ceylon cinnamon	*Cinnamomum verum* (also known as *C. zeylanicum*)
chaga mushroom	*Inonotus obliquus*
chai hu	*Bupleurum chinense, B. flavum*
chamomile	*Matricaria chamomilla, M. recutita*
chaparral	*Larrea tridentata*
chaste tree berry	*Vitex agnus-castus*
cherry	*Prunus* spp., *P. serotina*
chia	*Salvia hispanica*
chickweed	*Stellaria media*
China tree	*Azadirachta indica*
Chinese chaste tree	*Vitex negundo*
Chinese cinnamon	*Cinnamomum cassia*
Chinese foxglove	*Rehmannia glutinosa*
Chinese hibiscus	*Hibiscus rosa-sinensis*
Chinese knotweed	*Reynoutria multiflora* (also known as *Polygonum multiflorum*)
Chinese thoroughwax	*Bupleurum chinense, Bupleurum falcatum*
chuanxiong	*Ligusticum striatum* (also known as *L. chuanxiong, L. wallichii*)

cinnamon	*Cinnamomum verum* (also known as *C. zeylanicum*)
ciwujia	*Eleutherococcus senticosus* (also known as *Acanthopanax senticosus*)
cleavers	*Galium aparine*
cloves	*Syzygium aromaticum*
coconut	*Cocos nucifera*
coffee	*Coffea arabica*
coleus	*Coleus forskohlii* (also known as *Plectranthus forskohlii*)
comfrey	*Symphytum officinale*
common grape vine	*Vitis vinifera*
common hop	*Humulus lupulus*
common self-heal	*Prunella vulgaris*
coneflower	*Echinacea angustifolia, Echinacea purpurea*
cordyceps	*Cordyceps militaris, C. sinensis*
corn lily	*Veratrum viride, V. alba*
corydalis	*Corydalis yanhusuo*
cotton	*Gossypium hirsutum*
cow plant	*Gymnema sylvestre*
cow-itch	*Mucuna pruriens*
cowage	*Mucuna pruriens*
cramp bark	*Viburnum opulus*
creosote bush	*Larrea tridentata*
crow dipper	*Pinellia ternata*
curly dock	*Rumex crispus*
damask rose	*Rosa damascena*
damiana	*Turnera diffusa* (also known as *Turnera aphrodisiaca*)
dan shen	*Salvia miltiorrhiza*
dandelion	*Taraxacum officinale*
deadly nightshade	*Atropa belladonna*
Devil's club	*Oplopanax horridum*
di huang	*Rehmannia glutinosa*
dodder	*Cuscuta reflexa*
dog rose	*Rosa canina*
dong quai	*Angelica sinensis*

duan e huang lian	*Coptis chinensis*
East Indian globe thistle	*Sphaeranthus indicus*
elder	*Sambucus nigra*
elderberry	*Sambucus nigra*
elecampane	*Inula helenium*
eleuthero	*Eleutherococcus senticosus* (also known as *Acanthopanax senticosus*)
ergot	*Secale cornutum*
eucalyptus	*Eucalpytus globulus*
eyebright	*Euphrasia stricta*
false hellebore	*Veratrum viride, V. alba*
false jasmine	*Gelsemium sempervirens*
false unicorn	*Chamaelirium luteum* (also known as *Helonias*)
fennel	*Foeniculum vulgare*
fenugreek	*Trigonella foenum-graecum*
feverfew	*Tanacetum parthenium*
five flavor fruit	*Schisandra chinensis*
flax	*Linum usitatissimum*
flor de Jamaica	*Hibiscus sabdariffa*
fly agaric	*Amanita muscaria*
fo ti	*Reynoutria multiflora* (also known as *Polygonum multiflorum*)
forsythia	*Forsythia suspensa*
foxglove	*Digitalis purpurea*
frankincense	*Boswellia serrata*
fringe tree	*Chionanthus virginicus*
fu ling	*Poria cocos*
galangal	*Alpinia galanga, A. officinarum*
garlic	*Allium sativum*
gegen	*Pueraria montana* var. *lobata*
gentian	*Gentiana lutea*
giant dodder	*Cuscuta reflexa*
ginger	*Zingiber officinale*
ginkgo	*Ginkgo biloba*
ginseng	*Panax ginseng*

goat's rue	*Galega officinalis*
gobo root	*Arctium lappa*
goji	*Lycium barbarum, L. chinense*
goji berry	*Lycium barbarum, L. chinense*
golden ragwort	*Senecio aureus*
goldenrod	*Solidago chilensis* (also known as *S. repens*)
goldenseal	*Hydrastis canadensis*
goldthread	*Coptis chinensis, C. trifolia*
gotu kola	*Centella asiatica*
grapefruit	*Citrus paradisi*
gray mangrove	*Avicennia marina*
green tea	*Camellia sinensis*
gromwell	*Lithospermum officinale*
guarana	*Paullinia cupana*
guelder rose	*Viburnum opulus*
guggul	*Commiphora mukul, C. wightii*
guggulu	*Commiphora mukul, C. wightii*
gurmar	*Gymnema sylvestre*
gypsywort	*Lycopus europaeus*
haritaki	*Terminalia chebula*
hawthorn	*Crataegus laevigata, C. monogyna, C. oxyacantha*
he shou wu	*Reynoutria multiflora* (also known as *Polygonum multiflorum*)
heal-all	*Prunella vulgaris*
hedge hyssop	*Gratiola officinalis*
hemlock	*Conium maculatum*
hen of the woods	*Grifola frondosa*
hibiscus	*Hibiscus sabdariffa*
hijiki	*Sargassum fusiforme* (also known as *Hizikia fusiformis*)
hoelen mushroom	*Poria cocos*
holy basil	*Ocimum tenuiflorum* (also known as *O. sanctum*)
Honduran sarsaparilla	*Smilax ornata, S. regelii*
hops	*Humulus lupulus*

horny goatweed	*Epimedium brevicornu* (also known as *E. rotundatum*), *E. grandiflorum*, *E. koreanum*, *E. pubescens*
horse chestnut	*Aesculus hippocastanum*
horseradish	*Armoracia rusticana*
horsetail	*Equisetum arvense*, *E. hyemale*
huang qi	*Astragalus membranaceus*
huang qin	*Scutellaria baicalensis*
hyssop	*Hyssopus officinalis*
Indian frankincense	*Boswellia serrata*
Indian gooseberry	*Phyllanthus emblica*
Indian kudzu	*Pueraria tuberosa*
Indian snakeroot	*Rauvolfia serpentina*, *R. vomitoria*
Indian tobacco	*Lobelia inflata*
ispaghula	*Plantago ovata*
jaborandi	*Pilocarpus microphyllus*
Jamaican dogwood	*Piscidia piscipula* (also known as *P. erythrina*)
Jamaican sarsaparilla	*Smilax ornata*, *S. regelii*
jambul	*Syzygium cumini* (also known as *S. jambolanum*)
Japanese arrowroot	*Pueraria montana* var. *lobata*
Japanese wireweed	*Sargassum muticum*
Java plum	*Syzygium cumini* (also known as *S. jambolanum*)
Jerusalem artichoke	*Helianthus tuberosus*
jiaogulan	*Gynostemma pentaphyllum*
Job's tears	*Coix lacryma-jobi*
juniper	*Juniperus communis*
kava kava	*Piper methysticum*
kelp	*Fucus vesiculosus*
kelp	*Laminaria japonica*
khella	*Ammi visnaga*
Kindal tree	*Terminalia paniculata*
king of bitters	*Andrographis paniculata*
kino tree	*Pterocarpus marsupium*
Korean perilla	*Perilla frutescens*
kudzu	*P. montana* var. *chinensis* (also known as *P. thomsonii*), *Pueraria montana* var. *lobata*, *P. candollei* var. *mirifica*, *P. tuberosa*
lady's mantle	*Alchemilla vulgaris*
lavender	*Lavandula angustifolia* (also known as *L. officinalis*)
lebbeck tree	*Albizia lebbeck*
lemon balm	*Melissa officinalis*
lemon grass	*Cymbopogon citratus*
licorice	*Glycyrrhiza glabra*
ligusticum	*Ligusticum striatum* (also known as *L. chuanxiong*, *L. wallichii*)
lily of the valley	*Convallaria majalis*
lucerne	*Medicago sativa*
lucuma	*Pouteria lucuma*
luo han guo	*Siraitia grosvenorii* (also known as *Momordica grosvenorii*)
ma-kombu	*Laminaria japonica*
maca	*Lepidium meyenii*
macrotys	*Actaea racemosa* (also known as *Cimicifuga racemosa*)
magnolia	*Magnolia grandiflora*
magnolia vine	*Schisandra chinensis*
mahonia	*Berberis aquifolium* (also known as *Mahonia aquifolium*), *B. nervosa*
maidenhair tree	*Ginkgo biloba*
maitake mushroom	*Grifola frondosa*
mandarin	*Citrus reticulata*
mango	*Mangifera indica*
marshmallow	*Althaea officinalis*
mayapple	*Podophyllum peltatum*
meadowsweet	*Filipendula ulmaria*
milk thistle	*Silybum marianum*
milk vetch	*Astragalus membranaceus*
mimosa	*Albizia julibrissin*
mistletoe	*Viscum album*
monk fruit	*Siraitia grosvenorii* (also known as *Momordica grosvenorii*)

motherwort	*Leonurus cardiaca*
mountain germander	*Teucrium polium*
moutan	*Paeonia × suffruticosa*
mugwort	*Artemisia vulgaris*
muira puama	*Ptychopetalum olacoides*
mullein	*Verbascum thapsus*
mum	*Chrysanthemum morifolium*
mustard seed	*Sinapis alba* (also known as *Brassica alba*)
myrobalan	*Terminalia chebula*
myrrh	*Commiphora molmol, C. myrrha*
neem	*Azadirachta indica*
nettle	*Urtica dioica, U. urens*
New Jersey tea	*Ceanothus americanus*
night blooming cactus	*Selenicereus grandiflora*
North American ginseng	*Panax quinquefolius*
northern white cedar	*Thuja occidentalis*
notoginseng	*Panax notoginseng*
oats	*Avena sativa*
old man's beard	*Usnea barbata*
onion	*Allium cepa*
orange	*Citrus sinensis*
oregano	*Origanum vulgare*
oregano chiquito	*Hedeoma patens*
Oregon grape	*Berberis aquifolium* (also known as *Mahonia aquifolium*), *B. nervosa*
osthole	*Cnidium monnieri*
Pacific yew	*Taxus brevifolia*
parsley	*Petroselinum crispum*
partridge berry	*Mitchella repens*
pasque flower	*Anemone pulsatilla*
passionflower	*Passiflora incarnata*
pennyroyal	*Mentha pulegium*
pennywort	*Centella asiatica*
peony	*Paeonia × suffruticosa*

peppermint	*Mentha piperita*
perilla	*Perilla frutescens*
Persian oak	*Quercus brantii* (also known as *Q. persica*)
Persian sage	*Salvia rhytidea*
plantain	*Plantago* spp., *P. ovata*
poison hemlock	*Conium maculatum*
poke	*Phytolacca americana* (also known as *P. decandra*)
pokeroot	*Phytolacca americana* (also known as *P. decandra*)
pokeweed	*Phytolacca americana* (also known as *P. decandra*)
pomegranate	*Punica granatum*
pot marigold	*Calendula officinalis*
prickly pear	*Opuntia ficus-indica* (also known as *O. vulgaris*), *O. streptacantha*
psyllium	*Plantago* spp.
puccoon	*Sanguinaria canadensis*
pukeweed	*Lobelia inflata*
pumpkin	*Cucurbita pepo*
puncture vine	*Tribulus terrestris*
purple coneflower	*Echinacea purpurea*
purple gromwell	*Lithospermum erythrorhizon*
qing hao su	*Artemisia absinthium*
qinghao	*Artemisia annua*
rapeseed	*Brassica napus, B. juncea, B. rapa*
raspberry	*Rubus idaeus*
red algae	*Gelidium* spp., *Gracilaria* spp.
red clover	*Trifolium pratense*
red gromwell	*Lithospermum erythrorhizon*
red peony root	*Paeonia lactiflora*
red raspberry	*Rubus idaeus*
red root	*Ceanothus americanus*
red sage	*Salvia miltiorrhiza*
reishi	*Ganoderma lucidum*
reishi mushroom	*Ganoderma lucidum*
ren shen	*Panax ginseng*

rhodiola	*Rhodiola kirilowii, R. crenulata, R. quadrifida, R. rosea*
roseapple	*Syzygium cumini* (also known as *S. jambolanum*)
roselle	*Hibiscus sabdariffa*
rosemary	*Rosmarinus officinalis*
rue	*Ruta graveolens*
sacred bark	*Rhamnus purshiana*
safflower	*Carthamus tinctorius*
sage	*Salvia officinalis*
Saigon cinnamon	*Cinnamomum polyadelphum* (also known as *C. saigonicum*)
saiko	*Bupleurum falcatum*
salba	*Salvia hispanica*
san qi	*Panax notoginseng*
sarsaparilla	*Smilax ornata, S. regelii, Smilax* spp.
sassafras	*Sassafras* spp.
savin juniper	*Juniperus sabina*
saw palmetto	*Serenoa repens*
scouring rush	*Equisetum arvense, E. hyemale*
scute	*Scutellaria baicalensis*
sea buckthorn	*Hippophae rhamnoides*
shatavari	*Asparagus racemosus*
shengma	*Actaea cimicifuga* (also known as *Cimicifuga foetida*), *A. dahurica, A. heracleifolia, Actaea* spp.
shepherd's purse	*Capsella bursa-pastoris*
shiitake mushroom	*Lentinus edodes*
shiso	*Perilla frutescens*
Siberian ginseng	*Eleutherococcus senticosus* (also known as *Acanthopanax senticosus*)
silk tree	*Albizia julibrissin*
siris	*Albizia lebbeck*
Sitka valerian	*Valeriana sitchensis*
skullcap	*Scutellaria lateriflora*
slippery elm	*Ulmus fulva*
snow parsley	*Cnidium monnieri*
snowball bush	*Viburnum opulus*
southernwood	*Artemisia abrotanum*
soy	*Glycine max*
soybean	*Glycine max*
Spanish black radish	*Raphanus sativus* var. *niger*
spearmint	*Mentha spicata*
spurred rye	*Secale cornutum*
St. John's wort	*Hypericum perforatum*
St. Johnswort	*Hypericum perforatum*
stevia	*Stevia rebaudiana*
stinging nettle	*Urtica dioica, U. urens*
stonebreaker	*Phyllanthus niruri*
stoneroot	*Collinsonia canadensis*
stoneseed	*Lithospermum ruderale*
sugar destroyer	*Gymnema sylvestre*
suma	*Hebanthe eriana* (also known as *Pfaffia paniculata*)
sweet Annie	*Artemisia annua*
sweet leaf	*Stevia rebaudiana*
sweet wormwood	*Artemisia absinthium, A. annua*
Szechuan lovage	*Ligusticum striatum* (also known as *L. chuanxiong, L. wallichii*)
tamarind	*Tamarindus indica*
tea tree	*Melaleuca alternifolia*
thyme	*Thymus vulgaris*
tongkat ali	*Eurycoma longifolia*
tree moss	*Usnea barbata*
tree peony	*Paeonia × suffruticosa*
tropical almond	*Terminalia catappa*
tulsi	*Ocimum tenuiflorum* (also known as *O. sanctum*)
turkey corn	*Corydalis yanhusuo*
turmeric	*Curcuma longa*
valerian	*Valeriana officinalis*
vegetable mercury	*Iris versicolor, I. tenax*
velvet bean	*Mucuna pruriens*
vervain	*Verbena hastata*
vitex	*Vitex agnus-castus*

wakame	*Undaria pinnatifida*	wild yam	*Dioscorea villosa*
wax gourd	*Benincasa hispida*	witch hazel	*Hamamelis virginiana*
western red cedar	*Thuja plicata*	wolfberry	*Lycium barbarum, Lyciuim chinense*
white bryony	*Bryonia alba*	wolfsbane	*Aconitum napellus*
white button mushrooms	*Agaricus* spp.	wu wei zi	*Schisandra chinensis*
white kwao krua	*Pueraria candollei* var. *mirifica*	yacon	*Smallanthus sonchifolius*
		yarrow	*Achillea millefolium*
white mangrove	*Avicennia marina*	yellow gentian	*Gentiana lutea*
white mustard	*Sinapis alba* (also known as *Brassica alba*)	yellow jasmine	*Gelsemium sempervirens*
		yellow jessamine	*Gelsemium sempervirens*
white oak	*Quercus alba*	yellowdock	*Rumex crispus*
white peony root	*Paeonia lactiflora*	yin yang huo	*Epimedium brevicornu* (also known as *E. rotundatum*)
white willow	*Salix alba*		
wild geranium	*Geranium maculatum*	yohimbe	*Pausinystalia yohimbe*
wild iris	*Iris versicolor, I. tenax*	zhe bei mu	*Fritillaria thunbergii*

— GLOSSARY —
OF THERAPEUTIC TERMS

Abortifacient. An agent capable of promoting the expulsion of a developing fetus.

Absorbent. A drug that promotes the absorption of medicinal compounds.

Acidifier. An agent imparting acidity to body fluids, especially blood and urine.

Acute. A condition that has a new onset, comes on suddenly, and is relatively short-lasting in its entire duration.

Adaptogen. An agent that helps one to "adapt," a term stemming from research on herbs that improve resilience, immunity, and stress tolerance. The word has come to refer to herbs such as *Panax ginseng, Eleutherococcus, Withania,* and *Glycyrrhiza* that optimize HPA axis feedback loops and adrenal function.

Aerial parts. The parts of a plant that grow above ground.

Alkalinizer. An agent that increases the alkalinity of bodily fluids, especially the blood and urine.

Allopathic. A term applying to conventional, modern Western medicine. *Allo* refers to "opposite," and in this case, means to oppose pathology. For example: In cases of fever, an antipyretic is used; to treat inflammation, an anti-inflammatory is used; and to treat an infection, antimicrobials are used.

Alterative. An agent that favorably "alters" an individual's health. Alteratives stimulate digestive and absorptive functions while enhancing elimination of wastes. Alteratives are also traditionally said to "purify" the blood and optimize metabolic functions.

Amphoteric. An agent capable of both increasing or decreasing activity, such as estrogenic activity, in the body; a single compound that can both promote or reduce hormonal or other activity depending on the physiologic situation.

Analgesic. An agent that is pain-relieving.

Anaphrodisiac. An agent that diminishes sexual drive or function.

Anesthetic. An agent that diminishes pain and tactile sensations temporarily.

Anhydrotic. An agent that diminishes excessive sweating.

Anodyne. An agent that is pain-relieving.

Antacid. An agent that diminishes stomach acid.

Antagonist. An agent that opposes the action of some other medicine, usually a poison or toxic alkaloid.

Anthelmintic. An agent used to combat intestinal worms.

Antidote. A remedy to counteract the action of poisons or other strong actions.

Antiemetic. An agent that allays nausea and vomiting.

Antigalactogogue. An agent that diminishes lactation.

Antihemorrhagic. An agent that helps control excessive bleeding.

Anti-inflammatory. An agent that reduces inflammatory processes by a variety of mechanisms, reducing oxidative stress and protecting tissues from stress and damage.

Antilithic. An agent used to reduce the formation of stones and calculi in the body.

Antioxidant. An agent capable of accepting electrons or highly reactive molecules that could damage body membranes if left free in circulation.

Antiperiodic. An agent used to combat the periodic fevers of malaria.

Antiphlogistic. An agent used to reduce fever and inflammation.

Antiproliferative. An agent capable of reducing hormone-driven proliferation in hormone-sensitive tissues or of reducing other excessive stimulation of tissues (such as the endometrium or prostate) that can lead to enlargement or an increased risk of cancer.

Antipyretic. An agent used to reduce fever.

Antiscorbutic. An agent used to provide vitamin C and prevent or treat scurvy.

Antiseptic. An agent having antimicrobial capacity for the prevention of sepsis.

Antisialagogue. An agent capable of reducing salivation.

Antispasmodic. An agent capable of reducing painful spasms in muscles and hollow organs.

Antitussive. An agent used to diminish coughing.

Anxiolytic. An agent that reduces anxiety.

Aperient. A gentle nonirritating laxative.

Aphrodisiac. An agent used to stimulate the libido.

Aromatic. An agent with a strong fragrance to be inhaled or absorbed through the skin.

Astringent. An agent that dries, condenses, and shrinks inflamed or suppurative tissues.

Bitter. An agent that has a bitter flavor and is used to stimulate gastrointestinal tone and secretions. Bitters prepare mucosa for food, stimulate appetite, and enhance digestion.

Blood mover. A term from Traditional Chinese Medicine (TCM) used to refer to agents capable of improving circulation and relieving blood stagnation and tissue congestion.

Cardiotonic. An agent that improves heart function.

Carminative. An agent that reduces gas, bloating, flatulence, and associated pain.

Cathartic. A strong, potentially harsh laxative.

Caustic. An agent having a corrosive action on tissues.

Chi tonic. A term from TCM in which chi refers to the body's vital energy. Chi tonics are herbs purported to increase and support vitality, longevity, stamina, fertility, and other aspects of the vital force. Chi deficiency manifests as low energy and stamina, a weak voice, and coldness, as well as exercise intolerance, general fatigue, shortness of breath, and dizziness.

Cholagogue. An agent that increases gallbladder tone and the flow of bile from the gallbladder.

Choleretic. An agent that increases the production of bile.

Chronic. A condition that develops slowly over time and becomes persistent and sometimes permanent.

Corrigent. An agent that balances a harsh or strong action of another agent, a corrective.

Counterirritant. An agent that irritates local tissues to enhance blood flow to the area. Counterirritants are used to induce temporary hyperemia in chronic conditions in an attempt to relieve pain, promote healing, and reduce inflammation.

Dacryagogue. An agent that promotes the flow of tears (lacrimation).

Dampness. A term used in TCM and other energetic descriptions of physiologic tissue states that refers to fluid stagnation when evidenced by a coated tongue, chronic phlegm in the mucous membranes, fluid stagnation, and an increased tendency to opportunistic infections.

Deficient. Referring to low energy, low vitality, and poor-functioning tissues (herbal medicine). In Traditional Chinese Medicine, the term chi deficiency is used when the entire body is in a weakened state. The term may also be used to indicate a poorly functioning organ or biochemical state, such as digestive deficiency, circulatory deficiency, or metabolic deficiency.

Demulcent. A cooling, soothing, mucilagenous substance used internally or topically to emolliate abraded, inflamed, or irritated mucosal tissues.

Depurant. Any agent aimed at purifying, such as a liver depurant, a renal depurant, a blood depurant, and so on. Depurants have a purifying effect by promoting the elimination of wastes from the body.

Diaphoretic. An agent capable of inducing perspiration and often a temporary fever.

Diuretic. An agent that stimulates the production and flow of urine.

Ecbolic. An agent that stimulates childbirth (parturition).

Emetic. An agent that causes vomiting (emesis).

Emmenagogue. An agent that promotes menstrual flow.

Emollient. An agent that soothes and softens the skin and mucosal tissues.

Errhine. An agent that irritates the nasal mucosa and promotes sneezing and secretions.

Escharotic. Any caustic substances applied topically to diseased tissues to kill the cells and promote sloughing away. The word *eschar* means to cast off.

Essential oil. See Volatile oil.

Excess. Indicates a condition beyond normal range, such as too much heat, overstimulated bowel or muscle tone, or other situations of excess in various physiologic functions.

Excitant. An agent that causes excitation of nervous, circulatory, or motor functions; however, in Latin America, the term is more often used to refer to an aphrodisiac, or sexual excitant.

Exhilarant. An agent that causes excitation of psychic functions and promotes euphoria.

Exogenous estrogen. Exposure to estrogenic compounds that did not originate in the body, but influence physiology due to the ingestion of hormones in animal products or exposure to a variety of chemicals that bind to estrogen receptors in the body and have an estrogenic effect. Estrogens from outside the body.

Expectorant. An agent that promotes the flow of secretions from the respiratory tract.

Febrifuge. An agent used to bring down the temperature in cases of fever.

Fibrinolytic. An agent capable of breaking down fibrin, which may be deposited in vein and artery walls, as well as numerous tissues, in response to inflammatory processes.

GABAergic. GABA is the abbreviation for gamma amino butyric acid, one of the primary inhibitory neurotransmitters of the central nervous system, exerting a calming and relaxing affect. To match the terms *serotonergic* and *dopaminergic*, GABAergic is used to refer to promotion of GABA neurotransmission and signaling.

Galactogogue. An agent that stimulates lactation.

Hematic, hematinic. An agent that improves the quality of the blood, especially in cases of anemia, but may be used in other situations.

Hemostatic. An agent that reduces blood flow and promotes clotting in cases of hemorrhage, trauma, and internal bleeding.

Hepatic. An agent that improves the function of the liver.

Hepatotonic. An agent with a trophorestorative effect on the liver.

Hydragogue. An agent that promotes watery secretions.

Inotropic. An agent that supports ion flow in electrically active cardiac muscle and improves the contractile force of the heart, slowing and strengthening the heartbeat and improving circulation.

Irritant. An agent applied locally for the purposes of intentionally causing local hyperemia. See also Counterirritant.

Laxative. An agent that promotes a mild and painless evacuation of the bowels.

Lithotriptic. An agent aimed at dissolving calculi within the body.

Lipotropic. Literally translates as "fat mover," and used to refer to various alterative and cholagogue herbs, as well as substances such as choline that promote bile flow and biliary function, and thereby improve liver function, and hepatic clearance of lipids, carbohydrates, hormones, toxins, and chemicals.

Material dose. A term used by herbalists and alternative medicine practitioners to distinguish between a highly diluted or homeopathic preparation of a substance and the substance given in a more substantial or "material dose."

Miotic, myotic. An agent that causes the pupil to contract (miosis).

Mydriatic. An agent that promotes dilation of the pupil (mydriasis).

Narcotic. A drug that promotes stupor or sleep and is used to relieve pain or diminish consciousness.

Nervine. An agent having a tonifying effect on the nervous system, usually only used in the context of herbal medicine, with some herbs being referred to as nervine herbs.

Nutriant, nutrient, nutritive. An agent that enhances assimilation, metabolism, and nutrition.

Oxytocic. An agent that promotes uterine contractions and hastens childbirth.

Parturifacient. An agent that facilitates childbirth when taken during labor.

Partus preparator. An agent taken in the last months of pregnancy to tone the uterus and optimize labor and delivery.

Phytoestrogen. A steroid-like molecule occurring naturally in a plant and capable of exerting hormonal or other activity in animals via ligand activity at estrogen receptors.

Phytosterol. A steroid-like molecule occurring naturally in a plant and capable of exerting hormonal or other activity in animals.

Purgative. A strong laxative that may be irritating and cause cramping.

Refrigerant. An agent capable of imparting a cooling sensation when applied topically.

Revulsive. An agent used to enhance the blood flow to a particular body part (hand, foot) in order to draw it away from a congested, engorged area (head, uterus).

Rubefacient. An agent that promotes reddening or hyperemia of the tissues.

Sedative. An agent that calms in cases of nervousness, insomnia, and mania, and may be stronger and less tonifying than a nervine.

SERM. A selective estrogen response modifier, exerting effects at only particular estrogen receptors due to binding specific estrogen receptor types and subtypes, and variably in specific tissues.

Sialagogue, salivant. An agent that increases the flow of saliva.

Simple. A term used to refer to a single herb, not mixed with other herbs or used in a formula, but rather used as "a simple."

Specific. An agent thought to be of specific value for a collection of symptoms.

Sternutatory. An agent that promotes sneezing when inhaled.

Styptic. A strong astringent agent capable of reducing bleeding when applied topically.

Sudorific. An agent capable of inducing perspiration and regarded as being stronger than a diaphoretic.

Synergist. An agent that duplicates, enhances, or pulls together the action of a group of medicinal substances.

Taenicide. An agent that kills or weakens tapeworms.

Tonic. An agent that has a positive effect on the function of an organ or tissue and suggests an ability to restore normal function be it excess or deficient, atonic or hypertonic, overstimulated or understimulated. Tonic supports the optimal physiologic state.

Toxicity. A deranged, inflammatory, or otherwise corrupted or polluted biochemical state in the body.

Vasoconstrictor. An agent that constricts the blood vessels.

Vasodepressant. An agent that slows the pulse rate and lowers the pressure.

Vasodilator. An agent used to dilate the vasculature, usually used in cases of hypertension.

Vermifuge. An agent that promotes the expulsion of intestinal worms.

Vesicant. An agent that promotes blistering or vesication of the skin.

Volatile oil. An essential oil. Aromatic plants are high in volatile oils, so named due to the fact that they are small, light molecules that readily volatilize into the air, contributing to the aromatic quality. Volatile oils are often distilled out of aromatic plants such as mint, thyme, citrus, and numerous others and sold in small bottles to use in aromatherapy, to make body products, and for other purposes.

Xenoestrogen. A compound with an estrogenic effect on the tissue that originated from outside the body, such as an animal hormone, or a chemical having an estrogenic effect in the tissues.

— NOTES —

Introduction

1. Roni Jacobson, "Many Antidepressant Studies Found Tainted by Pharma Company Influence: A Review of Studies That Assess Clinical Antidepressants Shows Hidden Conflicts of Interest and Financial Ties to Corporate Drugmakers," *Scientific American*, October 21, 2015, https://www.scientificamerican.com/article/many-antidepressant-studies-found-tainted-by-pharma-company-influence/; Peter C. Gøtzsche et al., "Suicidality and Aggression During Antidepressant Treatment: Systematic Review and Meta-Analyses Based on Clinical Study Reports," *BMJ* 352 (2016): i65.

Chapter Two: Creating Herbal Formulas for Endocrine Conditions

1. C. M. Dutton et al., "Thyrotropin Receptor Expression in Adrenal, Kidney, and Thymus," *Thyroid: Official Journal of the American Thyroid Association* 7, no. 6 (1997): 879–84, https://doi.org/10.1089/thy.1997.7.879.
2. Jesús Morillo-Bernal et al., "Functional Expression of the Thyrotropin Receptor in C Cells: New Insights into Their Involvement in the Hypothalamic-Pituitary-Thyroid Axis," *Journal of Anatomy* 215, no. 2 (2009): 150–58, https://doi.org/10.1111/j.1469-7580.2009.01095.x.
3. Morillo-Bernal et al., "Functional expression of the Thyotropin Receptor in C Ce..s."
4. Antonio Mancini et al., "Thyroid Hormones, Oxidative Stress, and Inflammation," *Mediators of Inflammation* 2016 (2016): 6757154, https://doi.org/10.1155/2016/6757154.
5. Tânia M. Ortiga-Carvalho et al., "Thyroid Hormone Receptors and Resistance to Thyroid Hormone Disorders," *Nature Reviews Endocrinology* 10, no. 10 (2014): 582–91, https://doi.org/10.1038/nrendo.2014.143.
6. Shigenobu Nagataki, "The Average of Dietary Iodine Intake Due to the Ingestion of Seaweeds Is 1.2 mg/day in Japan," *Thyroid* 18, no. 6 (June 2008): 667–8, https://doi.org/10.1089/thy.2007.0379.
7. Robert Aquaron et al., "Bioavailability of Seaweed Iodine in Human Beings," *Cellular and Molecular Biology* 48, no. 5 (2002): 563–69, https://www.ncbi.nlm.nih.gov/pubmed/12146713.
8. X. Yan et al., "Fucoxanthin as the Major Antioxidant in Hijikia Fusiformis, a Common Edible Seaweed," *Bioscience, Biotechnology, and Biochemistry* 63, no. 3 (1999): 605–7, https://doi.org/10.1271/bbb.63.605.
9. A. Jiménez-Escrig et al., "Antioxidant Activity of Fresh and Processed Edible Seaweeds," *Journal of the Science of Food and Agriculture* 81, no. 5 (2001): 530–34, https://doi.org/10.1002/jsfa.842.
10. Albana Cumashi et al., "A Comparative Study of the Anti-Inflammatory, Anticoagulant, Antiangiogenic, and Antiadhesive Activities of Nine Different Fucoidans from Brown Seaweeds," *Glycobiology* 17, no. 5 (2007): 541–52, https://doi.org/10.1093/glycob/cwm014.
11. Hayato Maeda et al., "Fucoxanthinol, Metabolite of Fucoxanthin, Improves Obesity-Induced Inflammation in Adipocyte Cells," *Marine Drugs* 13, no. 8 (2015): 4799–4813, https://doi.org/10.3390/md13084799.
12. Hayato Maeda et al., "Fucoxanthin from Edible Seaweed, *Undaria pinnatifida*, Shows Antiobesity Effect through UCP1 Expression in White Adipose Tissues," *Biochemical and Biophysical Research Communications* 332, no. 2 (2005): 392–97, https://doi.org/10.1016/j.bbrc.2005.05.002; Hayato Maeda et al., "Seaweed Carotenoid, Fucoxanthin, as a Multi-Functional Nutrient," *Asia Pacific Journal of Clinical Nutrition* 17, Suppl 1 (2008): 196–99, https://www.ncbi.nlm.nih.gov/pubmed/18296336; Kh Muradian et al., "Fucoxanthin and Lipid Metabolism: A Minireview," *Nutrition, Metabolism, and Cardiovascular Diseases: NMCD* 25, no. 10 (2015): 891–97, https://doi.org/10.1016/j.numecd.2015.05.010.
13. Maria Alessandra Gammone and Nicolantonio D'Orazio, "Anti-Obesity Activity of the Marine Carotenoid Fucoxanthin," *Marine Drugs* 13, no. 4 (2015): 2196–2214, https://doi.org/10.3390/md13042196; Nicolantonio D'Orazio et al., "Fucoxantin: A Treasure from the Sea," *Marine Drugs* 10, no. 3 (2012): 604–16, https://doi.org/10.3390/md10030604.
14. Maria Alessandra Gammone and Nicolantonio D'Orazio, "Anti-Obesity Activity of the Marine Carotenoid Fucoxanthin," *Marine Drugs* 13, no. 4 (2015): 2196–2214, https://doi.org/10.3390/md13042196; Sho Nishikawa et al., "Fucoxanthin Promotes Translocation and Induction of Glucose Transporter 4 in Skeletal Muscles of Diabetic/Obese KK-A(y) Mice," *Phytomedicine: International Journal of Phytotherapy and Phytopharmacology* 19, no. 5 (2012): 389–94, https://doi.org/10.1016/j.phymed.2011.11.001.
15. Hayato Maeda, "Nutraceutical Effects of Fucoxanthin for Obesity and Diabetes Therapy: A Review," *Journal of Oleo Science* 64, no. 2 (2015): 125–32, https://doi.org/10.5650/jos.ess14226; Hayato Maeda et al., "Anti-Obesity and Anti-Diabetic Effects of Fucoxanthin on Diet-Induced Obesity Conditions in a Murine Model," *Molecular Medicine Reports* 2, no. 6 (2009): 897–902, https://doi.org/10.3892/mmr_00000189.

16. Xiaojie Hu et al., "Combination of Fucoxanthin and Conjugated Linoleic Acid Attenuates Body Weight Gain and Improves Lipid Metabolism in High-Fat Diet-Induced Obese Rats," *Archives of Biochemistry and Biophysics* 519, no. 1 (2012): 59–65, https://doi.org/10.1016/j.abb.2012.01.011; Seong-Il Kang et al., "*Petalonia binghamiae* Extract and Its Constituent Fucoxanthin Ameliorate High-Fat Diet-Induced Obesity by Activating AMP-Activated Protein Kinase," *Journal of Agricultural and Food Chemistry* 60, no. 13 (2012): 3389–95, https://doi.org/10.1021/jf2047652; Masashi Hosokawa et al., "Fucoxanthin Regulates Adipocytokine MRNA Expression in White Adipose Tissue of Diabetic/Obese KK-Ay Mice," *Archives of Biochemistry and Biophysics* 504, no. 1 (2010): 17–25, https://doi.org/10.1016/j.abb.2010.05.031; Kazuo Miyashita et al., "The Allenic Carotenoid Fucoxanthin, a Novel Marine Nutraceutical from Brown Seaweeds," *Journal of the Science of Food and Agriculture* 91, no. 7 (2011): 1166–74, https://doi.org/10.1002/jsfa.4353.

17. Seon-Min Jeon et al., "Fucoxanthin-Rich Seaweed Extract Suppresses Body Weight Gain and Improves Lipid Metabolism in High-Fat-Fed C57BL/6J Mice," *Biotechnology Journal* 5, no. 9 (2010): 961–69, https://doi.org/10.1002/biot.201000215.

18. Takayuki Tsukui et al., "Fucoxanthin and Fucoxanthinol Enhance the Amount of Docosahexaenoic Acid in the Liver of KKAy Obese/Diabetic Mice," *Journal of Agricultural and Food Chemistry* 55, no. 13 (2007): 5025–29, https://doi.org/10.1021/jf070110q.

19. Hayato Maeda et al., "Seaweed Carotenoid, Fucoxanthin, as a Multi-Functional Nutrient," *Asia Pacific Journal of Clinical Nutrition* 17 Suppl 1 (2008): 196–99, https://www.ncbi.nlm.nih.gov/pubmed/18296336.

20. Juan Peng et al., "Fucoxanthin, a Marine Carotenoid Present in Brown Seaweeds and Diatoms: Metabolism and Bioactivities Relevant to Human Health," *Marine Drugs* 9, no. 10 (2011): 1806–28, https://doi.org/10.3390/md9101806; Tomoko Okada et al., "Antiobesity Effects of Undaria Lipid Capsules Prepared with Scallop Phospholipids," *Journal of Food Science* 76, no. 1 (2011): H2-6, https://doi.org/10.1111/j.1750-3841.2010.01878.x.

21. Hayato Maeda et al., "Dietary Combination of Fucoxanthin and Fish Oil Attenuates the Weight Gain of White Adipose Tissue and Decreases Blood Glucose in Obese/Diabetic KK-Ay Mice," *Journal of Agricultural and Food Chemistry* 55, no. 19 (2007): 7701–6, https://doi.org/10.1021/jf071569n.

22. Carlos F. L. Gonçalves et al., "Flavonoids, Thyroid Iodide Uptake and Thyroid Cancer—A Review," *International Journal of Molecular Sciences* 18, no. 6 (2017), https://doi.org/10.3390/ijms18061247.

23. Catherine Cardot-Bauters and Jean-Louis Wémeau, "Autoimmune Thyroiditis," *La Revue Du Praticien* 64, no. 6 (2014): 835–38, https://www.ncbi.nlm.nih.gov/pubmed/25090773.

24. L. H. Duntas, "Environmental Factors and Thyroid Autoimmunity," *Annales d'Endocrinologie* 72, no. 2 (2011): 108–13, https://doi.org/10.1016/j.ando.2011.03.019.

25. L. Saranac et al., "Why Is the Thyroid So Prone to Autoimmune Disease?," *Hormone Research in Paediatrics* 75, no. 3 (2011): 157–65, https://doi.org/10.1159/000324442.

26. Yi Hou et al., "Meta-Analysis of the Correlation Between *Helicobacter pylori* Infection and Autoimmune Thyroid Diseases," *Oncotarget* 8, no. 70 (2017): 115691–700, https://doi.org/10.18632/oncotarget.22929.

27. Huijuan Liu et al., "Dihydroartemisinin Attenuates Autoimmune Thyroiditis by Inhibiting the CXCR3/PI3K/AKT/NF-KB Signaling Pathway," *Oncotarget* 8, no. 70 (2017): 115028–40, https://doi.org/10.18632/oncotarget.22854.

28. E. P. Kiseleva et al., "The Role of Components of Bifidobacterium and Lactobacillus in Pathogenesis and Serologic Diagnosis of Autoimmune Thyroid Diseases," *Beneficial Microbes* 2, no. 2 (2011): 139–54, https://doi.org/10.3920/BM2010.0011.

29. Mario Rotondi and Luca Chiovato, "The Chemokine System as a Therapeutic Target in Autoimmune Thyroid Diseases: A Focus on the Interferon-γ Inducible Chemokines and Their Receptor," *Current Pharmaceutical Design* 17, no. 29 (2011): 3202–16, https://www.ncbi.nlm.nih.gov/pubmed/21864266.

30. Fumina Doi et al., "Long-Term Outcome of Interferon-Alpha-Induced Autoimmune Thyroid Disorders in Chronic Hepatitis C," *Liver International: Official Journal of the International Association for the Study of the Liver* 25, no. 2 (2005): 242–46, https://doi.org/10.1111/j.1478-3231.2005.01089.x.

31. Elisabetta Caselli et al., "Virologic and Immunologic Evidence Supporting an Association between HHV-6 and Hashimoto's Thyroiditis," *PLoS Pathogens* 8, no. 10 (2012): e1002951, https://doi.org/10.1371/journal.ppat.1002951.

32. Rachel Desailloud and Didier Hober, "Viruses and Thyroiditis: An Update," *Virology Journal* 6 (2009): 5, https://doi.org/10.1186/1743-422X-6-5.

33. Raluca Trifănescu et al., "Autoimmune Thyroid Disease—A Continuous Spectrum," *Revue Roumaine de Médecine Interne (Romanian Journal of Internal Medicine)* 46, no. 4 (2008): 361–65, https://www.ncbi.nlm.nih.gov/pubmed/19480304.

34. Antoaneta B. Argatska and Boyan I. Nonchev, "Postpartum Thyroiditis," *Folia Medica* 56, no. 3 (2014): 145–51, https://www.ncbi.nlm.nih.gov/pubmed/25434070.

35. Maureen Groer and Cecilia Jevitt, "Symptoms and Signs Associated with Postpartum Thyroiditis," *Journal of Thyroid Research* 2014 (2014): 531969, https://doi.org/10.1155/2014/531969.

36. Allan Carlé et al., "Development of Autoimmune Overt Hypothyroidism Is Highly Associated with Live Births and Induced Abortions But Only in Premenopausal Women," *The Journal of Clinical Endocrinology and Metabolism* 99, no. 6 (2014): 2241–49, https://doi.org/10.1210/jc.2013-4474.

37. Flavia Di Bari et al., "Autoimmune Abnormalities of Postpartum Thyroid Diseases," *Frontiers in Endocrinology* 8 (2017): 166, https://doi.org/10.3389/fendo.2017.00166.

38. Anne Drutel et al., "Selenium and the Thyroid Gland: More Good News for Clinicians," *Clinical Endocrinology* 78, no. 2 (2013): 155–64, https://doi.org/10.1111/cen.12066.

39. Yanqiong Zhang et al., "Thyroid Hormone Synthesis: A Potential Target of a Chinese Herbal Formula Haizao Yuhu Decoction Acting on Iodine-Deficient Goiter," *Oncotarget* 7, no. 32 (2016): 51699–712, https://doi.org/10.18632/oncotarget.10329.

40. H. S. Parmar and A. Kar, "Antiperoxidative, Antithyroidal, Antihyperglycemic and Cardioprotective Role of Citrus Sinensis Peel Extract in Male Mice," *Phytotherapy Research: PTR* 22, no. 6 (2008): 791–95, https://doi.org/10.1002/ptr.2367.

41. Peter P. A. Smyth, "Role of Iodine in Antioxidant Defence in Thyroid and Breast Disease," *BioFactors* 19, no. 3–4 (2003): 121–30, https://www.ncbi.nlm.nih.gov/pubmed/14757962.

42. S. A. Cann et al., "Hypothesis: Iodine, Selenium and the Development of Breast Cancer," *Cancer Causes & Control: CCC* 11, no. 2 (2000): 121–27, https://www.ncbi.nlm.nih.gov/pubmed/10710195.

43. Csaba Balázs and Károly Rácz, "The Role of Selenium in Endocrine System Diseases," *Orvosi Hetilap* 154, no. 41 (2013): 1628–35, https://doi.org/10.1556/OH.2013.29723.

44. M. Auf'mkolk et al., "The Active Principles of Plant Extracts with Antithyrotropic Activity: Oxidation Products of Derivatives of 3,4-Dihydroxycinnamic Acid," *Endocrinology* 116, no. 5 (1985): 1677–86, https://doi.org/10.1210/endo-116-5-1677.

45. Debmalya Sanyal, "Spectrum of Hashimoto's Thyroiditis: Clinical, Biochemical & Cytomorphologic Profile," *The Indian Journal of Medical Research* 140, no. 6 (2014): 710–12, https://www.ncbi.nlm.nih.gov/pubmed/25758568.

46. Seyede Maryam Naghibi et al., "Carum Induced Hypothyroidism: An Interesting Observation and an Experiment," *Daru: Journal of Faculty of Pharmacy, Tehran University of Medical Sciences* 23 (2015): 5, https://doi.org/10.1186/s40199-015-0094-9.

47. Hamendra Singh Parmar and Anand Kar, "Protective Role of *Mangifera indica*, *Cucumis melo* and *Citrullus vulgaris* Peel Extracts in Chemically Induced Hypothyroidism," *Chemico-Biological Interactions* 177, no. 3 (2009): 254–58, https://doi.org/10.1016/j.cbi.2008.11.006.

48. Bronwyn A. Crawford et al., "Iodine Toxicity from Soy Milk and Seaweed Ingestion Is Associated with Serious Thyroid Dysfunction," *The Medical Journal of Australia* 193, no. 7 (2010): 413–15, https://www.ncbi.nlm.nih.gov/pubmed/20919974.

49. Jane Teas et al., "Seaweed and Soy: Companion Foods in Asian Cuisine and Their Effects on Thyroid Function in American Women," *Journal of Medicinal Food* 10, no. 1 (2007): 90–100, https://doi.org/10.1089/jmf.2005.056; Branka Sosić-Jurjević et al., "Suppressive Effects of Genistein and Daidzein on Pituitary-Thyroid Axis in Orchidectomized Middle-Aged Rats," *Experimental Biology and Medicine* 235, no. 5 (May 2010): 590–98, https://doi.org/10.1258/ebm.2009.009279.

50. Mark Messina and Geoffrey Redmond, "Effects of Soy Protein and Soybean Isoflavones on Thyroid Function in Healthy Adults and Hypothyroid Patients: A Review of the Relevant Literature," *Thyroid: Official Journal of the American Thyroid Association* 16, no. 3 (2006): 249–58, https://doi.org/10.1089/thy.2006.16.249.

51. Frank A. Simmen et al., "Soy Protein Diet Alters Expression of Hepatic Genes Regulating Fatty Acid and Thyroid Hormone Metabolism in the Male Rat," *The Journal of Nutritional Biochemistry* 21, no. 11 (2010): 1106–13, https://doi.org/10.1016/j.jnutbio.2009.09.008.

52. Gustavo C. Román et al., "Autism: Transient in Utero Hypothyroxinemia Related to Maternal Flavonoid Ingestion during Pregnancy and to Other Environmental Antithyroid Agents," *Journal of the Neurological Sciences* 262, no. 1–2 (2007): 15–26, https://doi.org/10.1016/j.jns.2007.06.023; Katarzyna Szkudelska and Leszek Nogowski, "Genistein—a Dietary Compound Inducing Hormonal and Metabolic Changes," *The Journal of Steroid Biochemistry and Molecular Biology* 105, no. 1–5 (2007): 37–45, https://doi.org/10.1016/j.jsbmb.2007.01.005.

53. Kylie Kavanagh et al., "High Isoflavone Soy Diet Increases Insulin Secretion without Decreasing Insulin Sensitivity in Premenopausal Nonhuman Primates," *Nutrition Research* 28, no. 6 (2008): 368–76, https://doi.org/10.1016/j.nutres.2008.03.011.

54. L. Hooper et al., "Effects of Soy Protein and Isoflavones on Circulating Hormone Concentrations in Pre- and Post-Menopausal Women: A Systematic Review and Meta-Analysis," *Human Reproduction Update* 15, no. 4 (2009): 423–40, https://doi.org/10.1093/humupd/dmp010.

55. Alessandra Bitto et al., "Genistein Aglycone Does Not Affect Thyroid Function: Results from a Three-Year, Randomized, Double-Blind, Placebo-Controlled Trial," *The Journal of Clinical Endocrinology and Metabolism* 95, no. 6 (2010): 3067–72, https://doi.org/10.1210/jc.2009-2779; Francene M. Steinberg et al., "Clinical Outcomes of a 2-y Soy Isoflavone Supplementation in Menopausal Women," *The American Journal of Clinical Nutrition* 93, no. 2 (2011): 356–67, https://doi.org/10.3945/ajcn.110.008359.

56. Richard Hampl et al., "Short-Term Effect of Soy Consumption on Thyroid Hormone Levels and Correlation with Phytoestrogen Level in Healthy Subjects," *Endocrine Regulations* 42, no. 2–3 (2008): 53–61, https://www.ncbi.nlm.nih.gov/pubmed/18624607.

57. Barbara L. Dillingham et al., "Soy Protein Isolates of Varied Isoflavone Content Do Not Influence Serum Thyroid Hormones in Healthy Young Men," *Thyroid: Official Journal of the American Thyroid Association* 17, no. 2 (2007): 131–37, https://doi.org/10.1089/thy.2006.0206.

58. Jaskanwal D. Sara et al., "Hypothyroidism Is Associated with Coronary Endothelial Dysfunction in Women," *Journal of the American Heart Association* 4, no. 8 (2015): e002225, https://doi.org/10.1161/JAHA.115.002225.

59. Y. B. Tripathi et al., "Thyroid Stimulating Action of Z-Guggulsterone Obtained from *Commiphora mukul*," *Planta Medica* 50, no. 1 (1984): 78–80, https://doi.org /10.1055/s-2007-969626; S. Panda and A. Kar, "Guggulu (*Commiphora mukul*) Potentially Ameliorates Hypothyroidism in Female Mice," *Phytotherapy Research: PTR* 19, no. 1 (2005): 78–80, https://doi.org/10.1002/ptr.1602; S. Panda and A. Kar, "Gugulu (*Commiphora mukul*) Induces Triiodothyronine Production: Possible Involvement of Lipid Peroxidation," *Life Sciences* 65, no. 12 (1999): PL137–141, https://www.ncbi.nlm.nih.gov/pubmed/10503949.

60. Jun Wu et al., "The Hypolipidemic Natural Product Guggulsterone Acts as an Antagonist of the Bile Acid Receptor," *Molecular Endocrinology* 16, no. 7 (2002): 1590–97, https://doi.org/10.1210/mend.16.7.0894.

61. Lise Anett Nohr et al., "Resin from the Mukul Myrrh Tree, Guggul, Can It Be Used for Treating Hypercholesterolemia? A Randomized, Controlled Study," *Complementary Therapies in Medicine* 17, no. 1 (2009): 16–22, https://doi.org /10.1016/j.ctim.2008.07.001; Bhavna Sharma et al., "Effects of Guggulsterone Isolated from *Commiphora mukul* in High Fat Diet Induced Diabetic Rats," *Food and Chemical Toxicology: An International Journal Published for the British Industrial Biological Research Association* 47, no. 10 (2009): 2631–39, https://doi.org/10.1016/j.fct.2009.07.021.

62. Y. B. Tripathi et al., "Thyroid Stimulating Action of Z-Guggulsterone Obtained from *Commiphora mukul*," *Planta Medica* 50, no. 1 (1984): 78–80, https://doi.org /10.1055/s-2007-969626.

63. Lise Anett Nohr et al., "Resin from the Mukul Myrrh Tree, Guggul, Can It Be Used for Treating Hypercholesterolemia? A Randomized, Controlled Study," *Complementary Therapies in Medicine* 17, no. 1 (2009): 16–22, https://doi.org/10.1016/j.ctim.2008.07.001.

64. Bhavna Sharma et al., "Effects of Guggulsterone Isolated from *Commiphora mukul* in High Fat Diet Induced Diabetic Rats," *Food and Chemical Toxicology: An International Journal Published for the British Industrial Biological Research Association* 47, no. 10 (2009): 2631–39, https://doi .org/10.1016/j.fct.2009.07.021.

65. Srujana Rayalam et al., "Anti-Obesity Effects of Xanthohumol plus Guggulsterone in 3T3-L1 Adipocytes," *Journal of Medicinal Food* 12, no. 4 (2009): 846–53, https://doi.org /10.1089/jmf.2008.0158.

66. Qing Gao et al., "Progress of Pathogenesis and Clinical Treatment of Hashimoto's Thyroiditis," *Zhongguo Zhong Yao Za Zhi (China Journal of Chinese Materia Medica)* 37, no. 20 (2012): 3003–6, https://www.ncbi.nlm.nih.gov /pubmed/23311142.

67. Paweł Zagrodzki and Elwira Przybylik-Mazurek, "Selenium and Hormone Interactions in Female Patients with Hashimoto Disease and Healthy Subjects," *Endocrine Research* 35, no. 1 (2010): 24–34, https://doi.org/10.3109 /07435800903551974.

68. C. C. Lei and C. C. Tang, "Successful Treatment of Postpartum Hypopituitarism with Decoction of *Radix glycyrrhizae* and *Radix ginseng*: Report of a Case," *Zhonghua Yi Xue Za Zhi* 11 (1973): 693–94, https://www.ncbi.nlm.nih .gov/pubmed/4203057.

69. Hayato Maeda et al., "Effect of Medium-Chain Triacylglycerols on Anti-Obesity Effect of Fucoxanthin," *Journal of Oleo Science* 56, no. 12 (2007): 615–21, https:// www.ncbi.nlm.nih.gov/pubmed/17992001.

70. Megan K. Horton et al., "Co-occurring Exposure to Perchlorate, Nitrate and Thiocyanate Alters Thyroid Function in Healthy Pregnant Women," *Environmental Research* 143 (2015): 1–9, https://doi.org/10.1016/j.envres.2015.09.013.

71. Briseis Aschebrook-Kilfoy et al., "Thyroid Cancer Risk and Dietary Nitrate and Nitrite Intake in the Shanghai Women's Health Study," *International Journal of Cancer* 132, no. 4 (2013): 897–904, https://doi.org/10.1002/ijc.27659.

72. Maki Inoue-Choi et al., "Nitrate and Nitrite Ingestion and Risk of Ovarian Cancer among Postmenopausal Women in Iowa," *International Journal of Cancer* 137, no. 1 (2015): 173–82, https://doi.org/10.1002/ijc.29365.

73. Peter Felker et al., "Concentrations of Thiocyanate and Goitrin in Human Plasma, Their Precursor Concentrations in Brassica Vegetables, and Associated Potential Risk for Hypothyroidism," *Nutrition Reviews* 74, no. 4 (2016): 248–58, https://doi.org/10.1093/nutrit/nuv110.

74. Jane V. Higdon et al., "Cruciferous Vegetables and Human Cancer Risk: Epidemiologic Evidence and Mechanistic Basis," *Pharmacological Research* 55, no. 3 (2007): 224–36, https://doi.org/10.1016/j.phrs.2007.01.009.

75. Jian-Min Yuan et al., "Clinical Trial of 2-Phenethyl Isothiocyanate as an Inhibitor of Metabolic Activation of a Tobacco-Specific Lung Carcinogen in Cigarette Smokers," *Cancer Prevention Research* 9, no. 5 (2016): 396–405, https://doi.org/10.1158/1940-6207.CAPR-15-0380.

76. Thérèse Truong et al., "Role of Dietary Iodine and Cruciferous Vegetables in Thyroid Cancer: A Countrywide Case-Control Study in New Caledonia," *Cancer Causes & Control: CCC* 21, no. 8 (2010): 1183–92, https://doi.org /10.1007/s10552-010-9545-2.

77. Ibrahim Taga et al., "Youth of West Cameroon Are at High Risk of Developing IDD Due to Low Dietary Iodine and High Dietary Thiocyanate," *African Health Sciences* 8, no. 4 (2008): 227–33, https://www.ncbi.nlm.nih.gov/pubmed/20589129.

78. G. Mejicanos et al., "Recent Advances in Canola Meal Utilization in Swine Nutrition," *Journal of Animal Science and Technology* 58 (2016): 7, https://doi.org/10.1186/s40781-016-0085-5.

79. Laurie C. Dolan et al., "Naturally Occurring Food Toxins," *Toxins* 2, no. 9 (2010): 2289–332, https://doi.org/10.3390 /toxins2092289.

80. Li Tang et al., "Total Isothiocyanate Yield from Raw Cruciferous Vegetables Commonly Consumed in the United States," *Journal of Functional Foods* 5, no. 4 (2013): 1996–2001, https://doi.org/10.1016/j.jff.2013.07.011.

81. Ralf J. Ludwig et al., "Mechanisms of Autoantibody-Induced Pathology," *Frontiers in Immunology* 8 (2017): 603, https://doi.org/10.3389/fimmu.2017.00603.

82. Miloš Zarković, "The Role of Oxidative Stress on the Pathogenesis of Graves' Disease," *Journal of Thyroid Research* 2012 (2012): 302537, https://doi.org/10.1155/2012/302537.

83. Y. S. Hussain et al., "Epidemiology, Management and Outcomes of Graves' Disease-Real Life Data," *Endocrine* 56, no. 3 (2017): 568–78, https://doi.org/10.1007/s12020-017-1306-5.

84. Malvinder S. Parmar, "Thyrotoxic Atrial Fibrillation," *MedGenMed: Medscape General Medicine* 7, no. 1 (2005): 74, https://www.ncbi.nlm.nih.gov/pubmed/16369379.

85. H. Yamamoto et al., "Caffeic Acid Oligomers in *Lithospermum erythrorhizon* Cell Suspension Cultures," *Phytochemistry* 53, no. 6 (2000): 651–57, https://www.ncbi.nlm.nih.gov/pubmed/10746877.

86. Dibakar Biswas et al., "Occurrence of Osteoporosis and Factors Determining Bone Mineral Loss in Young Adults with Graves' Disease," *The Indian Journal of Medical Research* 141, no. 3 (2015): 322–29, https://www.ncbi.nlm.nih.gov/pubmed/25963493.

87. Ashley Schaffer et al., "Recurrent Thyrotoxicosis Due to Both Graves' Disease and Hashimoto's Thyroiditis in the Same Three Patients," *Case Reports in Endocrinology* 2016 (2016): 6210493, https://doi.org/10.1155/2016/6210493.

88. Marcello Bagnasco et al., "Stress and Autoimmune Thyroid Diseases," *Neuroimmunomodulation* 13, no. 5–6 (2006): 309–17, https://doi.org/10.1159/000104859.

89. Sebastiano Bruno Solerte et al., "Defect of a Subpopulation of Natural Killer Immune Cells in Graves' Disease and Hashimoto's Thyroiditis: Normalizing Effect of Dehydroepiandrosterone Sulfate," *European Journal of Endocrinology* 152, no. 5 (2005): 703–12, https://doi.org/10.1530/eje.1.01906.

90. Seong-Mo Kim et al., "Antioxidant and Protective Effects of *Bupleurum falcatum* on the L-Thyroxine-Induced Hyperthyroidism in Rats," *Evidence-Based Complementary and Alternative Medicine: ECAM* 2012 (2012): 578497, https://doi.org/10.1155/2012/578497.

91. Yali Zhang et al., "Pingmu Decoction Induces Orbital Preadipocytes Apoptosis In Vitro," *Evidence-Based Complementary and Alternative Medicine: ECAM* 2017 (2017): 2109249, https://doi.org/10.1155/2017/2109249.

92. Hua Yang et al., "Clinical Efficacy of Yingliu Treatment for Graves Disease," *International Journal of Clinical and Experimental Medicine* 8, no. 4 (2015): 6145–53, https://www.ncbi.nlm.nih.gov/pubmed/26131218.

93. Keyvan Dastmalchi et al., "Chemical Composition and In Vitro Antioxidative Activity of a Lemon Balm (*Melissa officinalis L.*) Extract," *Food Science and Technology* 41, no. 3 (2008): 391–400, https://doi.org/10.1016/j.lwt.2007.03.007; A. P. Carnat et al., "The Aromatic and Polyphenolic Composition of Lemon Balm (*Melissa officinalis* L. subsp. *officinalis*) Tea," *Pharmaceutica Acta Helvetiae* 72, no. 5 (1998): 301–5, https://doi.org/10.1016/S0031-6865(97)00026-5.

94. Alvin Ibarra et al., "Importance of Extract Standardization and In Vitro/Ex Vivo Assay Selection for the Evaluation of Antioxidant Activity of Botanicals: A Case Study on Three *Rosmarinus officinalis* L. Extracts," *Journal of Medicinal Food* 13, no. 5 (2010): 1167–75, https://doi.org/10.1089/jmf.2009.0259.

95. Luo Fang et al., "Simultaneous Determination of Four Active Components in *Spica prunellae* by HPLC," *Zhongguo Zhong Yao Za Zhi (China Journal of Chinese Materia Medica)* 35, no. 5 (2010): 616–19, https://www.ncbi.nlm.nih.gov/pubmed/20506824.

96. Corinna Weitzel and Maike Petersen, "Enzymes of Phenylpropanoid Metabolism in the Important Medicinal Plant *Melissa officinalis* L.," *Planta* 232, no. 3 (2010): 731–42, https://doi.org/10.1007/s00425-010-1206-x.

97. Maike Petersen et al., "Evolution of Rosmarinic Acid Biosynthesis," *Phytochemistry* 70, no. 15–16 (2009): 1663–79, https://doi.org/10.1016/j.phytochem.2009.05.010.

98. Mouna Ben Farhat et al., "Variations in Essential Oil, Phenolic Compounds, and Antioxidant Activity of Tunisian Cultivated *Salvia officinalis* L.," *Journal of Agricultural and Food Chemistry* 57, no. 21 (2009): 10349–56, https://doi.org/10.1021/jf901877x.

99. M. Auf'mkolk et al., "Inhibition by Certain Plant Extracts of the Binding and Adenylate Cyclase Stimulatory Effect of Bovine Thyrotropin in Human Thyroid Membranes," *Endocrinology* 115, no. 2 (1984): 527–34, https://doi.org/10.1210/endo-115-2-527.

100. M. Auf'mkolk et al., "The Active Principles of Plant Extracts with Antithyrotropic Activity: Oxidation Products of Derivatives of 3,4-Dihydroxycinnamic Acid," *Endocrinology* 116, no. 5 (1985): 1677–86, https://doi.org/10.1210/endo-116-5-1677.

101. F. Santini et al., "In Vitro Assay of Thyroid Disruptors Affecting TSH-Stimulated Adenylate Cyclase Activity," *Journal of Endocrinological Investigation* 26, no. 10 (2003): 950–55, https://doi.org/10.1007/BF03348190; M. Auf'mkolk et al., "Extracts and Auto-Oxidized Constituents of Certain Plants Inhibit the Receptor-Binding and the Biological Activity of Graves' Immunoglobulins," *Endocrinology* 116, no. 5 (1985): 1687–93, https://doi.org/10.1210/endo-116-5-1687.

102. M. Auf'mkolk et al., "Antihormonal Effects of Plant Extracts: Iodothyronine Deiodinase of Rat Liver Is Inhibited by Extracts and Secondary Metabolites of Plants," *Hormones Et Metabolisme (Hormone and Metabolic Research)* 16, no. 4 (1984): 188–92, https://doi.org/10.1055/s-2007-1014739.

103. Christian Vonhoff et al., "Extract of *Lycopus europaeus* L. Reduces Cardiac Signs of Hyperthyroidism in Rats," *Life Sciences* 78, no. 10 (2006): 1063–70, https://doi.org/10.1016/j.lfs.2005.06.014.

104. A.-M. Beer et al., "*Lycopus europaeus* (Gypsywort): Effects on the Thyroidal Parameters and Symptoms Associated with

Thyroid Function," *Phytomedicine: International Journal of Phytotherapy and Phytopharmacology* 15, no. 1–2 (2008): 16–22, https://doi.org/10.1016/j.phymed.2007.11.001.

105. Sunanda Panda and Anand Kar, "Amelioration of L-Thyroxine-Induced Hyperthyroidism by Coumarin (1,2-Benzopyrone) in Female Rats," *Clinical and Experimental Pharmacology & Physiology* 34, no. 11 (2007): 1217–19, https://doi.org/10.1111/j.1440-1681.2007.04701.x.

106. Yuji Hiromatsu et al., "Graves' Ophthalmopathy: Epidemiology and Natural History," *Internal Medicine* 53, no. 5 (2014): 353–60, https://www.ncbi.nlm.nih.gov/pubmed/24583420.

107. Anne Drutel et al., "Selenium and the Thyroid Gland: More Good News for Clinicians," *Clinical Endocrinology* 78, no. 2 (2013): 155–64, https://doi.org/10.1111/cen.12066; Davide Nacamulli et al., "Influence of Physiological Dietary Selenium Supplementation on the Natural Course of Autoimmune Thyroiditis," *Clinical Endocrinology* 73, no. 4 (2010): 535–39, https://doi.org/10.1111/j.1365-2265.2009.03758.x.

108. R. Wang et al., "Salsalate and Salicylate Binding to and Their Displacement of Thyroxine from Thyroxine-Binding Globulin, Transthyrin, and Albumin," *Thyroid: Official Journal of the American Thyroid Association* 9, no. 4 (1999): 359–64, https://doi.org/10.1089/thy.1999.9.359.

109. Hong Li et al., "Icariin Inhibits AMPK-Dependent Autophagy and Adipogenesis in Adipocytes In Vitro and in a Model of Graves' Orbitopathy In Vivo," *Frontiers in Physiology* 8 (2017): 45, https://doi.org/10.3389/fphys.2017.00045.

110. H. Chang et al., "Pharmacology and Applications of Chinese Materia Medica," *Chinese University of Hong Kong* (1986): 16–29, https://doi.org/10.1142/9789814503143_0004; S. J. Fulder, "Ginseng and the Hypothalamic-Pituitary Control of Stress," *The American Journal of Chinese Medicine* 9, no. 2 (1981): 112–18, https://www.ncbi.nlm.nih.gov/pubmed/7345916.

111. E. V. Avakian et al., "Effect of *Panax ginseng* Extract on Energy Metabolism during Exercise in Rats," *Planta Medica* 50, no. 2 (1984): 151–54, https://doi.org/10.1055/s-2007-969657.

112. H. Chang et al., "Pharmacology and Applications of Chinese Materia Medica," *Chinese University of Hong Kong* (1986): 16–29, https://doi.org/10.1142/9789814503143_0004; S. J. Fulder, "Ginseng and the Hypothalamic-Pituitary Control of Stress," *The American Journal of Chinese Medicine* 9, no. 2 (1981): 112–18, https://www.ncbi.nlm.nih.gov/pubmed/7345916.

113. N. R. Farnsworth, *Economic and Medicinal Plant Research*, (Academic Press, 1985).

114. I. I. Brekhman and I. V. Dardymov, "New Substances of Plant Origin Which Increase Nonspecific Resistance," *Annual Review of Pharmacology* 9 (1969): 419–30, https://doi.org/10.1146/annurev.pa.09.040169.002223.

115. L. Davis and G. Kuttan, "Suppressive Effect of Cyclophosphamide-Induced Toxicity by *Withania somnifera* Extract in Mice," *Journal of Ethnopharmacology* 62, no. 3 (1998): 209–14, https://www.ncbi.nlm.nih.gov/pubmed/9849630.

116. Kirti Vaishnavi et al., "Differential Activities of the Two Closely Related Withanolides, Withaferin A and Withanone: Bioinformatics and Experimental Evidences," *PLoS One* 7, no. 9 (2012): e44419, https://doi.org/10.1371/journal.pone.0044419.

117. J. N. Dhuley, "Nootropic-Like Effect of Ashwagandha (*Withania somnifera* L.) in Mice," *Phytotherapy Research: PTR* 15, no. 6 (2001): 524–28, https://www.ncbi.nlm.nih.gov/pubmed/11536383.

118. A. Russo et al., "Indian Medicinal Plants as Antiradicals and DNA Cleavage Protectors," *Phytomedicine: International Journal of Phytotherapy and Phytopharmacology* 8, no. 2 (2001): 125–32, https://doi.org/10.1078/0944-7113-00021; S. D. Shukla et al., "Stress Induced Neuron Degeneration and Protective Effects of *Semecarpus anacardium Linn*. and *Withania somnifera Dunn*. in Hippocampus of Albino Rats: An Ultrastructural Study," *Indian Journal of Experimental Biology* 38, no. 10 (2000): 1007–13, https://www.ncbi.nlm.nih.gov/pubmed/11324152; J. N. Dhuley, "Effect of Ashwagandha on Lipid Peroxidation in Stress-Induced Animals," *Journal of Ethnopharmacology* 60, no. 2 (1998): 173–78, https://www.ncbi.nlm.nih.gov/pubmed/9582008.

119. J. N. Dhuley, "Effect of Some Indian Herbs on Macrophage Functions in Ochratoxin A Treated Mice," *Journal of Ethnopharmacology* 58, no. 1 (1997): 15–20, https://www.ncbi.nlm.nih.gov/pubmed/9324000; J. N. Dhuley, "Therapeutic Efficacy of Ashwagandha against Experimental Aspergillosis in Mice," *Immunopharmacology and Immunotoxicology* 20, no. 1 (1998): 191–98, https://doi.org/10.3109/08923979809034817.

120. A. Grandhi et al., "A Comparative Pharmacological Investigation of Ashwagandha and Ginseng," *Journal of Ethnopharmacology* 44, no. 3 (1994): 131–35, https://www.ncbi.nlm.nih.gov/pubmed/7898119.

121. S. K. Bhattacharya and S. K. Mitra, "Anxiolytic Activity of *Panax ginseng* Roots: An Experimental Study," *Journal of Ethnopharmacology* 34, no. 1 (1991): 87–92, https://www.ncbi.nlm.nih.gov/pubmed/1684404.

122. P. Ferrari et al., "The Role of the 11beta-Hydroxysteroid Dehydrogenase Type 2 in Human Hypertension," *Journal of Hypertension* 18, no. 3 (2000): 241–48, https://www.ncbi.nlm.nih.gov/pubmed/10726708.

123. Paal Methlie et al., "Grapefruit Juice and Licorice Increase Cortisol Availability in Patients with Addison's Disease," *European Journal of Endocrinology* 165, no. 5 (2011): 761–69, https://doi.org/10.1530/EJE-11-0518.

124. G. Valenti, "Neuroendocrine Hypothesis of Aging: The Role of Corticoadrenal Steroids," *Journal of Endocrinological Investigation* 27, no. 6 Suppl (2004): 62–63, https://www.ncbi.nlm.nih.gov/pubmed/15481804; E. Ferrari et al., "Cognitive and Affective Disorders in the Elderly: A Neuroendocrine Study," *Archives of Gerontology and*

Geriatrics. Supplement, no. 9 (2004): 171–82, https:// doi.org/10.1016/j.archger.2004.04.024.

125. E. Ferrari et al., "Neuroendocrine Features in Extreme Longevity," *Experimental Gerontology* 43, no. 2 (2008): 88–94, https://doi.org/10.1016/j.exger.2007.06.010; E. Ferrari et al., "Neuroendocrine Features in Extreme Longevity," *Experimental Gerontology* 43, no. 2 (2008): 88–94, https://doi.org/10.1016/j.exger.2007.06.010.

126. G. Valenti, "Neuroendocrine Hypothesis of Aging: The Role of Corticoadrenal Steroids," *Journal of Endocrinological Investigation* 27, no. 6 Suppl (2004): 62–63, https://www .ncbi.nlm.nih.gov/pubmed/15481804.

127. E. Ferrari et al., "Cognitive and Affective Disorders in the Elderly: A Neuroendocrine Study," *Archives of Gerontology and Geriatrics. Supplement*, no. 9 (2004): 171–82, https:// doi.org/10.1016/j.archger.2004.04.024; V. M. S. de Bruin et al., "Cortisol and Dehydroepiandosterone Sulfate Plasma Levels and Their Relationship to Aging, Cognitive Function, and Dementia," *Brain and Cognition* 50, no. 2 (2002): 316–23, https://www.ncbi.nlm.nih.gov/pubmed /12464198; Michael Ritsner et al., "Elevation of the Cortisol/Dehydroepiandrosterone Ratio in Schizophrenia Patients," *European Neuropsychopharmacology: The Journal of the European College of Neuropsychopharmacology* 14, no. 4 (2004): 267–73, https://doi.org/10.1016/j.euroneuro .2003.09.003; Kai G. Kahl et al., "Cortisol, the Cortisol-Dehydroepiandrosterone Ratio, and Pro-Inflammatory Cytokines in Patients with Current Major Depressive Disorder Comorbid with Borderline Personality Disorder," *Biological Psychiatry* 59, no. 7 (2006): 667–71, https:// doi.org/10.1016/j.biopsych.2005.08.001.

128. M. Palermo et al., "Urinary Free Cortisone and the Assessment of 11 Beta-Hydroxysteroid Dehydrogenase Activity in Man," *Clinical Endocrinology* 45, no. 5 (1996): 605–11, https://www.ncbi.nlm.nih.gov/pubmed/8977758; A. A. Al-Qarawi et al., "Liquorice (*Glycyrrhiza glabra*) and the Adrenal-Kidney-Pituitary Axis in Rats," *Food and Chemical Toxicology: An International Journal Published for the British Industrial Biological Research Association* 40, no. 10 (2002): 1525–27, https://www.ncbi.nlm.nih.gov/pubmed/12387318; M. Palermo et al., "Urinary Free Cortisone and the Assessment of 11 Beta-Hydroxysteroid Dehydrogenase Activity in Man," *Clinical Endocrinology* 45, no. 5 (1996): 605–11, https://www.ncbi.nlm.nih.gov/pubmed/8977758.

129. D. Vojnović Milutinović et al., "Hypothalamic-Pituitary-Adrenocortical Axis Hypersensitivity and Glucocorticoid Receptor Expression and Function in Women with Polycystic Ovary Syndrome," *Experimental and Clinical Endocrinology & Diabetes: Official Journal, German Society of Endocrinology and German Diabetes Association* 119, no. 10 (2011): 636–43, https://doi.org/10.1055/s-0031-1283122.

130. Yamei Huang et al., "Premenstrual Syndrome Is Associated with Blunted Cortisol Reactivity to the TSST," *Stress* 18, no. 2 (2015): 160–68, https://doi.org/10.3109/10253890.2014.999234.

131. M. T. Epstein et al., "Licorice Raises Urinary Cortisol in Man," *The Journal of Clinical Endocrinology and Metabolism* 47, no. 2 (1978): 397–400, https://doi.org/10.1210/jcem-47-2-397.

132. Farzaneh Zamansoltani et al., "Antiandrogenic Activities of *Glycyrrhiza glabra* in Male Rats," *International Journal of Andrology* 32, no. 4 (2009): 417–22, https://doi.org/10.1111 /j.1365-2605.2009.00944.x; Decio Armanini et al., "Licorice Reduces Serum Testosterone in Healthy Women," *Steroids* 69, no. 11–12 (2004): 763–66, https://doi.org/10.1016/j .steroids.2004.09.005.

133. Helga Agusta Sigurjonsdottir et al., "Liquorice in Moderate Doses Does Not Affect Sex Steroid Hormones of Biological Importance Although the Effect Differs between the Genders," *Hormone Research* 65, no. 2 (2006): 106–10, https://doi.org/10.1159/000091302; R. A. Josephs et al., "Liquorice Consumption and Salivary Testosterone Concentrations," *Lancet* 358, no. 9293 (2001): 1613–14, https://doi.org/10.1016/S0140-6736(01)06664-8.

134. H. Wagner et al., "Plant Adaptogens," *Phytomedicine: International Journal of Phytotherapy and Phytopharmacology* 1, no. 1 (1994): 63–76, https://doi.org /10.1016/S0944-7113(11)80025-5.

135. Ghulam Abbas et al., "Saponins: The Phytochemical with an Emerging Potential for Curing Clinical Depression," *Natural Product Research* 29, no. 4 (2015): 302–7, https:// doi.org/10.1080/14786419.2014.942661.

136. Paal Methlie et al., "Grapefruit Juice and Licorice Increase Cortisol Availability in Patients with Addison's Disease," *European Journal of Endocrinology* 165, no. 5 (2011): 761–69, https://doi.org/10.1530/EJE-11-0518.

137. M. T. Epstein et al., "Licorice Raises Urinary Cortisol in Man," *The Journal of Clinical Endocrinology and Metabolism* 47, no. 2 (1978): 397–400, https://doi.org /10.1210/jcem-47-2-397.

138. J. C. Bertoglio et al., "*Andrographis paniculata* Decreases Fatigue in Patients with Relapsing-Remitting Multiple Sclerosis: A 12-Month Double-Blind Placebo-Controlled Pilot Study," *BMC Neurology* 16 (2016): 77, https://doi.org /10.1186/s12883-016-0595-2.

139. Xuan Tang et al., "Arctigenin Efficiently Enhanced Sedentary Mice Treadmill Endurance," *PLoS One* 6, no. 8 (2011): e24224, https://doi.org/10.1371/journal.pone.0024224.

140. Chenchen Shi et al., "Anti-Inflammatory and Immunoregulatory Functions of Artemisinin and Its Derivatives," *Mediators of Inflammation* (2015): http:// dx.doi.org/10.1155/2015/435713.

141. Samira Alesaeidi and Sepide Miraj, "A Systematic Review of Anti-Malarial Properties, Immunosuppressive Properties, Anti-Inflammatory Properties, and Anti-Cancer Properties of *Artemisia annua*," *Electronic Physician* 8, no. 10 (2016): 3150–55, https://doi.org/10.19082/3150; Ying Li, "Qinghaosu (Artemisinin): Chemistry and Pharmacology," *Acta Pharmacologica Sinica* 33, no. 9 (2012): 1141–46, https://doi.org/10.1038/aps.2012.104.

142. Ryan Koenig et al., "Avenanthramide Supplementation Attenuates Exercise-Induced Inflammation in Postmenopausal Women," *Nutrition Journal* 13 (2014): 21, https://doi.org/10.1186/1475-2891-13-21.

143. Seong-Mo Kim et al., "Antioxidant and Protective Effects of *Bupleurum falcatum* on the L-Thyroxine-Induced Hyperthyroidism in Rats," *Evidence-Based Complementary and Alternative Medicine: ECAM* 2012 (2012): 578497, https://doi.org/10.1155/2012/578497.

144. Kim et al., "Antioxidant and Protective Effects of *Bupleurum falcatum*."

145. Okon Akpan Uduak et al., "Comparative Effect of *Citrus sinensis* and Carbimazole on Serum T4, T3 and TSH Levels," *Nigerian Medical Journal: Journal of the Nigeria Medical Association* 55, no. 3 (2014): 230–34, https://doi.org/10.4103/0300-1652.132049.

146. Shonteh Henderson et al., "Effects of *Coleus forskohlii* Supplementation on Body Composition and Hematological Profiles in Mildly Overweight Women," *Journal of the International Society of Sports Nutrition* 2 (2005): 54–62, https://doi.org/10.1186/1550-2783-2-2-54.

147. Hebbani Nagarajappa Shivaprasad et al., "Effect of *Coleus forskohlii* Extract on Cafeteria Diet-Induced Obesity in Rats," *Pharmacognosy Research* 6, no. 1 (2014): 42–45, https://doi.org/10.4103/0974-8490.122916.

148. Hayley L. Loftus et al., "*Coleus forskohlii* Extract Supplementation in Conjunction with a Hypocaloric Diet Reduces the Risk Factors of Metabolic Syndrome in Overweight and Obese Subjects: A Randomized Controlled Trial," *Nutrients* 7, no. 11 (2015): 9508–22, https://doi.org/10.3390/nu7115483.

149. F. Ahmad et al., "Insulin and Glucagon Releasing Activity of Coleonol (Forskolin) and Its Effect on Blood Glucose Level in Normal and Alloxan Diabetic Rats," *Acta Diabetologica Latina* 28, no. 1 (1991): 71–77, https://www.ncbi.nlm.nih.gov/pubmed/1650516.

150. A. K. Singh et al., "In Vitro Studies on Thyrogenic Effect of *Commiphora mukul* (Guggulu)," *Ancient Science of Life* 2, no. 1 (1982): 23–28, https://www.ncbi.nlm.nih.gov/pubmed/22556948.

151. Prerna Sarup et al., "Pharmacology and Phytochemistry of Oleo-Gum Resin of *Commiphora wightii* (Guggulu)," *Scientifica* 2015 (2015): 138039, https://doi.org/10.1155/2015/138039; S. Panda and A. Kar, "Gugulu (*Commiphora mukul*) Induces Triiodothyronine Production: Possible Involvement of Lipid Peroxidation," *Life Sciences* 65, no. 12 (1999): PL137–141, https://www.ncbi.nlm.nih.gov/pubmed/10503949.

152. Ramesh Bellamkonda et al., "Beneficiary Effect of *Commiphora mukul* Ethanolic Extract against High Fructose Diet Induced Abnormalities in Carbohydrate and Lipid Metabolism in Wistar Rats," *Journal of Traditional and Complementary Medicine* 8, no. 1 (2018): 203–11, https://doi.org/10.1016/j.jtcme.2017.05.007.

153. Francesco Di Pierro et al., "Preliminary Study about the Possible Glycemic Clinical Advantage in Using a Fixed Combination of *Berberis aristata* and *Silybum marianum* Standardized Extracts versus Only *Berberis aristata* in Patients with Type 2 Diabetes," *Clinical Pharmacology: Advances and Applications* 5 (2013): 167–74, https://doi.org/10.2147/CPAA.S54308.

154. Hui Li Tan et al., "*Rhizoma coptidis*: A Potential Cardiovascular Protective Agent," *Frontiers in Pharmacology* 7 (2016): 362, https://doi.org/10.3389/fphar.2016.00362.

155. Bhaktaprasad Gaire and Dongwook Lim, "Antidepressant Effects of *Radix et Caulis Acanthopanacis Santicosi* Extracts on Rat Models with Depression in Terms of Immobile Behavior," *Chung I Tsa Chih Ying Wen Pan (Journal of Traditional Chinese Medicine)* 34, no. 3 (2014): 317–23, https://www.ncbi.nlm.nih.gov/pubmed/24992759.

156. Maho Sumiyoshi and Yoshiyuki Kimura, "Effects Of *Eleutherococcus senticosus* Cortex on Recovery from the Forced Swimming Test and Fatty Acid β-Oxidation in the Liver and Skeletal Muscle of Mice," *The Natural Products Journal* 6, no. 1 (2016): 49–55, https://doi.org/10.2174/2210315506999151207145020.

157. Jie Han et al., "Effects of *Acanthopanax senticosus* Polysaccharide Supplementation on Growth Performance, Immunity, Blood Parameters and Expression of Pro-Inflammatory Cytokines Genes in Challenged Weaned Piglets," *Asian-Australasian Journal of Animal Sciences* 27, no. 7 (2014): 1035–43, https://doi.org/10.5713/ajas.2013.13659.

158. Jianhua Huang et al., "Epimedium Flavonoids Counteract the Side Effects of Glucocorticoids on Hypothalamic-Pituitary-Adrenal Axis," *Evidence-Based Complementary and Alternative Medicine: ECAM* 2013 (2013): 938425, https://doi.org/10.1155/2013/938425.

159. Zhisheng Wei et al., "Icariin Exerts Estrogen-Like Activity in Ameliorating EAE via Mediating Estrogen Receptor β, Modulating HPA Function and Glucocorticoid Receptor Expression," *American Journal of Translational Research* 8, no. 4 (2016): 1910–18, https://www.ncbi.nlm.nih.gov/pubmed/27186315.

160. Hyun Ku Kang et al., "Peripubertal Administration of Icariin and Icaritin Advances Pubertal Development in Female Rats," *Biomolecules & Therapeutics* 20, no. 2 (2012): 189–95, https://doi.org/10.4062/biomolther.2012.20.2.189.

161. Lingwen Kong et al., "BuShenYiQi Granule Inhibits Atopic Dermatitis via Improving Central and Skin Hypothalamic-Pituitary-Adrenal Axis Function," *PLoS One* 10, no. 2 (2015): e0116427, https://doi.org/10.1371/journal.pone.0116427.

162. Marcelo D. Catarino et al., "Fucaceae: A Source of Bioactive Phlorotannins," *International Journal of Molecular Sciences* 18, no. 6 (2017), https://doi.org/10.3390/ijms18061327.

163. Ulrike Grienke et al., "Accessing Biological Actions of Ganoderma Secondary Metabolites by in Silico Profiling," *Phytochemistry* 114 (2015): 114–24, https://doi.org/10.1016/j.phytochem.2014.10.010.

164. Carolina Arboleda et al., "Elimination of Bisphenol A and Triclosan Using the Enzymatic System of Autochthonous

Colombian Forest Fungi," *ISRN Biotechnology* 2013 (2013): 968241, https://doi.org/10.5402/2013/968241.

165. Mark Messina, "Soy and Health Update: Evaluation of the Clinical and Epidemiologic Literature," *Nutrients* 8, no. 12 (2016), https://doi.org/10.3390/nu8120754.

166. A. Kumagai et al., "Effect of Glycyrrhizin on Estrogen Action," *Endocrinologia Japonica* 14, no. 1 (1967): 34–38, https://www.ncbi.nlm.nih.gov/pubmed/6072411.

167. M. T. Epstein et al., "Effect of Eating Liquorice on the Renin-Angiotensin Aldosterone Axis in Normal Subjects," *British Medical Journal* 1, no. 6059 (1977): 488–90, https://www.ncbi.nlm.nih.gov/pubmed/837172.

168. Santosh Kumar Maurya et al., "Antidyslipidaemic Activity of *Glycyrrhiza glabra* in High Fructose Diet Induced Dsyslipidaemic Syrian Golden Hamsters," *Indian Journal of Clinical Biochemistry: IJCB* 24, no. 4 (2009): 404–9, https://doi.org/10.1007/s12291-009-0072-4.

169. Michael Dushkin et al., "Effects of *Rhaponticum car-thamoides* versus *Glycyrrhiza glabra* and *Punica granatum* Extracts on Metabolic Syndrome Signs in Rats," *BMC Complementary and Alternative Medicine* 14 (2014): 33, https://doi.org/10.1186/1472-6882-14-33.

170. Chia Hui Apphia Eu et al., "Glycyrrhizic Acid Improved Lipoprotein Lipase Expression, Insulin Sensitivity, Serum Lipid and Lipid Deposition in High-Fat Diet-Induced Obese Rats," *Lipids in Health and Disease* 9 (2010): 81, https://doi.org/10.1186/1476-511X-9-81.

171. Fai-Chu Wong et al., "Antioxidant, Metal Chelating, Anti-Glucosidase Activities and Phytochemical Analysis of Selected Tropical Medicinal Plants," *Iranian Journal of Pharmaceutical Research: IJPR* 13, no. 4 (2014): 1409–15, https://www.ncbi.nlm.nih.gov/pubmed/25587331.

172. Yan Dong et al., "Application of Traditional Chinese Medicine in Treatment of Atrial Fibrillation," *Evidence-Based Complementary and Alternative Medicine: ECAM* 2017 (2017): 1381732, https://doi.org/10.1155/2017/1381732.

173. H. Sourgens et al., "Effects of *Lithospermum officinale* and Related Plants on Hypophyseal and Thyroid Hormones in the Rat," *Pharmaceutical Biology* 24, no. 2 (1986): 53–63, https://doi.org/10.3109/13880208609083307.

174. Yue Zhang et al., "Shikonin Inhibits Migration and Invasion of Thyroid Cancer Cells by Downregulating DNMT1," *Medical Science Monitor: International Medical Journal of Experimental and Clinical Research* 24 (2018): 661–70, https://www.ncbi.nlm.nih.gov/pubmed/29389913.

175. Marhaba Hojahmat et al., "Lobeline Esters as Novel Ligands for Neuronal Nicotinic Acetylcholine Receptors and Neurotransmitter Transporters," *Bioorganic & Medicinal Chemistry* 18, no. 2 (2010): 640–49, https://doi.org/10.1016/j.bmc.2009.12.002.

176. Christopher L. German et al., "Regulation of the Dopamine and Vesicular Monoamine Transporters: Pharmacological Targets and Implications for Disease," *Pharmacological Reviews* 67, no. 4 (2015): 1005–24, https://doi.org/10.1124/pr.114.010397.

177. Shyamsunder R. Joolakanti et al., "Lobelane Analogues Containing 4-Hydroxy and 4-(2-Fluoroethoxy) Aromatic Substituents: Potent and Selective Inhibitors of [(3)H]Dopamine Uptake at the Vesicular Monoamine Transporter-2," *Bioorganic & Medicinal Chemistry Letters* 26, no. 10 (2016): 2422–27, https://doi.org/10.1016/j.bmcl.2016.03.119.

178. S. Kleemann and H. Winterhoff, "Rosmarinic Acid and Freeze-Dried Extract (FDE) of *Lycopus virginicus* Are Able to Inhibit Forsklin-Induced Activation of Adenylate Cyclase in Cultured Rat Thyroid Cells," *Planta Med* 56, no. 6 (1990): 683, https://doi.org/10.1055/s-2006-961356.

179. Christian Vonhoff et al., "Extract of *Lycopus europaeus* L. Reduces Cardiac Signs of Hyperthyroidism in Rats," *Life Sciences* 78, no. 10 (2006): 1063–70, https://doi.org/10.1016/j.lfs.2005.06.014.

180. H. Winterhoff et al., "Endocrine Effects of *Lycopus europaeus* L. Following Oral Application," *Arzneimittel-Forschung* 44, no. 1 (1994): 41–45, https://www.ncbi.nlm.nih.gov/pubmed/8135877.

181. F. Santini et al., "In Vitro Assay of Thyroid Disruptors Affecting TSH-Stimulated Adenylate Cyclase Activity," *Journal of Endocrinological Investigation* 26, no. 10 (2003): 950–55, https://doi.org/10.1007/BF03348190.

182. Siyavash Joukar and Haleh Asadipour, "Evaluation of *Melissa officinalis* (Lemon Balm) Effects on Heart Electrical System," *Research in Cardiovascular Medicine* 4, no. 2 (2015): e27013, https://doi.org/10.5812/cardiovascmed.4(2)2015.27013.

183. Fatemeh Alijaniha et al., "Heart Palpitation Relief with *Melissa officinalis* Leaf Extract: Double Blind, Randomized, Placebo Controlled Trial of Efficacy and Safety," *Journal of Ethnopharmacology* 164 (2015): 378–84, https://doi.org/10.1016/j.jep.2015.02.007.

184. Keyvan Dastmalchi et al., "Chemical Composition and In Vitro Antioxidative Activity of a Lemon Balm (*Melissa officinalis* L.) Extract," *Food Science and Technology* 41, no. 3 (2008): 391–400, https://doi.org/10.1016/j.lwt.2007.03.007; A. P. Carnat et al., "The Aromatic and Polyphenolic Composition of Lemon Balm (*Melissa officinalis* L. subsp. *officinalis*) Tea," *Pharmaceutica Acta Helvetiae* 72, no. 5 (1998): 301–5, https://doi.org/10.1016/S0031-6865(97)00026-5.

185. Abolfazl Shakeri et al., "*Melissa officinalis* L.—A Review of Its Traditional Uses, Phytochemistry and Pharmacology," *Journal of Ethnopharmacology* 188 (2016): 204–28, https://doi.org/10.1016/j.jep.2016.05.010.

186. Hyeong-Geug Kim et al., "Antifatigue Effects of *Panax ginseng* C.A. Meyer: A Randomised, Double-Blind, Placebo-Controlled Trial," *PLoS One* 8, no. 4 (2013): e61271, https://doi.org/10.1371/journal.pone.0061271.

187. Rong Di and Nilgun E. Tumer, "Pokeweed Antiviral Protein: Its Cytotoxicity Mechanism and Applications in Plant Disease Resistance," *Toxins* 7, no. 3 (2015): 755–72, https://doi.org/10.3390/toxins7030755.

188. Artem V. Domashevskiy and Dixie J. Goss, "Pokeweed Antiviral Protein, a Ribosome Inactivating Protein: Activity, Inhibition and Prospects," *Toxins* 7, no. 2 (2015): 274–98, https://doi.org/10.3390/toxins7020274.

189. Fatih M. Uckun et al., "CNS Activity of Pokeweed Anti-Viral Protein (PAP) in Mice Infected with Lymphocytic Choriomeningitis Virus (LCMV)," *BMC Infectious Diseases* 5 (2005): 9, https://doi.org/10.1186/1471-2334-5-9.

190. D. E. Myers et al., "Large Scale Manufacturing of TXU (Anti-CD7)-Pokeweed Antiviral Protein (PAP) Immunoconjugate for Clinical Trials," *Leukemia & Lymphoma* 27, no. 3–4 (1997): 275–302, https://doi.org/10.3109/10428199709059683; F. M. Uckun et al., "In Vivo Efficacy of B43 (Anti-CD19)-Pokeweed Antiviral Protein Immunotoxin against Human Pre-B Cell Acute Lymphoblastic Leukemia in Mice with Severe Combined Immunodeficiency," *Blood* 79, no. 9 (1992): 2201–14, https://www.ncbi.nlm.nih.gov/pubmed/1373967.

191. José M. Zubeldia et al., "Exploring New Applications for *Rhodiola rosea*: Can We Improve the Quality of Life of Patients with Short-Term Hypothyroidism Induced by Hormone Withdrawal?," *Journal of Medicinal Food* 13, no. 6 (2010): 1287–92, https://doi.org/10.1089/jmf.2009.0286.

192. Nan Xia et al., "*Schisandra chinensis* and *Rhodiola rosea* Exert an Anti-Stress Effect on the HPA Axis and Reduce Hypothalamic c-Fos Expression in Rats Subjected to Repeated Stress," *Experimental and Therapeutic Medicine* 11, no. 1 (2016): 353–59, https://doi.org/10.3892/etm.2015.2882.

193. Shui-Jin Yang et al., "Antidepressant-Like Effects of Salidroside on Olfactory Bulbectomy-Induced pro-Inflammatory Cytokine Production and Hyperactivity of HPA Axis in Rats," *Pharmacology, Biochemistry, and Behavior* 124 (2014): 451–57, https://doi.org/10.1016/j.pbb.2014.07.015.

194. Somayeh Javidanpour et al., "The Cardioprotective Effect of Rosmarinic Acid on Acute Myocardial Infarction and Genes Involved in Ca2+homeostasis," *Free Radical Research* 51, no. 11–12 (2017): 911–23, https://doi.org/10.1080/10715762.2017.1390227.

195. Mesfin Yimam et al., "A Botanical Composition from *Morus alba*, *Ilex paraguariensis*, and *Rosmarinus officinalis* for Body Weight Management," *Journal of Medicinal Food* 20, no. 11 (2017): 1100–1112, https://doi.org/10.1089/jmf.2017.0002.

196. Margot Loussouarn et al., "Carnosic Acid and Carnosol, Two Major Antioxidants of Rosemary, Act through Different Mechanisms," *Plant Physiology* 175, no. 3 (2017): 1381–94, https://doi.org/10.1104/pp.17.01183.

197. Kai Kang et al., "Carnosic Acid Slows Photoreceptor Degeneration in the Pde6b(Rd10) Mouse Model of Retinitis Pigmentosa," *Scientific Reports* 6 (2016): 22632, https://doi.org/10.1038/srep22632.

198. Ya-Yu Chen et al., "Anti-Fibrotic Effect of Rosmarinic Acid on Inhibition of Pterygium Epithelial Cells," *International Journal of Ophthalmology* 11, no. 2 (2018): 189–95, https://doi.org/10.18240/ijo.2018.02.02; Thing-Fong Tzeng et al., "Antioxidant-Rich Extract from Plantaginis Semen Ameliorates Diabetic Retinal Injury in a Streptozotocin-Induced Diabetic Rat Model," *Nutrients* 8, no. 9 (2016), https://doi.org/10.3390/nu8090572.

199. Yun Hwan Kang et al., "Antiobesity Effects of the Water-Soluble Fraction of the Ethanol Extract of *Smilax china* L. Leaf in 3T3-L1 Adipocytes," *Nutrition Research and Practice* 9, no. 6 (2015): 606–12, https://doi.org/10.4162/nrp.2015.9.6.606.

200. Hwan Kang et al., "Antiobesity Effects of the Water-Soluble Fraction of the Ethanol Extract of Smilax China L. Leaf in 3T3-L1 Adipocytes."

201. Hyo Young Jung et al., "Valerenic Acid Protects Against Physical and Psychological Stress by Reducing the Turnover of Serotonin and Norepinephrine in Mouse Hippocampus-Amygdala Region," *Journal of Medicinal Food* 18, no. 12 (2015): 1333–39, https://doi.org/10.1089/jmf.2014.3412.

202. Heng-Wen Chen et al., "Chemical Components and Cardiovascular Activities of *Valeriana* Spp," *Evidence-Based Complementary and Alternative Medicine: ECAM* 2015 (2015): 947619, https://doi.org/10.1155/2015/947619.

203. Qi Ye et al., "Casticin, a Flavonoid Isolated from *Vitex rotundifolia*, Inhibits Prolactin Release In Vivo and In Vitro," *Acta Pharmacologica Sinica* 31, no. 12 (2010): 1564–68, https://doi.org/10.1038/aps.2010.178.

Chapter Three: Creating Herbal Formulas for Metabolic Conditions

1. Roberto Miatello et al., "Mechanisms of Cardiovascular Changes in an Experimental Model of Syndrome X and Pharmacological Intervention on the Renin-Angiotensin-System," *Current Vascular Pharmacology* 2, no. 4 (2004): 371–7, https://doi.org/info:doi/10.2174/1570161043385510.

2. D. S. Sartorelli et al., "Dietary Fructose, Fruits, Fruit Juices and Glucose Tolerance Status in Japanese-Brazilians," *Nutrition, Metabolism, and Cardiovascular Diseases: NMCD* 19, no. 2 (2009): 77–83, https://doi.org/10.1016/j.numecd.2008.04.004.

3. F. Panza et al., "Metabolic Syndrome, Mild Cognitive Impairment, and Dementia," *Current Alzheimer Research* 8, no. 5 (2011): 492–509, https://www.ncbi.nlm.nih.gov/pubmed/21605050.

4. Huroki Fujii et al., "Impact of Dietary Fiber Intake on Glycemic Control, Cardiovascular Risk Factors and Chronic Kidney Disease in Japanese Patients with Type 2 Diabetes Mellitus: The Fukuoka Diabetes Registry," *Nutrition Journal* 12 (2013): 159, https://doi.org/10.1186/1475-2891-12-159.

5. Johnson W. McRorie, "Evidence-Based Approach to Fiber Supplements and Clinically Meaningful Health Benefits, Part 2: What to Look for and How to Recommend an Effective Fiber Therapy," *Nutrition Today* 50, no. 2 (2015): 90–97, https://doi.org/10.1097/NT.0000000000000089.

6. Bettina Nowotny et al., "Low-Energy Diets Differing in Fibre, Red Meat and Coffee Intake Equally Improve Insulin Sensitivity in Type 2 Diabetes: A Randomised Feasibility Trial," *Diabetologia* 58, no. 2 (2015): 255–64, https://doi.org/10.1007/s00125-014-3457-8.

7. Anja Bosy-Westphal and Manfred J. Müller, "Impact of Carbohydrates on Weight Regain," *Current Opinion in Clinical Nutrition and Metabolic Care* 18, no. 4 (2015): 389–94, https://doi.org/10.1097/MCO.0000000000000193.

8. Kristina H. Jackson et al., "Effects of Whole and Refined Grains in a Weight-Loss Diet on Markers of Metabolic Syndrome in Individuals with Increased Waist Circumference: A Randomized Controlled-Feeding Trial," *The American Journal of Clinical Nutrition* 100, no. 2 (2014): 577–86, https://doi.org/10.3945/ajcn.113.078048; Di Dong et al., "Consumption of Specific Foods and Beverages and Excess Weight Gain Among Children and Adolescents," *Health Affairs (Project Hope)* 34, no. 11 (2015): 1940–48, https://doi.org/10.1377/hlthaff.2015.0434; Michael McCarthy, "High Fibre Diet May Be Good Alternative to Complex Weight Loss Regimen, US Study Finds," *BMJ (Clinical Research Ed.)* 350 (2015): h965, https://www.ncbi.nlm.nih.gov/pubmed/25698063.

9. Yunsheng Ma et al., "Single-Component versus Multicomponent Dietary Goals for the Metabolic Syndrome: A Randomized Trial," *Annals of Internal Medicine* 162, no. 4 (2015): 248–57, https://doi.org/10.7326/M14-0611.

10. Anna Maria Fulghesu et al., "N-Acetyl-Cysteine Treatment Improves Insulin Sensitivity in Women with Polycystic Ovary Syndrome," *Fertility and Sterility* 77, no. 6 (2002): 1128–35, https://www.ncbi.nlm.nih.gov/pubmed/12057717.

11. Christian K. Roberts et al., "HMG-CoA Reductase, Cholesterol 7alpha-Hydroxylase, LDL Receptor, SR-B1, and ACAT in Diet-Induced Syndrome X," *Kidney International* 66, no. 4 (2004): 1503–11, https://www.ncbi.nlm.nih.gov/pubmed/1.1523-1755.2004.00914.x.

12. George V. Vahouny and David Kritchevsky, eds., *Dietary Fiber in Health and Disease* (New York: Plenum Press 1982).

13. Najma Zaheer Baquer et al., "Metabolic and Molecular Action of *Trigonella foenum-graecum* (Fenugreek) and Trace Metals in Experimental Diabetic Tissues," *Journal of Biosciences* 36, no. 2 (2011): 383–96, https://www.ncbi.nlm.nih.gov/pubmed/21654091.

14. S. Kannappan and C. V. Anuradha, "Insulin Sensitizing Actions of Fenugreek Seed Polyphenols, Quercetin & Metformin in a Rat Model," *The Indian Journal of Medical Research* 129, no. 4 (2009): 401–8, https://www.ncbi.nlm.nih.gov/pubmed/19535835.

15. Radha Moorthy et al., "Anti-Hyperglycemic Compound (GII) from Fenugreek (*Trigonella foenum-graecum Linn.*) Seeds, Its Purification and Effect in Diabetes Mellitus," *Indian Journal of Experimental Biology* 48, no. 11 (2010): 1111–18, https://www.ncbi.nlm.nih.gov/pubmed/21117451.

16. Fu-rong Lu et al., "Clinical Observation on *Trigonella foenum-graecum* L. Total Saponins in Combination with Sulfonylureas in the Treatment of Type 2 Diabetes Mellitus," *Chinese Journal of Integrative Medicine* 14, no. 1 (2008): 56–60, https://doi.org/10.1007/s11655-007-9005.

17. Jack N. Losso et al., "Fenugreek Bread: A Treatment for Diabetes Mellitus," *Journal of Medicinal Food* 12, no. 5 (2009): 1046–49, https://doi.org/10.1089/jmf.2008.0199.

18. H. Leclerc, "Le *Galega officinalis* L.," *Presse Med (Paris)* 36, (1928): 1634; G. Parturier and G. Hugonot, "Le Galega dans le Traitement du Diabete," *Presse Med (Paris)* 43, no. 14 (1935): 258–60.

19. K. Pufahl and K. Schreiber, "Isolation of a New Guanidine Derivative from Goat's Rue, *Galega officinalis* L," *Experientia* 17 (1961): 302–3, https://www.ncbi.nlm.nih.gov/pubmed/13738405.

20. M. Sendrail et al., "Experimental Study of the Action of Plant Drugs with a Glycopenic Effect on the Cytological Structure of the Insular Pancreas," *La Semaine des Hopitaux: Organe Fonde par l'Association d'Enseignement Medical des Hopitaux de Paris* 37 (1961): 389–98, https://www.ncbi.nlm.nih.gov/pubmed/13750069.

21. V. Bettini et al., "Vasodilator and Inhibitory Effects of *Vaccinium myrtillus* on the Contractile Responses of Coronary Artery Segments to Acetylcholine: Role of the Prostacyclins and of the Endothelium-Derived Relaxing Factor," *Fitoter* 42, no. 1 (1991): 15–28; P. Morazzoni and M. J. Magistretti, "Effects of *Vaccinium myrtillus* Anthocyanosides on Prostacyclin-Like Activity in Rat Arterial Tissue," *Fitoterapia* 57 (1986): 11–14.

22. A. Colantuoni et al., "Effects of *Vaccinium myrtillus* Anthocyanosides on Arterial Vasomotion," *Arzneimittel-Forschung* 41, no. 9 (1991): 905–9, https://www.ncbi.nlm.nih.gov/pubmed/1796918.

23. B. Bever and G. R. Zahnd, "Plants with Oral Hypoglycaemic Action," *Quarterly Journal of Crude Drug Research* 17, no. 3–4 (1979): 139–96, https://doi.org/10.3109/13880207909065167.

24. K. K. Sharma et al., "Antihyperglycemic Effect of Onion: Effect on Fasting Blood Sugar and Induced Hyperglycemia in Man," *The Indian Journal of Medical Research* 65, no. 3 (1977): 422–29, https://www.ncbi.nlm.nih.gov/pubmed/336527.

25. A. A. Silver and John C. Krantz Jr., "The Effect of the Ingestion of Burdock Root on Normal and Diabetic Individuals: A Preliminary Report," *Annals of Internal Medicine* 5, no. 3 (1931): 274, https://doi.org/10.7326/0003-4819-5-3-274; J. J. Rumessen et al., "Fructans of Jerusalem Artichokes: Intestinal Transport, Absorption, Fermentation, and Influence on Blood Glucose, Insulin, and C-Peptide Responses in Healthy Subjects," *The American Journal of Clinical Nutrition* 52, no. 4 (1990): 675–81, https://www.ncbi.nlm.nih.gov/pubmed/2206038.

26. H. C. Simpson et al., "A High Carbohydrate Leguminous Fibre Diet Improves All Aspects of Diabetic Control," *Lancet* 1, no. 8210 (1981): 1–5, https://www.ncbi.nlm.nih.gov/pubmed/6109047.

27. Simpson et al., "A High Carbohydrate Leguminous Fibre Diet"; R. J. Marles and N. R. Farnsworth, "Antidiabetic Plants and Their Active Constituents," *Phytomedicine: International Journal of Phytotherapy and Phytopharmacology* 2, no. 2 (1995): 137–89, https://doi.org/10.1016/S0944-7113(11)80059-0; D. J. Jenkins et al., "Diabetic Diets: High Carbohydrate Combined with High Fiber," *The American*

Journal of Clinical Nutrition 33, no. 8 (1980): 1729–33, https://www.ncbi.nlm.nih.gov/pubmed/6250394.

28. Yinrun Ding et al., "The Mechanisms Underlying the Hypolipidaemic Effects of *Grifola frondosa* in the Liver of Rats," *Frontiers in Microbiology* 7 (2016): 1186, https://doi.org/10.3389/fmicb.2016.01186.

29. Hyun Yang et al., "*Lentinus edodes* Promotes Fat Removal in Hypercholesterolemic Mice," *Experimental and Therapeutic Medicine* 6, no. 6 (2013): 1409–13, https://doi.org/10.3892/etm.2013.1333.

30. Yu-Sheng Wu et al., "*Ganoderma lucidum* Beta 1,3/1,6 Glucan as an Immunomodulator in Inflammation Induced by a High-Cholesterol Diet," *BMC Complementary and Alternative Medicine* 16, no. 1 (2016): 500, https://doi.org/10.1186/s12906-016-1476-3.

31. Yen-Jung Chou et al., "Renal Protective Effects of Low Molecular Weight of Inonotus Obliquus Polysaccharide (LIOP) on HFD/STZ-Induced Nephropathy in Mice," *International Journal of Molecular Sciences* 17, no. 9 (2016), https://doi.org/10.3390/ijms17091535.

32. I. Abete et al., "Obesity and Metabolic Syndrome: Potential Benefit from Specific Nutritional Components," *Nutrition, Metabolism, and Cardiovascular Diseases: NMCD* 21, Suppl. 2 (2011): B1–15, https://doi.org/10.1016/j.numecd.2011.05.001.

33. Myung-Sook Choi et al., "Metabolic Response of Soy Pinitol on Lipid-Lowering, Antioxidant and Hepatoprotective Action in Hamsters Fed High Fat and High Cholesterol Diet," *Molecular Nutrition & Food Research* 53, no. 6 (2009): 751–59, https://doi.org/10.1002/mnfr.200800241.

34. Chia Hui Apphia Eu et al., "Glycyrrhizic Acid Improved Lipoprotein Lipase Expression, Insulin Sensitivity, Serum Lipid and Lipid Deposition in High-Fat Diet-Induced Obese Rats," *Lipids in Health and Disease* 9 (2010): 81, https://doi.org/10.1186/1476-511X-9-81.

35. Kaku Nakagawa et al., "Licorice Flavonoids Suppress Abdominal Fat Accumulation and Increase in Blood Glucose Level in Obese Diabetic KK-A(y) Mice," *Biological & Pharmaceutical Bulletin* 27, no. 11 (2004): 1775–78, https://www.ncbi.nlm.nih.gov/pubmed/15516721.

36. Chia Hui Apphia Eu et al., "Glycyrrhizic Acid Improved Lipoprotein Lipase Expression."

37. Hazem Hasan Kataya et al., "Effect of Licorice Extract on the Complications of Diabetes Nephropathy in Rats," *Drug and Chemical Toxicology* 34, no. 2 (2011): 101–8, https://doi.org/10.3109/01480545.2010.510524.

38. Debleena Dey et al., "A Lupinoside Prevented Fatty Acid Induced Inhibition of Insulin Sensitivity in 3T3 L1 Adipocytes," *Molecular and Cellular Biochemistry* 300, no. 1–2 (2007): 149–57, https://doi.org/10.1007/s11010-006-9378-1; Hong-yan Bai et al., "Influence of *Pueraria thomsonii* on Insulin Resistance Induced by Dexamethasone," *Zhongguo Zhong Yao Za Zhi = Zhongguo Zhongyao Zazhi (China Journal of Chinese Materia Medica)* 29, no. 4 (2004): 356–58, https://www.ncbi.nlm.nih.gov/pubmed/15706877.

39. Richard A. Anderson et al., "Cinnamon Extract Lowers Glucose, Insulin and Cholesterol in People with Elevated Serum Glucose," *Journal of Traditional and Complementary Medicine* 6, no. 4 (2016): 332–36, https://doi.org/10.1016/j.jtcme.2015.03.005.

40. R. Boniface et al., "Pharmacological Properties of Myrtillus Anthocyanosides: Correlation with Results of Treatment of Diabetic Microangiopathy," *Flavonoids and Bioflavonoids* 23 (1985): 293–301; G. Lagrue et al., "Pathology of the Microcirculation in Diabetes and Alterations of the Biosynthesis of Intracellular Matrix Molecules," *Frontiers of Matrix Biology* 7 (1970): 324–35; A. Scharrer and M. Ober, "Anthocyanosides in the Treatment of Retinopathies," *Klinische Monatsblatter fur Augenheilkunde* 178, no. 5 (1981): 386–89, https://doi.org/10.1055/s-2008-1057228.

41. G. Mussgnug and J. Alemany, "Studies on Peripheral Arterial Blood Circulation Disorders. XV. On the Problems of Conservative Therapy of Obliterating Peripheral Blood Circulation Disorders Demonstrated on Tincture and Extract from *Ginkgo biloba* L.," *Arzneimittel-Forschung* 18, no. 5 (1968): 543–50, https://www.ncbi.nlm.nih.gov/pubmed/5755881; H. P. Gau, "On the Treatment of Cerebral and Peripheral Arterial Obliterating Vascular Disorders with Tebonin," *Arztliche Forschung* 23, no. 3 (1969): 103–11, https://www.ncbi.nlm.nih.gov/pubmed/5819754.

42. J.-C. Stoclet and V. Schini-Kerth, "Dietary Flavonoids and Human Health," *Annales Pharmaceutiques Francaises* 69, no. 2 (2011): 78–90, https://doi.org/10.1016/j.pharma.2010.11.004.

43. Nozomu Matsunaga et al., "Bilberry and Its Main Constituents Have Neuroprotective Effects against Retinal Neuronal Damage In Vitro and In Vivo," *Molecular Nutrition & Food Research* 53, no. 7 (2009): 869–77, https://doi.org/10.1002/mnfr.200800394.

44. Nan Yao et al., "Protective Effects of Bilberry (*Vaccinium myrtillus* L.) Extract against Endotoxin-Induced Uveitis in Mice," *Journal of Agricultural and Food Chemistry* 58, no. 8 (2010): 4731–36, https://doi.org/10.1021/jf904572a.

45. Mary F.-F. Chong et al., "Fruit Polyphenols and CVD Risk: A Review of Human Intervention Studies," *The British Journal of Nutrition* 104, Suppl. 3 (2010): S28–39, https://doi.org/10.1017/S0007114510003922.

46. A. Kadar et al., "Influence of Anthocyanoside Treatment on the Cholesterol-Induced Atherosclerosis in the Rabbit," *Paroi Arterielle* 5, no. 4 (1979): 187–205, https://www.ncbi.nlm.nih.gov/pubmed/554974.

47. Seiji Miyake et al., "Vision Preservation during Retinal Inflammation by Anthocyanin-Rich Bilberry Extract: Cellular and Molecular Mechanism," *Laboratory Investigation: A Journal of Technical Methods and Pathology* 92, no. 1 (2012): 102–9, https://doi.org/10.1038/labinvest.2011.132; Hideto Osada et al., "Neuroprotective Effect of Bilberry Extract in a Murine Model of Photo-Stressed Retina," *PLoS One* 12, no. 6 (2017): e0178627, https://doi.org/10.1371/journal.pone.0178627.

48. Yong Wang et al., "Retinoprotective Effects of Bilberry Anthocyanins via Antioxidant, Anti-Inflammatory, and Anti-Apoptotic Mechanisms in a Visible Light-Induced Retinal Degeneration Model in Pigmented Rabbits," *Molecules (Basel, Switzerland)* 20, no. 12 (2015): 22395–410, https://doi.org/10.3390/molecules201219785.

49. Junghyun Kim et al., "*Vaccinium myrtillus* Extract Prevents or Delays the Onset of Diabetes-Induced Blood-Retinal Barrier Breakdown," *International Journal of Food Sciences and Nutrition* 66, no. 2 (2015): 236–42, https://doi.org/10.3109/09637486.2014.979319.

50. Chia Hui Apphia Eu et al., "Glycyrrhizic Acid Improved Lipoprotein Lipase Expression, Insulin Sensitivity, Serum Lipid and Lipid Deposition in High-Fat Diet-Induced Obese Rats," *Lipids in Health and Disease* 9 (2010): 81, https://doi.org/10.1186/1476-511X-9-81.

51. Hazem Hasan Kataya et al., "Effect of Licorice Extract on the Complications of Diabetes Nephropathy in Rats," *Drug and Chemical Toxicology* 34, no. 2 (2011): 101–8, https://doi.org/10.3109/01480545.2010.510524.

52. Debleena Dey et al., "A Lupinoside Prevented Fatty Acid Induced Inhibition of Insulin Sensitivity in 3T3 L1 Adipocytes," *Molecular and Cellular Biochemistry* 300, no. 1–2 (2007): 149–57, https://doi.org/10.1007/s11010-006-9378-1.

53. Hong-yan Bai et al., "Influence of *Pueraria thomsonii* on Insulin Resistance Induced by Dexamethasone," *Zhongguo Zhong Yao Za Zhi = Zhongguo Zhongyao Zazhi (China Journal of Chinese Materia Medica)* 29, no. 4 (2004): 356–58, https://www.ncbi.nlm.nih.gov/pubmed/15706877.

54. A. Al-Romaiyan et al., "A Novel *Gymnema sylvestre* Extract Stimulates Insulin Secretion from Human Islets In Vivo and In Vitro," *Phytotherapy Research: PTR* 24, no. 9 (2010): 1370–76, https://doi.org/10.1002/ptr.3125.

55. Vinay Kumar et al., "Evaluation of Antiobesity and Cardioprotective Effect of *Gymnema sylvestre* Extract in Murine Model," *Indian Journal of Pharmacology* 44, no. 5 (2012): 607–13, https://doi.org/10.4103/0253-7613.100387; Vinay Kumar et al., "Evaluation of Antiobesity and Cardioprotective Effect of *Gymnema sylvestre* Extract in Murine Model," *Indian Journal of Pharmacology* 44, no. 5 (2012): 607–13, https://doi.org/10.4103/0253-7613.100387; V. Kumar et al., "Protective Effect of *Gymnema sylvestre* Ethanol Extract on High Fat Diet-Induced Obese Diabetic Wistar Rats," *Indian Journal of Pharmaceutical Sciences* 76, no. 4 (2014): 315–22, https://www.ncbi.nlm.nih.gov/pubmed/25284929; V. Kumar et al., "Anti-Obesity Effect of *Gymnema sylvestre* Extract on High Fat Diet-Induced Obesity in Wistar Rats," *Drug Research* 63, no. 12 (2013): 625–32, https://doi.org/10.1055/s-0033-1349852.

56. Smriti Nanda Kumar et al., "An Open Label Study on the Supplementation of *Gymnema sylvestre* in Type 2 Diabetics," *Journal of Dietary Supplements* 7, no. 3 (2010): 273–82, https://doi.org/10.3109/19390211.2010.505901.

57. K. Baskaran et al., "Antidiabetic Effect of a Leaf Extract from *Gymnema sylvestre* in Non-Insulin-Dependent Diabetes Mellitus Patients," *Journal of Ethnopharmacology* 30, no. 3 (1990): 295–300, https://www.ncbi.nlm.nih.gov/pubmed/2259217; E. R. Shanmugasundaram et al., "Use of *Gymnema sylvestre* Leaf Extract in the Control of Blood Glucose in Insulin-Dependent Diabetes Mellitus," *Journal of Ethnopharmacology* 30, no. 3 (1990): 281–94, https://www.ncbi.nlm.nih.gov/pubmed/2259216.

58. Bhuvaneswari Ch et al., "Abiotic Elicitation of Gymnemic Acid in the Suspension Cultures of *Gymnema sylvestre*," *World Journal of Microbiology & Biotechnology* 28, no. 2 (2012): 741–47, https://doi.org/10.1007/s11274-011-0870-8; Giovanni Di Fabio et al., "Triterpenoids from *Gymnema sylvestre* and Their Pharmacological Activities," *Molecules* 19, no. 8 (2014): 10956–81, https://doi.org/10.3390/molecules190810956.

59. Parijat Kanetkar et al., "*Gymnema sylvestre*: A Memoir," *Journal of Clinical Biochemistry and Nutrition* 41, no. 2 (2007): 77–81, https://doi.org/10.3164/jcbn.2007010.

60. E. R. Shanmugasundaram et al., "Use of *Gymnema sylvestre* Leaf Extract in the Control of Blood Glucose in Insulin-Dependent Diabetes Mellitus," *Journal of Ethnopharmacology* 30, no. 3 (1990): 281–94, https://www.ncbi.nlm.nih.gov/pubmed/2259216; K. Baskaran et al., "Antidiabetic Effect of a Leaf Extract from *Gymnema sylvestre* in Non-Insulin-Dependent Diabetes Mellitus Patients," *Journal of Ethnopharmacology* 30, no. 3 (1990): 295–300, https://www.ncbi.nlm.nih.gov/pubmed/2259217; A. Al-Romaiyan et al., "A Novel Extract of *Gymnema sylvestre* Improves Glucose Tolerance In Vivo and Stimulates Insulin Secretion and Synthesis In Vitro," *Phytotherapy Research: PTR* 27, no. 7 (2013): 1006–11, https://doi.org/10.1002/ptr.4815.

61. E. R. Shanmugasundaram et al., "Possible Regeneration of the Islets of Langerhans in Streptozotocin-Diabetic Rats Given *Gymnema sylvestre* Leaf Extracts," *Journal of Ethnopharmacology* 30, no. 3 (1990): 265–79, https://www.ncbi.nlm.nih.gov/pubmed/2259215; A. Bakrudeen Ali Ahmed et al., "In Vitro Callus and In Vivo Leaf Extract of *Gymnema sylvestre* Stimulate β-Cells Regeneration and Anti-Diabetic Activity in Wistar Rats," *Phytomedicine: International Journal of Phytotherapy and Phytopharmacology* 17, no. 13 (2010): 1033–39, https://doi.org/10.1016/j.phymed.2010.03.019; G. L. Snigur et al., "Structural Alterations in Pancreatic Islets in Streptozotocin-Induced Diabetic Rats Treated with Bioactive Additive on the Basis of *Gymnema sylvestre*," *Morfologiia* 133, no. 1 (2008): 60–64, https://www.ncbi.nlm.nih.gov/pubmed/19069418.

62. V. Krecman et al., "Silymarin Inhibits the Development of Diet-Induced Hypercholesterolemia in Rats," *Planta Medica* 64, no. 2 (1998): 138–42, https://www.ncbi.nlm.nih.gov/pubmed/9525106.

63. E. Shaker et al., "Silymarin, the Antioxidant Component and *Silybum marianum* Extracts Prevent Liver Damage," *Food and Chemical Toxicology: An International Journal Published for*

the British Industrial Biological Research Association 48, no. 3 (2010): 803–6, https://doi.org/10.1016/j.fct.2009.12.011; Shu Yun Zhu et al., "*Silybum marianum* Oil Attenuates Oxidative Stress and Ameliorates Mitochondrial Dysfunction in Mice Treated with D-Galactose," *Pharmacognosy Magazine* 10, Suppl. 1 (2014): S92–99, https://doi.org/10.4103/0973-1296.127353.

64. V. Krecman et al., "Silymarin Inhibits the Development of Diet-Induced Hypercholesterolemia in Rats," *Planta Medica* 64, no. 2 (1998): 138–42, https://www.ncbi.nlm.nih.gov/pubmed/9525106.

65. N. Skottová and V. Krecman, "Silymarin as a Potential Hypocholesterolaemic Drug," *Physiological Research* 47, no. 1 (1998): 1–7, https://www.ncbi.nlm.nih.gov/pubmed/9708694; Fulvio Cacciapuoti et al., "Silymarin in Non Alcoholic Fatty Liver Disease," *World Journal of Hepatology* 5, no. 3 (2013): 109–13, https://doi.org/10.4254/wjh.v5.i3.109.

66. Fulvio Cacciapuoti et al., "Silymarin in Non Alcoholic Fatty Liver Disease," *World Journal of Hepatology* 5, no. 3 (2013): 109–13, https://doi.org/10.4254/wjh.v5.i3.109; Nina Skottová et al., "Phenolics-Rich Extracts from *Silybum marianum* and *Prunella vulgaris* Reduce a High-Sucrose Diet Induced Oxidative Stress in Hereditary Hypertriglyceridemic Rats," *Pharmacological Research* 50, no. 2 (2004): 123–30, https://doi.org/10.1016/j.phrs.2003.12.013.

67. Zhiguo Zhang et al., "Berberine Activates Thermogenesis in White and Brown Adipose Tissue," *Nature Communications* 5 (2014): 5493, https://doi.org/10.1038/ncomms6493.

68. Yueshan Hu et al., "Metformin and Berberine Prevent Olanzapine-Induced Weight Gain in Rats," *PLoS One* 9, no. 3 (2014): e93310, https://doi.org/10.1371/journal.pone.0093310.

69. Yun S. Lee et al., "Berberine, a Natural Plant Product, Activates AMP-Activated Protein Kinase with Beneficial Metabolic Effects in Diabetic and Insulin-Resistant States," *Diabetes* 55, no. 8 (2006): 2256–64, https://doi.org/10.2337/db06-0006; Jing Yang et al., "Berberine Improves Insulin Sensitivity by Inhibiting Fat Store and Adjusting Adipokines Profile in Human Preadipocytes and Metabolic Syndrome Patients," *Evidence-Based Complementary and Alternative Medicine: ECAM* 2012 (2012): 363845, https://doi.org/10.1155/2012/363845.

70. Jing Yang et al., "Berberine Improves Insulin Sensitivity by Inhibiting Fat Store and Adjusting Adipokines Profile in Human Preadipocytes and Metabolic Syndrome Patients," *Evidence-Based Complementary and Alternative Medicine: ECAM* 2012 (2012): 363845, https://doi.org/10.1155/2012/363845.

71. Yuan An et al., "The Use of Berberine for Women with Polycystic Ovary Syndrome Undergoing IVF Treatment," *Clinical Endocrinology* 80, no. 3 (2014): 425–31, https://doi.org/10.1111/cen.12294.

72. Yueshan Hu et al., "Lipid-Lowering Effect of Berberine in Human Subjects and Rats," *Phytomedicine: International Journal of Phytotherapy and Phytopharmacology* 19, no. 10 (2012): 861–67, https://doi.org/10.1016/j.phymed.2012.05.009.

73. Francesco Di Pierro et al., "Pilot Study on the Additive Effects of Berberine and Oral Type 2 Diabetes Agents for Patients with Suboptimal Glycemic Control," *Diabetes, Metabolic Syndrome and Obesity: Targets and Therapy* 5 (2012): 213–17, https://doi.org/10.2147/DMSO.S33718.

74. Francesco Di Pierro et al., "Clinical Role of a Fixed Combination of Standardized *Berberis aristata* and *Silybum marianum* Extracts in Diabetic and Hypercholesterolemic Patients Intolerant to Statins," *Diabetes, Metabolic Syndrome and Obesity: Targets and Therapy* 8 (2015): 89–96, https://doi.org/10.2147/DMSO.S78877.

75. Chang Chu et al., "Proliposomes for Oral Delivery of Dehydrosilymarin: Preparation and Evaluation In Vitro and In Vivo," *Acta Pharmacologica Sinica* 32, no. 7 (2011): 973–80, https://doi.org/10.1038/aps.2011.25.

76. Francesco Di Pierro et al., "Preliminary Study about the Possible Glycemic Clinical Advantage in Using a Fixed Combination of *Berberis aristata* and *Silybum marianum* Standardized Extracts versus Only *Berberis aristata* in Patients with Type 2 Diabetes," *Clinical Pharmacology: Advances and Applications* 5 (2013): 167–74, https://doi.org/10.2147/CPAA.S54308.

77. J. Cunnick and D. Takemoto, "Bitter Melon (*Mormordica charantia*)," *Journal Naturopathic Medicine* 4, no. 1 (1993): 16–21; Y. Srivastava et al., "Antidiabetic and Adaptogenic Properties of *Momordica charantia* Extract: An Experimental and Clinical Evaluation," *Phytotherapy Research* 7, no. 4 (1993): 285–289, http://onlinelibrary.wiley.com/doi/10.1002/ptr.2650070405/abstract; J. Welihinda et al., "Effect of *Momordica charantia* on the Glucose Tolerance in Maturity Onset Diabetes," *Journal of Ethnopharmacology* 17, no. 3 (1986): 277–82, https://www.ncbi.nlm.nih.gov/pubmed/3807390; T. B. Ng et al., "Insulin-like Molecules in *Momordica charantia* Seeds," *Journal of Ethnopharmacology* 15, no. 1 (1986): 107–17, https://www.ncbi.nlm.nih.gov/pubmed/3520153; B. A. Leatherdale et al., "Improvement in Glucose Tolerance Due to *Momordica charantia* (Karela)," *British Medical Journal (Clinical Research Ed.)* 282, no. 6279 (1981): 1823–24, https://www.ncbi.nlm.nih.gov/pubmed/6786635; J. Welihinda et al., "The Insulin-Releasing Activity of the Tropical Plant *Momordica charantia*," *Acta Biologica et Medica Germanica* 41, no. 12 (1982): 1229–40, https://www.ncbi.nlm.nih.gov/pubmed/6765165.

78. E. R. Shanmugasundaram et al., "Use of *Gymnema sylvestre* Leaf Extract in the Control of Blood Glucose in Insulin-Dependent Diabetes Mellitus," *Journal of Ethnopharmacology* 30, no. 3 (1990): 281–94, https://www.ncbi.nlm.nih.gov/pubmed/2259216; K. Baskaran et al., "Antidiabetic Effect of a Leaf Extract from *Gymnema sylvestre* in Non-Insulin-Dependent Diabetes Mellitus Patients," *Journal of Ethnopharmacology* 30, no. 3 (1990): 295–300, https://www.ncbi.nlm.nih.gov/pubmed/2259217.

79. E. R. Shanmugasundaram et al., "Possible Regeneration of the Islets of Langerhans in Streptozotocin-Diabetic

Rats Given *Gymnema sylvestre* Leaf Extracts," *Journal of Ethnopharmacology* 30, no. 3 (1990): 265–79, https://www.ncbi.nlm.nih.gov/pubmed/2259215.

80. Pragya Tiwari et al., "Phytochemical and Pharmacological Properties of *Gymnema sylvestre*: An Important Medicinal Plant," *BioMed Research International* 2014 (2014): 830285, https://doi.org/10.1155/2014/830285.

81. Giovanni Di Fabio et al., "C-4 Gem-Dimethylated Oleanes of *Gymnema sylvestre* and Their Pharmacological Activities," *Molecules* 18, no. 12 (2013): 14892–919, https://doi.org/10.3390/molecules181214892.

82. R. J. Marles and N. R. Farnsworth, "Antidiabetic Plants and Their Active Constituents," *Phytomedicine: International Journal of Phytotherapy and Phytopharmacology* 2, no. 2 (1995): 137–89, https://doi.org/10.1016/S0944-7113(11)80059-0.

83. Z. Madar et al., "Glucose-Lowering Effect of Fenugreek in Non-Insulin Dependent Diabetics," *European Journal of Clinical Nutrition* 42, no. 1 (1988): 51–54, https://www.ncbi.nlm.nih.gov/pubmed/3286242; R. D. Sharma, "Effect of Fenugreek Seeds and Leaves on Blood Glucose and Serum Insulin Responses in Human Subjects," *Nutrition Research* 6, no. 12 (1986): 1353–64, https://www.sciencedirect.com/science/article/pii/S0271531786800203; Z. Madar et al., "Glucose-Lowering Effect of Fenugreek in Non-Insulin Dependent Diabetics," *European Journal of Clinical Nutrition* 42, no. 1 (1988): 51–54, https://www.ncbi.nlm.nih.gov/pubmed/3286242.

84. R. D. Sharma et al., "Effect of Fenugreek Seeds on Blood Glucose and Serum Lipids in Type I Diabetes," *European Journal of Clinical Nutrition* 44, no. 4 (1990): 301–6, https://www.ncbi.nlm.nih.gov/pubmed/2194788; G. Ribes et al., "Effects of Fenugreek Seeds on Endocrine Pancreatic Secretions in Dogs," *Annals of Nutrition & Metabolism* 28, no. 1 (1984): 37–43, https://www.ncbi.nlm.nih.gov/pubmed/6703649; M. A. Riyad et al., "Effect of Fenugreek and Lupine Seeds on the Development of Experimental Diabetes in Rats," *Planta Medica* 54, no. 4 (1988): 286–90, https://doi.org/10.1055/s-2006-962434; M. A. Ajabnoor and A. K. Tilmisany, "Effect of *Trigonella foenum-graecum* on Blood Glucose Levels in Normal and Alloxan-Diabetic Mice," *Journal of Ethnopharmacology* 22, no. 1 (1988): 45–49, https://www.ncbi.nlm.nih.gov/pubmed/3352284.

85. B. K. Chakravarthy et al., "Functional Beta Cell Regeneration in Rats by Epicatechin," *Lancet* 2 (1981): 749–60.

86. B. K. Chakravarthy et al., "Functional Beta Cell Regeneration in the Islets of Pancreas in Alloxan Induced Diabetic Rats by (-)-Epicatechin," *Life Sciences* 31, no. 24 (1982): 2693–97, https://www.ncbi.nlm.nih.gov/pubmed/6759833.

87. Gail Nunlee-Bland et al., "Vitamin D Deficiency and Insulin Resistance in Obese African-American Adolescents," *Journal of Pediatric Endocrinology & Metabolism: JPEM* 24, no. 1–2 (2011): 29–33, https://www.ncbi.nlm.nih.gov/pubmed/21528812; Han Seok Choi et al., "Low Serum Vitamin D Is Associated with High Risk of Diabetes in Korean Adults," *The Journal of Nutrition* 141, no. 8 (2011): 1524–28, https://doi.org/10.3945/jn.111.139121.

88. Shin Terada et al., "Dietary Intake of Medium- and Long-Chain Triacylglycerols Prevents the Progression of Hyperglycemia in Diabetic Ob/Ob Mice," *Journal of Oleo Science* 64, no. 6 (2015): 683–88, https://doi.org/10.5650/jos.ess14287.

89. Sabri Ahmed Rial et al., "Gut Microbiota and Metabolic Health: The Potential Beneficial Effects of a Medium Chain Triglyceride Diet in Obese Individuals," *Nutrients* 8, no. 5 (2016), https://doi.org/10.3390/nu8050281.

90. Jonathan Thevenet et al., "Medium-Chain Fatty Acids Inhibit Mitochondrial Metabolism in Astrocytes Promoting Astrocyte-Neuron Lactate and Ketone Body Shuttle Systems," *FASEB Journal: Official Publication of the Federation of American Societies for Experimental Biology* 30, no. 5 (2016): 1913–26, https://doi.org/10.1096/fj.201500182.

91. Ying Chen et al., "Evaluating Pharmacological Effects of Two Major Components of Shuangdan Oral Liquid: Role of Danshensu and Paeonol in Diabetic Nephropathy Rat," *Biomolecules & Therapeutics* 24, no. 5 (2016): 536–42, https://doi.org/10.4062/biomolther.2015.191.

92. S. K. Goyal et al., "Stevia (*Stevia rebaudiana*) a Bio-Sweetener: A Review," *International Journal of Food Sciences and Nutrition* 61, no. 1 (2010): 1–10, https://doi.org/10.3109/09637480903193049.

93. J.-C. Chang et al., "Increase of Insulin Sensitivity by Stevioside in Fructose-Rich Chow-Fed Rats," *Hormon-Und Stoffwechselforschung = Hormones et Metabolisme (Hormone and Metabolic Research)* 37, no. 10 (2005): 610–16, https://doi.org/10.1055/s-2005-870528; Himanshu Misra et al., "Antidiabetic Activity of Medium-Polar Extract from the Leaves of *Stevia rebaudiana Bert.* (Bertoni) on Alloxan-Induced Diabetic Rats," *Journal of Pharmacy & Bioallied Sciences* 3, no. 2 (2011): 242–48, https://doi.org/10.4103/0975-7406.80779.

94. Himanshu Misra et al., "Antidiabetic Activity of Medium-Polar Extract from the Leaves of *Stevia rebaudiana Bert.* (Bertoni) on Alloxan-Induced Diabetic Rats," *Journal of Pharmacy & Bioallied Sciences* 3, no. 2 (2011): 242–48, https://doi.org/10.4103/0975-7406.80779.

95. Catherine Ulbricht et al., "An Evidence-Based Systematic Review of Stevia by the Natural Standard Research Collaboration," *Cardiovascular & Hematological Agents in Medicinal Chemistry* 8, no. 2 (2010): 113–27, https://www.ncbi.nlm.nih.gov/pubmed/20370653; P. Chan et al., "A Double-Blind Placebo-Controlled Study of the Effectiveness and Tolerability of Oral Stevioside in Human Hypertension," *British Journal of Clinical Pharmacology* 50, no. 3 (2000): 215–20, https://www.ncbi.nlm.nih.gov/pubmed/10971305.

96. R. S. Kujur et al., "Antidiabetic Activity and Phytochemical Screening of Crude Extract of *Stevia rebaudiana* in Alloxan-Induced Diabetic Rats," *Pharmacognosy Research* 2, no. 4 (2010): 258–63, https://doi.org/10.4103/0974-8490.69128.

97. Jeong-Eun Park and Youn-Soo Cha, "*Stevia rebaudiana Bertoni* Extract Supplementation Improves Lipid and Carnitine Profiles in C57BL/6J Mice Fed a High-Fat Diet," *Journal of the Science of Food and Agriculture* 90, no. 7 (2010): 1099–1105, https://doi.org/10.1002/jsfa.3906.

98. Stig Eric Underbjerg Dyrskog et al., "Preventive Effects of a Soy-Based Diet Supplemented with Stevioside on the Development of the Metabolic Syndrome and Type 2 Diabetes in Zucker Diabetic Fatty Rats," *Metabolism: Clinical and Experimental* 54, no. 9 (2005): 1181–88, https://doi.org/10.1016/j.metabol.2005.03.026; Tso-Hsiao Chen et al., "Mechanism of the Hypoglycemic Effect of Stevioside, a Glycoside of *Stevia rebaudiana*," *Planta Medica* 71, no. 2 (2005): 108–13, https://doi.org/10.1055/s-2005-837775; P. B. Jeppesen et al., "Antihyperglycemic and Blood Pressure-Reducing Effects of Stevioside in the Diabetic Goto-Kakizaki Rat," *Metabolism: Clinical and Experimental* 52, no. 3 (2003): 372–78, https://doi.org/10.1053/meta.2003.50058.

99. Yuan Chen et al., "Characterization of the Organic Component of Low-Molecular-Weight Chromium-Binding Substance and Its Binding of Chromium," *The Journal of Nutrition* 141, no. 7 (2011): 1225–32, https://doi.org/10.3945/jn.111.139147.

100. John B. Vincent, "Recent Advances in the Nutritional Biochemistry of Trivalent Chromium," *The Proceedings of the Nutrition Society* 63, no. 1 (2004): 41–47, https://doi.org/10.1079/PNS2003315.

101. Dr. Duke's Phytochemical and Ethnobotanical Databases, National Agricultural Library, https://phytochem.nal.usda.gov/phytochem/search.

102. Adriana G. Chicco et al., "Dietary Chia Seed (*Salvia hispanica* L.) Rich in Alpha-Linolenic Acid Improves Adiposity and Normalises Hypertriacylglycerolaemia and Insulin Resistance in Dyslipaemic Rats," *The British Journal of Nutrition* 101, no. 1 (2009): 41–50, https://doi.org/10.1017/S000711450899053X.

103. Andrea S. Rossi et al., "Dietary Chia Seed Induced Changes in Hepatic Transcription Factors and Their Target Lipogenic and Oxidative Enzyme Activities in Dyslipidaemic Insulin-Resistant Rats," *The British Journal of Nutrition* 109, no. 9 (2013): 1617–27, https://doi.org/10.1017/S0007114512003558; M. R. Ferreira et al., "Dietary Salba (*Salvia hispanica* L.) Ameliorates the Adipose Tissue Dysfunction of Dyslipemic Insulin-Resistant Rats through Mechanisms Involving Oxidative Stress, Inflammatory Cytokines and Peroxisome Proliferator-Activated Receptor γ," *European Journal of Nutrition* 57, no. 1 (2018): 83–94, https://doi.org/10.1007/s00394-016-1299-5.

104. M. E. Oliva et al., "Dietary Salba (*Salvia hispanica* L.) Seed Rich in α-Linolenic Acid Improves Adipose Tissue Dysfunction and the Altered Skeletal Muscle Glucose and Lipid Metabolism in Dyslipidemic Insulin-Resistant Rats," *Prostaglandins, Leukotrienes, and Essential Fatty Acids* 89, no. 5 (2013): 279–89, https://doi.org/10.1016/j.plefa.2013.09.010.

105. Agustina Creus et al., "Dietary Salba (*Salvia hispanica* L.) Improves the Altered Metabolic Fate of Glucose and Reduces Increased Collagen Deposition in the Heart of Insulin-Resistant Rats," *Prostaglandins, Leukotrienes, and Essential Fatty Acids* 121 (2017): 30–39, https://doi.org/10.1016/j.plefa.2017.06.002.

106. Agustina Creus et al., "Mechanisms Involved in the Improvement of Lipotoxicity and Impaired Lipid Metabolism by Dietary α-Linolenic Acid Rich *Salvia hispanica* L. (Salba) Seed in the Heart of Dyslipemic Insulin-Resistant Rats," *Journal of Clinical Medicine* 5, no. 2 (2016): 18, https://doi.org/10.3390/jcm5020018.

107. H. Satoh et al., "Yacon Diet (*Smallanthus sonchifolius*, Asteraceae) Improves Hepatic Insulin Resistance via Reducing Trb3 Expression in Zucker Fa/Fa Rats," *Nutrition & Diabetes* 3 (2013): e70, https://doi.org/10.1038/nutd.2013.11.

108. Ni Wayan Arya Utami et al., "Comparison of Yacon (*Smallanthus sonchifolius*) Tuber with Commercialized Fructo-Oligosaccharides (FOS) in Terms of Physiology, Fermentation Products and Intestinal Microbial Communities in Rats," *Bioscience of Microbiota, Food and Health* 32, no. 4 (2013): 167–78, https://doi.org/10.12938/bmfh.32.167.

109. Brunno F. R. Caetano et al., "Yacon (*Smallanthus sonchifolius*) as a Food Supplement: Health-Promoting Benefits of Fructooligosaccharides," *Nutrients* 8, no. 7 (2016), https://doi.org/10.3390/nu8070436.

110. Vlasios Goulas et al., "Functional Components of Carob Fruit: Linking the Chemical and Biological Space," *International Journal of Molecular Sciences* 17, no. 11 (2016): 1875, https://doi.org/10.3390/ijms17111875.

111. G. Baños et al., "Medicinal Agents in the Metabolic Syndrome," *Cardiovascular & Hematological Agents in Medicinal Chemistry* 6, no. 4 (2008): 237–52, https://www.ncbi.nlm.nih.gov/pubmed/18855636; William T. Cefalu et al., "Efficacy of Dietary Supplementation with Botanicals on Carbohydrate Metabolism in Humans," *Endocrine, Metabolic & Immune Disorders Drug Targets* 8, no. 2 (2008): 78–81, https://www.ncbi.nlm.nih.gov/pubmed/18537692.

112. Sahng-Wook Hahm et al., "*Opuntia humifusa* Stems Lower Blood Glucose and Cholesterol Levels in Streptozotocin-Induced Diabetic Rats," *Nutrition Research* 31, no. 6 (2011): 479–87, https://doi.org/10.1016/j.nutres.2011.05.002.

113. Francisco Javier Alarcon-Aguilar et al., "Hypoglycemic Activity of Two Polysaccharides Isolated from *Opuntia ficus-indica* and *O. streptacantha*," *Proceedings of the Western Pharmacology Society* 46 (2003): 139–42, https://www.ncbi.nlm.nih.gov/pubmed/14699912.

114. R. Ibañez-Camacho and R. Roman-Ramos, "Hypoglycemic Effect of *Opuntia cactus*," *Archivos de Investigacion Medica* 10, no. 4 (1979): 223–30, https://www.ncbi.nlm.nih.gov/pubmed/539865.

115. Elodie Linarès et al., "The Effect of NeOpuntia on Blood Lipid Parameters—Risk Factors for the Metabolic Syndrome (Syndrome X)," *Advances in Therapy* 24, no. 5 (2007): 1115–25, https://www.ncbi.nlm.nih.gov/pubmed/18029338; Chun-yan Li et al., "Regulative Effect of Opuntia Powder on

Blood Lipids in Rats and Its Mechanism," *Zhongguo Zhong Yao Za Zhi = Zhongguo Zhongyao Zazhi (China Journal of Chinese Materia Medica)* 30, no. 9 (2005): 694–96, https://www.ncbi.nlm.nih.gov/pubmed/16075737; Phil-Sun Oh and Kye-Taek Lim, "Glycoprotein (90 KDa) Isolated from *Opuntia ficus-indica* var. *saboten* MAKINO Lowers Plasma Lipid Level through Scavenging of Intracellular Radicals in Triton WR-1339-Induced Mice," *Biological & Pharmaceutical Bulletin* 29, no. 7 (2006): 1391–96, https://www.ncbi.nlm.nih.gov/pubmed/16819175.

116. Phil-Sun Oh and Kye-Taek Lim, "Glycoprotein (90 KDa) Isolated from *Opuntia ficus-indica* var. *saboten* MAKINO Lowers Plasma Lipid Level through Scavenging of Intracellular Radicals in Triton WR-1339-Induced Mice," *Biological & Pharmaceutical Bulletin* 29, no. 7 (2006): 1391–96, https://www.ncbi.nlm.nih.gov/pubmed/16819175.

117. Monia Ennouri et al., "Evaluation of Some Biological Parameters of *Opuntia ficus-indica*. 2. Influence of Seed Supplemented Diet on Rats," *Bioresource Technology* 97, no. 16 (2006): 2136–40, https://doi.org/10.1016/j.biortech.2005.09.031.

118. Ennouri et al., "Evaluation of Some Biological Parameters of *Opuntia ficus-indica*"; Rubeena Saleem et al., "Hypotensive Activity, Toxicology and Histopathology of Opuntioside-I and Methanolic Extract of *Opuntia dillenii*," *Biological & Pharmaceutical Bulletin* 28, no. 10 (2005): 1844–51, https://www.ncbi.nlm.nih.gov/pubmed/16204933.

119. Elena Castellanos-Santiago and Elhadi M. Yahia, "Identification and Quantification of Betalains from the Fruits of 10 Mexican Prickly Pear Cultivars by High-Performance Liquid Chromatography and Electrospray Ionization Mass Spectrometry," *Journal of Agricultural and Food Chemistry* 56, no. 14 (2008): 5758–64, https://doi.org/10.1021/jf800362t.

120. Florian C. Stintzing et al., "Color, Betalain Pattern, and Antioxidant Properties of Cactus Pear (*Opuntia* spp.) Clones," *Journal of Agricultural and Food Chemistry* 53, no. 2 (2005): 442–51, https://doi.org/10.1021/jf048751y; Luisa Tesoriere et al., "Supplementation with Cactus Pear (*Opuntia ficus-indica*) Fruit Decreases Oxidative Stress in Healthy Humans: A Comparative Study with Vitamin C," *The American Journal of Clinical Nutrition* 80, no. 2 (2004): 391–95, https://www.ncbi.nlm.nih.gov/pubmed/15277160.

121. Ming Hong Lee et al., "Inhibition of Nitric Oxide Synthase Expression in Activated Microglia and Peroxynitrite Scavenging Activity by *Opuntia ficus indica* var. *saboten*," *Phytotherapy Research: PTR* 20, no. 9 (2006): 742–47, https://doi.org/10.1002/ptr.1942.

122. C. Gentile et al., "Antioxidant Betalains from Cactus Pear (*Opuntia ficus-indica*) Inhibit Endothelial ICAM-1 Expression," *Annals of the New York Academy of Sciences* 1028 (2004): 481–86, https://doi.org/10.1196/annals.1322.057; Jae Youl Cho et al., "Radical Scavenging and Anti-Inflammatory Activity of Extracts from *Opuntia humifusa* raf," *The Journal of Pharmacy and Pharmacology* 58, no. 1 (2006): 113–19, https://doi.org/10.1211/jpp.58.1.0014.

123. Najla Hfaiedh et al., "Protective Effect of Cactus (*Opuntia ficus indica*) Cladode Extract upon Nickel-Induced Toxicity in Rats," *Food and Chemical Toxicology: An International Journal Published for the British Industrial Biological Research Association* 46, no. 12 (2008): 3759–63, https://doi.org/10.1016/j.fct.2008.09.059; Lazhar Zourgui et al., "Cactus (*Opuntia ficus-indica*) Cladodes Prevent Oxidative Damage Induced by the Mycotoxin Zearalenone in Balb/C Mice," *Food and Chemical Toxicology: An International Journal Published for the British Industrial Biological Research Association* 46, no. 5 (2008): 1817–24, https://doi.org/10.1016/j.fct.2008.01.023; Saida Ncibi et al., "*Opuntia ficus indica* Extract Protects against Chlorpyrifos-Induced Damage on Mice Liver," *Food and Chemical Toxicology: An International Journal Published for the British Industrial Biological Research Association* 46, no. 2 (2008): 797–802, https://doi.org/10.1016/j.fct.2007.08.047; E. M. Galati et al., "*Opuntia ficus indica* (l.) Mill. Fruit Juice Protects Liver from Carbon Tetrachloride-Induced Injury," *Phytotherapy Research: PTR* 19, no. 9 (2005): 796–800, https://doi.org/10.1002/ptr.1741.

124. Ning Zhang et al., "Astragaloside IV Improves Metabolic Syndrome and Endothelium Dysfunction in Fructose-Fed Rats," *Molecules* 16, no. 5 (2011): 3896–3907, https://doi.org/10.3390/molecules16053896; Xiaoxing Yin et al., "The Antioxidative Effects of *Astragalus saponin I* Protect against Development of Early Diabetic Nephropathy," *Journal of Pharmacological Sciences* 101, no. 2 (2006): 166–73, https://www.ncbi.nlm.nih.gov/pubmed/16766854.

125. Ming-en Xu et al., "Effects of Astragaloside IV on Pathogenesis of Metabolic Syndrome in Vitro," *Acta Pharmacologica Sinica* 27, no. 2 (2006): 229–36, https://doi.org/10.1111/j.1745-7254.2006.00243.x.

126. Feng Zou et al., "Astragalus Polysaccharides Alleviates Glucose Toxicity and Restores Glucose Homeostasis in Diabetic States via Activation of AMPK," *Acta Pharmacologica Sinica* 30, no. 12 (2009): 1607–15, https://doi.org/10.1038/aps.2009.168.

127. Min Liu et al., "Astragalus Polysaccharide Improves Insulin Sensitivity in KKAy Mice: Regulation of PKB/GLUT4 Signaling in Skeletal Muscle," *Journal of Ethnopharmacology* 127, no. 1 (2010): 32–37, https://doi.org/10.1016/j.jep.2009.09.055.

128. Menglei Chao et al., "Improving Insulin Resistance with Traditional Chinese Medicine in Type 2 Diabetic Patients," *Endocrine* 36, no. 2 (2009): 268–74, https://doi.org/10.1007/s12020-009-9222-y.

129. Da-Duo Liu et al., "Effects of *Siraitia grosvenorii* Fruits Extracts on Physical Fatigue in Mice," *Iranian Journal of Pharmaceutical Research: IJPR* 12, no. 1 (2013): 115–21, https://www.ncbi.nlm.nih.gov/pubmed/24250579.

130. Naoki Harada et al., "Mogrol Derived from *Siraitia grosvenorii* Mogrosides Suppresses 3T3-L1 Adipocyte Differentiation by Reducing CAMP-Response Element-Binding Protein Phosphorylation and Increasing AMP-Activated Protein Kinase Phosphorylation," *PLoS One* 11, no. 9 (2016): e0162252, https://doi.org/10.1371/journal.pone.0162252.

131. Q. Xu et al., "Antioxidant Effect of Mogrosides against Oxidative Stress Induced by Palmitic Acid in Mouse Insulinoma NIT-1 Cells," *Revista Brasileira de Pesquisas Medicas e Biologicas (Brazilian Journal of Medical and Biological Research)* 46, no. 11 (2013): 949–55, https://doi.org/10.1590/1414-431X20133163.

132. A. M. Gray and P. R. Flatt, "Pancreatic and Extra-Pancreatic Effects of the Traditional Anti-Diabetic Plant, *Medicago sativa* (Lucerne)," *The British Journal of Nutrition* 78, no. 2 (1997): 325–34, https://www.ncbi.nlm.nih.gov/pubmed/9301421.

133. M. S. Kurzer and X. Xu, "Dietary Phytoestrogens," *Annual Review of Nutrition* 17 (1997): 353–81, https://doi.org /10.1146/annurev.nutr.17.1.353; Francesco Branca and Stefano Lorenzetti, "Health Effects of Phytoestrogens," *Forum of Nutrition*, no. 57 (2005): 100–111, https://www .ncbi.nlm.nih.gov/pubmed/15702593.

134. Ireneusz Kapusta et al., "Triterpene Saponins from Barrel Medic (*Medicago truncatula*) Aerial Parts," *Journal of Agricultural and Food Chemistry* 53, no. 6 (2005): 2164–70, https://doi.org/10.1021/jf048178i.

135. Xin-ping Liang et al., "Effects of Alfalfa Saponin Extract on MRNA Expression of Ldlr, LXRα, and FXR in BRL Cells," *Journal of Zhejiang University. Science. B* 16, no. 6 (2015): 479–86, https://doi.org/10.1631/jzus.B1400343; Yinghua Shi et al., "The Regulation of Alfalfa Saponin Extract on Key Genes Involved in Hepatic Cholesterol Metabolism in Hyperlipidemic Rats," *PLoS One* 9, no. 2 (2014): e88282, https://doi.org/10.1371/journal.pone.0088282; Ireneusz Kapusta et al., "Triterpene Saponins from Barrel Medic (*Medicago truncatula*) Aerial Parts," *Journal of Agricultural and Food Chemistry* 53, no. 6 (2005): 2164–70, https://doi .org/10.1021/jf048178i.

136. Mohsen Hosseini et al., "Inhibitory Potential of Pure Isoflavonoids, Red Clover, and Alfalfa Extracts on Hemoglobin Glycosylation," *ARYA Atherosclerosis* 11, no. 2 (2015): 133–38, https://www.ncbi.nlm.nih.gov/pubmed/26405442.

137. A. M. Gray and P. R. Flatt, "Pancreatic and Extra-Pancreatic Effects of the Traditional Anti-Diabetic Plant, *Medicago sativa* (Lucerne)," *The British Journal of Nutrition* 78, no. 2 (1997): 325–34, https://www.ncbi.nlm.nih.gov/pubmed/9301421.

138. P. Kishore et al., "Xylitol Prevents NEFA-Induced Insulin Resistance in Rats," *Diabetologia* 55, no. 6 (2012): 1808–12, https://doi.org/10.1007/s00125-012-2527-z.

139. Kikuko Amo et al., "Effects of Xylitol on Metabolic Parameters and Visceral Fat Accumulation," *Journal of Clinical Biochemistry and Nutrition* 49, no. 1 (2011): 1–7, https://doi.org/10.3164/jcbn.10-111.

140. Min-Jung Kang et al., "Hypoglycemic Effects of Welsh Onion in an Animal Model of Diabetes Mellitus," *Nutrition Research and Practice* 4, no. 6 (2010): 486–91, https://doi .org/10.4162/nrp.2010.4.6.486.

141. Farnaz Jafarpour-Sadegh et al., "Consumption of Fresh Yellow Onion Ameliorates Hyperglycemia and Insulin Resistance in Breast Cancer Patients During Doxorubicin-Based Chemotherapy: A Randomized Controlled Clinical Trial," *Integrative Cancer Therapies* 16, no. 3 (2017): 276–89, https://doi.org/10.1177/1534735416656915.

142. Ning Zhang et al., "Astragaloside IV Improves Metabolic Syndrome and Endothelium Dysfunction in Fructose-Fed Rats," *Molecules (Basel, Switzerland)* 16, no. 5 (2011): 3896–3907, https://doi.org/10.3390/molecules16053896; Xiaoxing Yin et al., "The Antioxidative Effects of Astragalus Saponin I Protect against Development of Early Diabetic Nephropathy," *Journal of Pharmacological Sciences* 101, no. 2 (2006): 166–73, https://www.ncbi.nlm.nih.gov/pubmed/16766854.

143. Ming-en Xu et al., "Effects of Astragaloside IV on Pathogenesis of Metabolic Syndrome in Vitro," *Acta Pharmacologica Sinica* 27, no. 2 (2006): 229–36, https://doi .org/10.1111/j.1745-7254.2006.00243.x.

144. Feng Zou et al., "Astragalus Polysaccharides Alleviates Glucose Toxicity and Restores Glucose Homeostasis in Diabetic States via Activation of AMPK," *Acta Pharmacologica Sinica* 30, no. 12 (2009): 1607–15, https:// doi.org/10.1038/aps.2009.168.

145. Min Liu et al., "Astragalus Polysaccharide Improves Insulin Sensitivity in KKAy Mice: Regulation of PKB/ GLUT4 Signaling in Skeletal Muscle," *Journal of Ethnopharmacology* 127, no. 1 (2010): 32–37, https://doi.org /10.1016/j.jep.2009.09.055.

146. Menglei Chao et al., "Improving Insulin Resistance with Traditional Chinese Medicine in Type 2 Diabetic Patients," *Endocrine* 36, no. 2 (2009): 268–74, https://doi.org/10.1007 /s12020-009-9222-y.

147. Vlasios Goulas et al., "Functional Components of Carob Fruit: Linking the Chemical and Biological Space," *International Journal of Molecular Sciences* 17, no. 11 (2016): 1875, https://doi.org/10.3390/ijms17111875.

148. E. B. C. Lima et al., "*Cocos nucifera* (L.) (Arecaceae): A Phytochemical and Pharmacological Review," *Revista Brasileira de Pesquisas Medicas e Biologicas (Brazilian Journal of Medical and Biological Research)* 48, no. 11 (2015): 953–64, https://doi.org/10.1590/1414-431X20154773.

149. Laurence Eyres et al., "Coconut Oil Consumption and Cardiovascular Risk Factors in Humans," *Nutrition Reviews* 74, no. 4 (2016): 267–80, https://doi.org/10.1093/nutrit/nuw002.

150. Hayley L. Loftus et al., "*Coleus forskohlii* Extract Supplementation in Conjunction with a Hypocaloric Diet Reduces the Risk Factors of Metabolic Syndrome in Overweight and Obese Subjects: A Randomized Controlled Trial," *Nutrients* 7, no. 11 (2015): 9508–22, https://doi.org /10.3390/nu7115483.

151. Ramesh Bellamkonda et al., "Beneficiary Effect of *Commiphora mukulethanolic* Extract against High Fructose Diet Induced Abnormalities in Carbohydrate and Lipid Metabolism in Wistar Rats," *Journal of Traditional and Complementary Medicine* 8, no. 1 (2018): 203–11, https://doi .org/10.1016/j.jtcme.2017.05.007.

152. Francesco Di Pierro et al., "Preliminary Study about the Possible Glycemic Clinical Advantage in Using a Fixed

Combination of *Berberis aristata* and *Silybum marianum* Standardized Extracts versus Only *Berberis aristata* in Patients with Type 2 Diabetes," *Clinical Pharmacology: Advances and Applications* 5 (2013): 167–74, https://doi.org/10.2147/CPAA.S54308.

153. Hui Li Tan et al., "*Rhizoma coptidis*: A Potential Cardiovascular Protective Agent," *Frontiers in Pharmacology* 7 (2016): 362, https://doi.org/10.3389/fphar.2016.00362.

154. Maryem Ben Salem et al., "Protective Effects of *Cynara scolymus* Leaves Extract on Metabolic Disorders and Oxidative Stress in Alloxan-Diabetic Rats," *BMC Complementary and Alternative Medicine* 17, no. 1 (2017): 328, https://doi.org/10.1186/s12906-017-1835-8.

155. Vajiheh Rangboo et al., "The Effect of Artichoke Leaf Extract on Alanine Aminotransferase and Aspartate Aminotransferase in the Patients with Nonalcoholic Steatohepatitis," *International Journal of Hepatology* 2016 (2016): 4030476, https://doi.org/10.1155/2016/4030476.

156. Carsten Gründemann et al., "*Equisetum arvense* (Common Horsetail) Modulates the Function of Inflammatory Immunocompetent Cells," *BMC Complementary and Alternative Medicine* 14 (2014): 283, https://doi.org/10.1186/1472-6882-14-283.

157. Azam Asgharikhatooni et al., "The Effect of *Equisetum arvense* (Horse Tail) Ointment on Wound Healing and Pain Intensity after Episiotomy: A Randomized Placebo-Controlled Trial," *Iranian Red Crescent Medical Journal* 17, no. 3 (2015): e25637, https://doi.org/10.5812/ircmj.25637.

158. Shamkant B. Badgujar et al., "*Foeniculum vulgare* Mill: A Review of Its Botany, Phytochemistry, Pharmacology, Contemporary Application, and Toxicology," *BioMed Research International* 2014 (2014): 842674, https://doi.org/10.1155/2014/842674.

159. JiYoung Bae et al., "Fennel (*Foeniculum vulgare*) and Fenugreek (*Trigonella foenum-graecum*) Tea Drinking Suppresses Subjective Short-Term Appetite in Overweight Women," *Clinical Nutrition Research* 4, no. 3 (2015): 168–74, https://doi.org/10.7762/cnr.2015.4.3.168.

160. Mariia Nagalievska et al., "*Galega officinalis* Extract Regulate the Diabetes Mellitus Related Violations of Proliferation, Functions and Apoptosis of Leukocytes," *BMC Complementary and Alternative Medicine* 18, no. 1 (2018): 4, https://doi.org/10.1186/s12906-017-2079-3.

161. M. H. Mooney et al., "Mechanisms Underlying the Metabolic Actions of Galegine That Contribute to Weight Loss in Mice," *British Journal of Pharmacology* 153, no. 8 (2008): 1669–77, https://doi.org/10.1038/bjp.2008.37.

162. Santosh Kumar Maurya et al., "Antidyslipidaemic Activity of *Glycyrrhiza glabra* in High Fructose Diet Induced Dyslipidaemic Syrian Golden Hamsters," *Indian Journal of Clinical Biochemistry: IJCB* 24, no. 4 (2009): 404–9, https://doi.org/10.1007/s12291-009-0072-4.

163. Michael Dushkin et al., "Effects of *Rhaponticum carthamoides* versus *Glycyrrhiza glabra* and *Punica granatum* Extracts on Metabolic Syndrome Signs in Rats," *BMC Complementary and Alternative Medicine* 14 (2014): 33, https://doi.org/10.1186/1472-6882-14-33.

164. Nazli Namazi et al., "The Effect of Dried *Glycyrrhiza glabra* L. Extract on Obesity Management with Regard to PPAR-Γ2 (Pro12Ala) Gene Polymorphism in Obese Subjects Following an Energy Restricted Diet," *Advanced Pharmaceutical Bulletin* 7, no. 2 (2017): 221–28, https://doi.org/10.15171/apb.2017.027.

165. Chia Hui Apphia Eu et al., "Glycyrrhizic Acid Improved Lipoprotein Lipase Expression, Insulin Sensitivity, Serum Lipid and Lipid Deposition in High-Fat Diet-Induced Obese Rats," *Lipids in Health and Disease* 9 (2010): 81, https://doi.org/10.1186/1476-511X-9-81.

166. V. T. T. Huyen et al., "*Gynostemma pentaphyllum* Tea Improves Insulin Sensitivity in Type 2 Diabetic Patients," *Journal of Nutrition and Metabolism* 2013 (2013): 765383, https://doi.org/10.1155/2013/765383.

167. V. T. T. Huyen et al., "Antidiabetic Effects of Add-On *Gynostemma pentaphyllum* Extract Therapy with Sulfonylureas in Type 2 Diabetic Patients," *Evidence-Based Complementary and Alternative Medicine: ECAM* 2012 (2012): 452313, https://doi.org/10.1155/2012/452313.

168. Miao Wang et al., "Metabonomics Study of the Therapeutic Mechanism of *Gynostemma pentaphyllum* and Atorvastatin for Hyperlipidemia in Rats," *PLoS One* 8, no. 11 (2013): e78731, https://doi.org/10.1371/journal.pone.0078731.

169. Ezarul Faradianna Lokman et al., "Evaluation of Antidiabetic Effects of the Traditional Medicinal Plant *Gynostemma pentaphyllum* and the Possible Mechanisms of Insulin Release," *Evidence-Based Complementary and Alternative Medicine: ECAM* 2015 (2015): 120572, https://doi.org/10.1155/2015/120572.

170. B. T. Diwakar et al., "Bio-Availability and Metabolism of n-3 Fatty Acid Rich Garden Cress (*Lepidium sativum*) Seed Oil in Albino Rats," *Prostaglandins, Leukotrienes, and Essential Fatty Acids* 78, no. 2 (2008): 123–30, https://doi.org/10.1016/j.plefa.2007.12.001.

171. Mhamed Maghrani et al., "Antihypertensive Effect of *Lepidium sativum* L. in Spontaneously Hypertensive Rats," *Journal of Ethnopharmacology* 100, no. 1–2 (2005): 193–97, https://doi.org/10.1016/j.jep.2005.02.024.

172. Katerina Valentová et al., "Maca (*Lepidium meyenii*) and Yacon (*Smallanthus sonchifolius*) in Combination with Silymarin as Food Supplements: In Vivo Safety Assessment," *Food and Chemical Toxicology: An International Journal Published for the British Industrial Biological Research Association* 46, no. 3 (2008): 1006–13, https://doi.org/10.1016/j.fct.2007.10.031.

173. Rostislav Vecera et al., "The Influence of Maca (*Lepidium meyenii*) on Antioxidant Status, Lipid and Glucose Metabolism in Rat," *Plant Foods for Human Nutrition (Dordrecht, Netherlands)* 62, no. 2 (2007): 59–63, https://doi.org/10.1007/s11130-007-0042-z; M. Eddouks et al., "Study of the Hypoglycaemic Activity of *Lepidium sativum* L. Aqueous Extract in Normal and Diabetic Rats," *Journal of Ethnopharmacology* 97, no. 2 (2005): 391–95, https://doi.org/10.1016/j.jep.2004.11.030.

174. Mohamed Eddouks and Mhamed Maghrani, "Effect of *Lepidium sativum* L. on Renal Glucose Reabsorption and Urinary TGF-Beta 1 Levels in Diabetic Rats," *Phytotherapy Research: PTR* 22, no. 1 (2008): 1–5, https://doi.org/10.1002/ptr.2101.

175. M. S. Kurzer and X. Xu, "Dietary Phytoestrogens," *Annual Review of Nutrition* 17 (1997): 353–81, https://doi.org/10.1146/annurev.nutr.17.1.353; Francesco Branca and Stefano Lorenzetti, "Health Effects of Phytoestrogens," *Forum of Nutrition*, no. 57 (2005): 100–111, https://www.ncbi.nlm.nih.gov/pubmed/15702593.

176. G. A. Bhat et al., "GLP-I Secretion in Healthy and Diabetic Wistar Rats in Response to Aqueous Extract of *Momordica charantia*," *BMC Complementary and Alternative Medicine* 18, no. 1 (2018): 162; Baby Joseph and D. Jini, "Antidiabetic Effects of *Momordica charantia* (Bitter Melon) and Its Medicinal Potency," *Asian Pacific Journal of Tropical Disease* 3, no. 2 (2013): 93–102.

177. Sushil Kumar Middha et al., "A Review on Antihyperglycemic and Antihepatoprotective Activity of Eco-Friendly *Punica granatum* Peel Waste," *Evidence-Based Complementary and Alternative Medicine: ECAM* 2013 (2013): 656172, https://doi.org/10.1155/2013/656172.

178. Yun Hwan Kang et al., "Antiobesity Effects of the Water-Soluble Fraction of the Ethanol Extract of *Smilax china* L. Leaf in 3T3-L1 Adipocytes," *Nutrition Research and Practice* 9, no. 6 (2015): 606–12, https://doi.org/10.4162/nrp.2015.9.6.606.

179. Kang et al., "Antiobesity Effects."

180. Parvinnesh S. Kumar et al., "Anticancer Potential of *Syzygium aromaticum* L. in MCF-7 Human Breast Cancer Cell Lines," *Pharmacognosy Research* 6, no. 4 (2014): 350–54, https://doi.org/10.4103/0974-8490.138291.

181. Stephen Adeniyi Adefegha and Ganiyu Oboh, "In Vitro Inhibition Activity of Polyphenol-Rich Extracts from *Syzygium aromaticum* (L.) Merr. & Perry (Clove) Buds against Carbohydrate Hydrolyzing Enzymes Linked to Type 2 Diabetes and Fe(2+)-Induced Lipid Peroxidation in Rat Pancreas," *Asian Pacific Journal of Tropical Biomedicine* 2, no. 10 (2012): 774–81, https://doi.org/10.1016/S2221-1691(12)60228-7.

182. Andile Khathi et al., "Effects of *Syzygium aromaticum*-Derived Triterpenes on Postprandial Blood Glucose in Streptozotocin-Induced Diabetic Rats Following Carbohydrate Challenge," *PLoS One* 8, no. 11 (2013): e81632, https://doi.org/10.1371/journal.pone.0081632.

183. Anayt Ulla et al., "Supplementation of *Syzygium cumini* Seed Powder Prevented Obesity, Glucose Intolerance, Hyperlipidemia and Oxidative Stress in High Carbohydrate High Fat Diet Induced Obese Rats," *BMC Complementary and Alternative Medicine* 17 (2017): 289, https://doi.org/10.1186/s12906-017-1799-8.

184. Masahito Takikawa et al., "Dietary Anthocyanin-Rich Bilberry Extract Ameliorates Hyperglycemia and Insulin Sensitivity via Activation of AMP-Activated Protein Kinase in Diabetic Mice," *The Journal of Nutrition* 140, no. 3 (2010): 527–33, https://doi.org/10.3945/jn.109.118216.

185. H.-M. Lehtonen et al., "Different Berries and Berry Fractions Have Various but Slightly Positive Effects on the Associated Variables of Metabolic Diseases on Overweight and Obese Women," *European Journal of Clinical Nutrition* 65, no. 3 (2011): 394–401, https://doi.org/10.1038/ejcn.2010.268.

186. R. Boniface, "Pharmacological Properties of Myrtillus Anthocyanosides: Correlation with Results of Treatment of Diabetic Microangiopathy," *Study Org Chem* 23 (1986): 293–301.

187. Yiming Li et al., "Preventive and Protective Properties of *Zingiber officinale* (Ginger) in Diabetes Mellitus, Diabetic Complications, and Associated Lipid and Other Metabolic Disorders: A Brief Review," *Evidence-Based Complementary and Alternative Medicine: ECAM* 2012 (2012): 516870, https://doi.org/10.1155/2012/516870.

188. Nafiseh Khandouzi et al., "The Effects of Ginger on Fasting Blood Sugar, Hemoglobin A1c, Apolipoprotein B, Apolipoprotein a-I and Malondialdehyde in Type 2 Diabetic Patients," *Iranian Journal of Pharmaceutical Research: IJPR* 14, no. 1 (2015): 131–40, https://www.ncbi.nlm.nih.gov/pubmed/25561919.

Chapter Four: Creating Herbal Formulas for Reproductive Endocrine Conditions

1. Li Li et al., "Non-Monotonic Dose-Response Relationship in Steroid Hormone Receptor-Mediated Gene Expression," *Journal of Molecular Endocrinology* 38, no. 5 (2007): 569–85, https://doi.org/10.1677/JME-07-0003; L. Payen et al., "The Drug Efflux Pump MRP2: Regulation of Expression in Physiopathological Situations and by Endogenous and Exogenous Compounds," *Cell Biology and Toxicology* 18, no. 4 (2002): 221–33, https://www.ncbi.nlm.nih.gov/pubmed/12206135.

2. Tze-Chen Hsieh and Joseph M. Wu, "Targeting CWR22Rv1 Prostate Cancer Cell Proliferation and Gene Expression by Combinations of the Phytochemicals EGCG, Genistein and Quercetin," *Anticancer Research* 29, no. 10 (2009): 4025–32, https://www.ncbi.nlm.nih.gov/pubmed/19846946; Anja Bliedtner et al., "Effects of Genistein and Estrogen Receptor Subtype-Specific Agonists in ArKO Mice Following Different Administration Routes," *Molecular and Cellular Endocrinology* 314, no. 1 (2010): 41–52, https://doi.org/10.1016/j.mce.2009.07.032.

3. Marie-Pierre Lézé et al., "2- and 3-[(Aryl)(Azolyl)Methyl] Indoles as Potential Non-Steroidal Aromatase Inhibitors," *Journal of Enzyme Inhibition and Medicinal Chemistry* 19, no. 6 (2004): 549–57, https://doi.org/10.1080/1475636040004631; J. Drsata, "Enzyme Inhibition in the Drug Therapy of Benign Prostatic Hyperplasia," *Casopis Lekaru Ceskych* 141, no. 20 (2002): 630–35, https://www.ncbi.nlm.nih.gov/pubmed/12515038.

4. D. Gansser and G. Spiteller, "Aromatase Inhibitors from *Urtica dioica* Roots," *Planta Medica* 61, no. 2 (1995): 138–40, https://doi.org/10.1055/s-2006-958033.

5. H. J. Schneider et al., "Treatment of Benign Prostatic Hyperplasia. Results of a Treatment Study with the Phytogenic Combination of Sabal Extract WS 1473 and Urtica Extract WS 1031 in Urologic Specialty Practices," *Fortschritte der Medizin* 113, no. 3 (1995): 37–40, https://www.ncbi.nlm.nih.gov/pubmed/7534258.

6. Christine Johnston and Lawrence Corey, "Current Concepts for Genital Herpes Simplex Virus Infection: Diagnostics and Pathogenesis of Genital Tract Shedding," *Clinical Microbiology Reviews* 29, no. 1 (2016): 149–61, https://doi.org/10.1128/CMR.00043-15.

7. Zhangsheng Yang et al., "The Chlamydia-Secreted Protease CPAF Promotes Chlamydial Survival in the Mouse Lower Genital Tract," *Infection and Immunity* 84, no. 9 (2016): 2697–2702, https://doi.org/10.1128/IAI.00280-16.

8. Fangfang Liu et al., "Concurrence of Oral and Genital Human Papillomavirus Infection in Healthy Men: A Population-Based Cross-Sectional Study in Rural China," *Scientific Reports* 5 (2015): 15637, https://doi.org/10.1038/srep15637.

9. Parvin Bahadoran et al., "Investigating the Therapeutic Effect of Vaginal Cream Containing Garlic and Thyme Compared to Clotrimazole Cream for the Treatment of Mycotic Vaginitis," *Iranian Journal of Nursing and Midwifery Research* 15, no. Suppl 1 (2010): 343–49, https://www.ncbi.nlm.nih.gov/pubmed/22069409.

10. Mansoure Masoudi et al., "Comparison between the Efficacy of Metronidazole Vaginal Gel and *Berberis vulgaris* (*Berberis vulgaris*) Combined with Metronidazole Gel Alone in the Treatment of Bacterial Vaginosis," *Electronic Physician* 8, no. 8 (2016): 2818–27, https://doi.org/10.19082/2818.

11. Sanjiveeni Dhamgaye et al., "Molecular Mechanisms of Action of Herbal Antifungal Alkaloid Berberine, in *Candida albicans*," *PLoS One* 9, no. 8 (2014): e104554, https://doi.org/10.1371/journal.pone.0104554.

12. G. M. El-Sherbiny and E. T. El Sherbiny, "The Effect of *Commiphora molmol* (Myrrh) in Treatment of Trichomoniasis Vaginalis Infection," *Iranian Red Crescent Medical Journal* 13, no. 7 (2011): 480–86, https://www.ncbi.nlm.nih.gov/pubmed/22737515.

13. El-Sherbiny and El Sherbiny, "The Effect of *Commiphora molmol* (Myrrh)."

14. Solmaz Hassani et al., "Effects of Different Extracts of *Eucalyptus camaldulensis* on Trichomonas Vaginalis Parasite in Culture Medium," *Advanced Biomedical Research* 2 (2013): 47, https://doi.org/10.4103/2277-9175.114187.

15. Fereshteh Behmanesh et al., "Antifungal Effect of Lavender Essential Oil (*Lavandula angustifolia*) and Clotrimazole on *Candida albicans*: An In Vitro Study," *Scientifica* 2015 (2015): 261397, https://doi.org/10.1155/2015/261397.

16. Wei Zou et al., "The Anti-Inflammatory Effect of *Andrographis paniculata* (Burm. f.) Nees on Pelvic Inflammatory Disease in Rats through Down-Regulation of the NF-KB Pathway," *BMC Complementary and Alternative Medicine* 16, no. 1 (2016): 483, https://doi.org/10.1186/s12906-016-1466-5.

17. H. Y. Xiang et al., "Pharmacodynamic Studies on the Effect of Jingangteng Dispersible Tablets for Chronic Pelvic Inflammatory Disease," *Journal of Chinese Medicinal Materials* 30, no. 4 (2007): 449–452, http://caod.oriprobe.com/articles/12367818/jin_gang_teng_fen_san_pian_dui_da_shu_man_xing_pen_qiang_yan_de_yao_xi.htm.

18. Mu-Kai Chen et al., "Effects of *Salvia miltiorrhiza* on Chlamydia Trachomatis Mice of Salpingitis," *Zhongguo Zhong Yao Za Zhi = Zhongguo Zhongyao Zazhi (China Journal of Chinese Materia Medica)* 32, no. 6 (2007): 523–25, https://www.ncbi.nlm.nih.gov/pubmed/17552161.

19. Ziyu Hua et al., "Andrographolide Inhibits Intracellular Chlamydia Trachomatis Multiplication and Reduces Secretion of Proinflammatory Mediators Produced by Human Epithelial Cells," *Pathogens and Disease* 73, no. 1 (2015): 1–11, https://doi.org/10.1093/femspd/ftu022.

20. Yeonsu Oh et al., "Anti-Inflammatory Effects of the Natural Compounds *Cortex phellodendri* and *Humulus japonicus* on Pelvic Inflammatory Disease in Mice," *International Journal of Medical Sciences* 14, no. 8 (2017): 729–34, https://doi.org/10.7150/ijms.19616.

21. A. Minson, "Interactions of Herpes Simplex Viruses with the Host Cell," *Biochemical Society Transactions* 22, no. 2 (1994): 298–301.

22. S. Dewhurst, "Herpesviruses: Lecture 1," University of Rochester Medical Center, 1997.

23. J. F. Jones et al., "Evidence for Active Epstein-Barr Virus Infection in Patients with Persistent, Unexplained Illnesses: Elevated Anti-Early Antigen Antibodies," *Annals of Internal Medicine* 102, no. 1 (1985): 1–7, https://www.ncbi.nlm.nih.gov/pubmed/2578266; S. E. Straus et al., "Persisting Illness and Fatigue in Adults with Evidence of Epstein-Barr Virus Infection," *Annals of Internal Medicine* 102, no. 1 (1985): 7–16, https://www.ncbi.nlm.nih.gov/pubmed/2578268; R. E. DuBois et al., "Chronic Mononucleosis Syndrome," *Southern Medical Journal* 77, no. 11 (1984): 1376–82, https://www.ncbi.nlm.nih.gov/pubmed/6093268; M. Tobi et al., "Prolonged Atypical Illness Associated with Serological Evidence of Persistent Epstein-Barr Virus Infection," *Lancet* 1, no. 8263 (1982): 61–64, https://www.ncbi.nlm.nih.gov/pubmed/6119490.

24. P. Sen and S. E. Barton, "Genital Herpes and Its Management," *BMJ* 334, no. 7602 (2007): 1048–52, https://doi.org/10.1136/bmj.39189.504306.55.

25. Sen and Barton, "Genital Herpes and Its Management."

26. Mehdi Ansari et al., "In Vitro Evaluation of Anti-Herpes Simplex-1 Activity of Three Standardized Medicinal Plants from Lamiaceae," *Ancient Science of Life* 34, no. 1 (2014): 33–38, https://doi.org/10.4103/0257-7941.150777.

27. J. A. Scott, "On the Biochemical Similarities of Ascorbic Acid and Interferon," *Journal of Theoretical Biology* 98, no. 2 (1982): 235–38, https://www.ncbi.nlm.nih.gov/pubmed/6184576.

28. G. T. Terezhalmy et al., "The Use of Water-Soluble Bioflavonoid-Ascorbic Acid Complex in the Treatment of Recurrent Herpes Labialis," *Oral Surgery, Oral Medicine, and Oral Pathology* 45, no. 1 (1978): 56–62, https://www.ncbi.nlm.nih.gov/pubmed/339141.

29. M. J. Bunk et al., "Relationship of Cytotoxic Activity of Natural Killer Cells to Growth Rates and Serum Zinc Levels of Female RIII Mice Fed Zinc," *Nutrition and Cancer* 10, no. 1–2 (1987): 79–87, https://doi.org/10.1080/01635588709513942; E. Katz and E. Margalith, "Inhibition of Vaccinia Virus Maturation by Zinc Chloride," *Antimicrobial Agents and Chemotherapy* 19, no. 2 (1981): 213–17, https://www.ncbi.nlm.nih.gov/pubmed/7347557.

30. R. S. Griffith et al., "A Multicentered Study of Lysine Therapy in Herpes Simplex Infection," *Dermatologica* 156, no. 5 (1978): 257–67, https://www.ncbi.nlm.nih.gov/pubmed/640102.

31. Akram Astani et al., "Attachment and Penetration of Acyclovir-Resistant Herpes Simplex Virus Are Inhibited by *Melissa officinalis* Extract," *Phytotherapy Research: PTR* 28, no. 10 (2014): 1547–52, https://doi.org/10.1002/ptr.5166.

32. Akram Astani et al., "*Melissa officinalis* Extract Inhibits Attachment of Herpes Simplex Virus In Vitro," *Chemotherapy* 58, no. 1 (2012): 70–77, https://doi.org/10.1159/000335590.

33. G. Mazzanti et al., "Inhibitory Activity of *Melissa officinalis* L. Extract on Herpes Simplex Virus Type 2 Replication," *Natural Product Research* 22, no. 16 (2008): 1433–40, https://doi.org/10.1080/14786410802075939.

34. A. D. Klein and N. S. Penneys "*Aloe vera*," *JAMER* 18 (1988): 714–19; James Duke, *Handbook of Medicinal Herbs* (Boca Raton, FL: CRC Press, 1985), 31.

35. H. R. McDaniel, "An Increase in Circulating Monocyte/Macrophages Is Induced by Acemannan," *American Journal of Clinical Pathology* 94 (1990): 516–17.

36. M. C. Kemp et al., "In Vitro Evaluation of the Antiviral Effects of Acemannan on the Replication and Pathogenesis of HIV-1 and Other Enveloped Viruses: Modification of the Processing of Glycoprotein Glycoprotein Precursors," *Antiviral Research* 13 (1990): 83, http://dx.doi.org/10.1016/0166-3542(90)90156-2; J. B. Kahlon et al., "In Vitro Evaluation of the Synergistic Antiviral Effects of Acemannan in Combination with Azidothymidine and Acyclovir," *Molecular Biotherapy* 3, no. 4 (1991): 214–23, https://www.ncbi.nlm.nih.gov/pubmed/1662957.

37. J. A. Singer, "Randomized Placebo Controlled Trial of Oral Acemannon as an Adjuctive to Anti-Retroviral Therapy," *International AIDS* 9, no. 1 (1993): 494.

38. Mohammad-Taghi Moradi et al., "A Review Study on the Effect of Iranian Herbal Medicines against In Vitro Replication of Herpes Simplex Virus," *Avicenna Journal of Phytomedicine* 6, no. 5 (2016): 506–15, https://www.ncbi.nlm.nih.gov/pubmed/27761420.

39. Kyungtaek Im et al., "Ginseng, the Natural Effectual Antiviral: Protective Effects of Korean Red Ginseng against Viral Infection," *Journal of Ginseng Research* 40, no. 4 (2016): 309–14, https://doi.org/10.1016/j.jgr.2015.09.002.

40. R. Pompei et al., "Antiviral Activity of Glycyrrhizic Acid," *Experientia* 36, no. 3 (1980): 304, https://www.ginsengres.com/article/S1226-8453(15)00084-6/fulltext; M. Partridge and D. E. Poswillo, "Topical Carbenoxolone Sodium in the Management of Herpes Simplex Infection," *The British Journal of Oral & Maxillofacial Surgery* 22, no. 2 (1984): 138–45, https://www.ncbi.nlm.nih.gov/pubmed/6585224; G. W. Csonka and D. A. Tyrrell, "Treatment of Herpes Genitalis with Carbenoxolone and Cicloxolone Creams: A Double Blind Placebo Controlled Clinical Trial," *The British Journal of Venereal Diseases* 60, no. 3 (1984): 178–81, https://www.ncbi.nlm.nih.gov/pubmed/6375805.

41. D. Poswillo and M. Partridge, "Management of Recurrent Aphthous Ulcers. A Trial of Carbenoxolone Sodium Mouthwash," *British Dental Journal* 157, no. 2 (1984): 55–57, https://doi.org/10.1038/sj.bdj.4805417; R. Pompei, "Activity of *Glycyrrhiza glabra* Extracts and Glycyrrhizic Acid on Virus Growth and Infectivity," *Rivista di Farmacologia e Terapia* 10 (1979): 281–84.

42. L. S. Kucera and E. C. Herrmann, "Antiviral Substances in Plants of the Mint Family (Labiatae). I. Tannin of *Melissa officinalis*," *Proceedings of the Society for Experimental Biology and Medicine. Society for Experimental Biology and Medicine* 124, no. 3 (1967): 865–69, https://www.ncbi.nlm.nih.gov/pubmed/4290277; J. David Phillipson et al., *Herbal Drugs and Phytopharmaceuticals: A Handbook for Practice on a Scientific Basis* (Centurion, South Africa: Medpharm Scientific Publishers, 1994), 332.

43. Dean Bruenton, *Pharmacognosy Medical Plants* (Paris: Lauoisier, 1995), 430–31; R. F. Weiss, *Herbal Medication* (Beaconsfield, UK: Beaconsfield, 1988), 125.

44. Silvia Geuenich et al., "Aqueous Extracts from Peppermint, Sage and Lemon Balm Leaves Display Potent Anti-HIV-1 Activity by Increasing the Virion Density," *Retrovirology* 5 (2008): 27, https://doi.org/10.1186/1742-4690-5-27.

45. Akram Astani et al., "Screening for Antiviral Activities of Isolated Compounds from Essential Oils," *Evidence-Based Complementary and Alternative Medicine: ECAM* 2011 (2011): 253643, https://doi.org/10.1093/ecam/nep187.

46. A. Bennett et al., "Aspirin-Induced Gastric Mucosal Damage in Rats: Cimetidine and Degly-Cyhrrhizinated Liquorice Together Give Greater Protection Than Low Doses of Either Drug Alone," *Journal of Pharmacy and Pharmacology* 32, no. 1 (1980): 151; S. Okabe, *Oyo Yakuri* 18 (1979): 469.

47. Vaibhav Tiwari et al., "In Vitro Antiviral Activity of Neem (*Azardirachta indica* L.) Bark Extract against Herpes Simplex Virus Type-1 Infection," *Phytotherapy Research: PTR* 24, no. 8 (2010): 1132–40, https://doi.org/10.1002/ptr.3085.

48. "STD Facts—Human Papillomavirus (HPV)," November 16, 2017. https://www.cdc.gov/std/hpv/stdfact-hpv.htm.

49. N. Munoz et al., "The Epidemiology of Human Papilloma Virus and Cervical Cancer," *IARC Scientific Publications* 119 (1992), http://publications.iarc.fr/Book-And-Report-Series/Iarc-Scientific-Publications/The-Epidemiology-Of

-Human-Papillomavirus-And-Cervical-Cancer-1992; D. A. Galloway, "Human Papillomavirus Vaccines: A Warty Problem," *Infections Agents and Disease* 3 (1997): 187–93.

50. F. Xavier Bosch et al., "Comprehensive Control of Human Papillomavirus Infections and Related Diseases," *Vaccine* 31 Suppl 7 (2013): H1–31, https://doi.org/10.1016/j.vaccine.2013.10.003; Jeong-Min Kim et al., "Efficacy of 5% Imiquimod Cream on Vulvar Intraepithelial Neoplasia in Korea: Pilot Study," *Annals of Dermatology* 27, no. 1 (2015): 66–70, https://doi.org/10.5021/ad.2015.27.1.66.

51. V. Graham et al., "Phase II Trial of Beta-All-Trans-Retinoic Acid for Cervical Intraepithelial Neoplasia Delivered via a Collagen Sponge and Cervical Cap," *The Western Journal of Medicine* 145, no. 2 (1986): 192–95, https://www.ncbi.nlm.nih.gov/pubmed/3765597.

52. Lynne Hampson et al., "A Single-Arm, Proof-of-Concept Trial of Lopimune (Lopinavir/Ritonavir) as a Treatment for HPV-Related Pre-Invasive Cervical Disease," *PLoS One* 11, no. 1 (2016): e0147917, https://doi.org/10.1371/journal.pone.0147917.

53. Yu-Ping Dang et al., "Curcumin Improves the Paclitaxel-Induced Apoptosis of HPV-Positive Human Cervical Cancer Cells via the NF-KB-P53-Caspase-3 Pathway," *Experimental and Therapeutic Medicine* 9, no. 4 (2015): 1470–76, https://doi.org/10.3892/etm.2015.2240.

54. Pooja Chandrakant Thacker and Devarajan Karunagaran, "Curcumin and Emodin Down-Regulate TGF-β Signaling Pathway in Human Cervical Cancer Cells," *PLoS One* 10, no. 3 (2015): e0120045, https://doi.org/10.1371/journal.pone.0120045.

55. Hamidreza Ardalani et al., "Podophyllotoxin: A Novel Potential Natural Anticancer Agent," *Avicenna Journal of Phytomedicine* 7, no. 4 (2017): 285–94, https://www.ncbi.nlm.nih.gov/pubmed/28884079.

56. Kimberly Windstar et al., "Escharotic Treatment for ECC-Positive CIN3 in Childbearing Years: A Case Report," *Integrative Medicine* 13, no. 2 (2014): 43–49, https://www.ncbi.nlm.nih.gov/pubmed/26770091.

57. R. Joseph et al., "Successful Treatment of Verruca Vulgaris with *Thuja occidentalis* in a Renal Allograft Recipient," *Indian Journal of Nephrology* 23, no. 5 (2013): 362–64, https://doi.org/10.4103/0971-4065.116316.

58. Sanower Hossain et al., "*Andrographis paniculata* (Burm. f.) Wall. Ex Nees: A Review of Ethnobotany, Phytochemistry, and Pharmacology," *The Scientific World Journal* 2014 (2014): 274905, https://doi.org/10.1155/2014/274905.

59. S. Fangkham et al., "The Effect of Andrographolide on Human Papillomavirus Type 16 (HPV16) Positive Cervical Cancer Cells (SiHa)," *International Journal of Infectious Diseases* 16 (2012): e80, https://doi.org/10.1016/j.ijid.2012.05.192.

60. J. Steinmann et al., "Anti-Infective Properties of Epigallocatechin-3-Gallate (EGCG), a Component of Green Tea," *British Journal of Pharmacology* 168, no. 5 (2013): 1059–73, https://doi.org/10.1111/bph.12009.

61. Jian-Ming Lü et al., "Molecular Mechanisms and Clinical Applications of Nordihydroguaiaretic Acid (NDGA) and Its Derivatives: An Update," *Medical Science Monitor: International Medical Journal of Experimental and Clinical Research* 16, no. 5 (2010): RA93–100, https://www.ncbi.nlm.nih.gov/pubmed/20424564.

62. Sutapa Mahata et al., "Berberine Modulates AP-1 Activity to Suppress HPV Transcription and Downstream Signaling to Induce Growth Arrest and Apoptosis in Cervical Cancer Cells," *Molecular Cancer* 10 (2011): 39, https://doi.org/10.1186/1476-4598-10-39.

63. Andrew Croaker et al., "*Sanguinaria canadensis*: Traditional Medicine, Phytochemical Composition, Biological Activities and Current Uses," *International Journal of Molecular Sciences* 17, no. 9 (2016), https://doi.org/10.3390/ijms17091414.

64. Kimberly Windstar et al., "Escharotic Treatment for ECC-Positive CIN3 in Childbearing Years: A Case Report," *Integrative Medicine* 13, no. 2 (2014): 43–49, https://www.ncbi.nlm.nih.gov/pubmed/26770091.

65. Parisa Yavarikia et al., "Comparing the Effect of Mefenamic Acid and *Vitex agnus* on Intrauterine Device Induced Bleeding," *Journal of Caring Sciences* 2, no. 3 (2013): 245–54, https://doi.org/10.5681/jcs.2013.030.

66. Kyoung-Sun Park et al., "A Review of In Vitro and In Vivo Studies on the Efficacy of Herbal Medicines for Primary Dysmenorrhea," *Evidence-Based Complementary and Alternative Medicine: ECAM* 2014 (2014): 296860, https://doi.org/10.1155/2014/296860.

67. Y. Q. Hua et al., "Danggui-Shaoyao-San, a Traditional Chinese Prescription, Suppresses PGF2alpha Production in Endometrial Epithelial Cells by Inhibiting COX-2 Expression and Activity," *Phytomedicine: International Journal of Phytotherapy and Phytopharmacology* 15, no. 12 (2008): 1046–52, https://doi.org/10.1016/j.phymed.2008.06.010.

68. Molouk Jaafarpour et al., "The Effect of Cinnamon on Menstrual Bleeding and Systemic Symptoms with Primary Dysmenorrhea," *Iranian Red Crescent Medical Journal* 17, no. 4 (2015): e27032, https://doi.org/10.5812/ircmj.17(4)2015.27032.

69. Birgit M. Dietz et al., "Botanicals and Their Bioactive Phytochemicals for Women's Health," *Pharmacological Reviews* 68, no. 4 (2016): 1026–73, https://doi.org/10.1124/pr.115.010843.

70. Khadijeh Onsory et al., "Hormone Receptor-Related Gene Polymorphisms and Prostate Cancer Risk in North Indian Population," *Molecular and Cellular Biochemistry* 314, no. 1–2 (2008): 25–35, https://doi.org/10.1007/s11010-008-9761-1.

71. Georges Pelletier, "Expression of Steroidogenic Enzymes and Sex-Steroid Receptors in Human Prostate," *Best Practice & Research. Clinical Endocrinology & Metabolism* 22, no. 2 (2008): 223–28, https://doi.org/10.1016/j.beem.2008.02.004.

72. Musiliyu A. Musa et al., "A Review of Coumarin Derivatives in Pharmacotherapy of Breast Cancer," *Current Medicinal Chemistry* 15, no. 26 (2008): 2664–79, https://www.ncbi.nlm.nih.gov/pubmed/18991629.

73. S. Paoletta et al., "Screening of Herbal Constituents for Aromatase Inhibitory Activity," *Bioorganic & Medicinal Chemistry* 16, no. 18 (2008): 8466–70, https://doi.org/10.1016/j.bmc.2008.08.034.

74. Zahra Ghodsi and Maryam Asltoghiri, "The Effect of Fennel on Pain Quality, Symptoms, and Menstrual Duration in Primary Dysmenorrhea," *Journal of Pediatric and Adolescent Gynecology* 27, no. 5 (2014): 283–86, https://doi.org/10.1016/j.jpag.2013.12.003.

75. G. Bhuvanalakshmi et al., "Breast Cancer Stem-Like Cells Are Inhibited by Diosgenin, a Steroidal Saponin, by the Attenuation of the Wnt β-Catenin Signaling via the Wnt Antagonist Secreted Frizzled Related Protein-4," *Frontiers in Pharmacology* 8 (2017): 124, https://doi.org/10.3389/fphar.2017.00124.

76. Pin-Shern Chen et al., "Diosgenin, a Steroidal Saponin, Inhibits Migration and Invasion of Human Prostate Cancer PC-3 Cells by Reducing Matrix Metalloproteinases Expression," *PLoS One* 6, no. 5 (2011): e20164, https://doi.org/10.1371/journal.pone.0020164.

77. Mingjie Shen et al., "Observation of the Influences of Diosgenin on Aging Ovarian Reserve and Function in a Mouse Model," *European Journal of Medical Research* 22, no. 1 (2017): 42, https://doi.org/10.1186/s40001-017-0285-6.

78. A. Kumagai et al., "Effect of Glycyrrhizin on Estrogen Action," *Endocrinologia Japonica* 14, no. 1 (1967): 34–38, https://www.ncbi.nlm.nih.gov/pubmed/6072411.

79. T. Yaginuma et al., "Effect of Traditional Herbal Medicine on Serum Testosterone Levels and Its Induction of Regular Ovulation in Hyperandrogenic and Oligomenorrheic Women," *Nihon Sanka Fujinka Gakkai Zasshi* 34, no. 7 (1982): 939–44, https://www.ncbi.nlm.nih.gov/pubmed/7108310.

80. T. Yaginuma et al., "Effect of Traditional Herbal Medicine on Serum Testosterone Levels."

81. Susan Arentz et al., "Herbal Medicine for the Management of Polycystic Ovary Syndrome (PCOS) and Associated Oligo/Amenorrhoea and Hyperandrogenism; a Review of the Laboratory Evidence for Effects with Corroborative Clinical Findings," *BMC Complementary and Alternative Medicine* 14 (2014): 511, https://doi.org/10.1186/1472-6882-14-511.

82. Qi Ye et al., "Casticin, a Flavonoid Isolated from *Vitex rotundifolia*, Inhibits Prolactin Release In Vivo and In Vitro," *Acta Pharmacologica Sinica* 31, no. 12 (2010): 1564–68, https://doi.org/10.1038/aps.2010.178.

83. Michaela Döll, "The Premenstrual Syndrome: Effectiveness of *Vitex agnus-castus*," *Medizinische Monatsschrift für Pharmazeuten* 32, no. 5 (2009): 186–91, https://www.ncbi.nlm.nih.gov/pubmed/19469189.

84. A. R. Carmichael, "Can *Vitex agnus-castus* Be Used for the Treatment of Mastalgia? What Is the Current Evidence?" *Evidence-Based Complementary and Alternative Medicine: ECAM* 5, no. 3 (2008): 247–50, https://doi.org/10.1093/ecam/nem074; N. A. Ibrahim et al., "Gynecological Efficacy and Chemical Investigation of *Vitex agnus-castus* L. Fruits Growing in Egypt," *Natural Product Research* 22, no. 6 (2008): 537–46, https://doi.org/10.1080/14786410701592612; Y. Hu et al., "Anti-Nociceptive and Anti-Hyperprolactinemia Activities of *Fructus viticis* and Its Effective Fractions and Chemical Constituents," *Phytomedicine: International Journal of Phytotherapy and Phytopharmacology* 14, no. 10 (2007): 668–74, https://doi.org/10.1016/j.phymed.2007.01.008.

85. A. Milewicz et al., "*Vitex agnus-castus* Extract in the Treatment of Luteal Phase Defects Due to Latent Hyperprolactinemia. Results of a Randomized Placebo-Controlled Double-Blind Study," *Arzneimittel-Forschung* 43, no. 7 (1993): 752–56, https://www.ncbi.nlm.nih.gov/pubmed/8369008.

86. Molouk Jaafarpour et al., "Comparative Effect of Cinnamon and Ibuprofen for Treatment of Primary Dysmenorrhea: A Randomized Double-Blind Clinical Trial," *Journal of Clinical and Diagnostic Research: JCDR* 9, no. 4 (2015): QC04–07, https://doi.org/10.7860/JCDR/2015/12084.5783.

87. Marzieh Akbarzadeh et al., "Effect of *Melissa officinalis* Capsule on the Intensity of Premenstrual Syndrome Symptoms in High School Girl Students," *Nursing and Midwifery Studies* 4, no. 2 (2015): e27001, https://doi.org/10.17795/nmsjournal27001.

88. Samira Khayat et al., "Effect of Treatment with Ginger on the Severity of Premenstrual Syndrome Symptoms," *ISRN Obstetrics and Gynecology* 2014 (2014): 792708, https://doi.org/10.1155/2014/792708.

89. Francesco Branca and Stefano Lorenzetti, "Health Effects of Phytoestrogens," *Forum of Nutrition*, no. 57 (2005): 100–111, https://www.ncbi.nlm.nih.gov/pubmed/15702593; S. S. Strom et al., "Phytoestrogen Intake and Prostate Cancer: A Case-Control Study Using a New Database," *Nutrition and Cancer* 33, no. 1 (1999): 20–25, https://doi.org/10.1080/01635589909514743; Wayne A. Fritz et al., "Genistein Alters Growth But Is Not Toxic to the Rat Prostate," *The Journal of Nutrition* 132, no. 10 (2002): 3007–11, https://doi.org/10.1093/jn/131.10.3007.

90. Kristian Almstrup et al., "Dual Effects of Phytoestrogens Result in U-Shaped Dose-Response Curves," *Environmental Health Perspectives* 110, no. 8 (2002): 743–48, https://www.ncbi.nlm.nih.gov/pubmed/12153753.

91. Nan Zhang et al., "Effect of M-Phase Kinase Phosphorylations on Type 1 Inositol 1,4,5-Trisphosphate Receptor-Mediated Ca2+ Responses in Mouse Eggs," *Cell Calcium* 58, no. 5 (2015): 476–88, https://doi.org/10.1016/j.ceca.2015.07.004.

92. E. Papaleo et al., "Contribution of Myo-Inositol to Reproduction," *European Journal of Obstetrics, Gynecology, and Reproductive Biology* 147, no. 2 (2009): 120–23, https://doi.org/10.1016/j.ejogrb.2009.09.008.

93. V. Unfer et al., "Effects of Myo-Inositol in Women with PCOS: A Systematic Review of Randomized Controlled Trials," *Gynecological Endocrinology: The Official Journal of the*

International Society of Gynecological Endocrinology 28, no. 7 (2012): 509–15, https://doi.org/10.3109/09513590.201 1.650660; Francesca Caprio et al., "Myo-Inositol Therapy for Poor-Responders during IVF: A Prospective Controlled Observational Trial," *Journal of Ovarian Research* 8 (2015): 37, https://doi.org/10.1186/s13048-015-0167-x.

94. Osamu Negishi et al., "Content of Methylated Inositols in Familiar Edible Plants," *Journal of Agricultural and Food Chemistry* 63, no. 10 (2015): 2683–88, https://doi.org /10.1021/jf5041367.

95. Ana Isabel Ruiz-Matute et al., "A GC Method for Simultaneous Analysis of Bornesitol, Other Polyalcohols and Sugars in Coffee and Its Substitutes," *Journal of Separation Science* 30, no. 4 (2007): 557–62, https://www .ncbi.nlm.nih.gov/pubmed/17444224.

96. Osamu Negishi et al., "Content of Methylated Inositols in Familiar Edible Plants."

97. Nedim Tetik et al., "Determination of D-Pinitol in Carob Syrup," *International Journal of Food Sciences and Nutrition* 62, no. 6 (2011): 572–76, https://doi.org/10.3109/09637486 .2011.560564.

98. Celia Bañuls et al., "Chronic Consumption of an Inositol-Enriched Carob Extract Improves Postprandial Glycaemia and Insulin Sensitivity in Healthy Subjects: A Randomized Controlled Trial," *Clinical Nutrition* 35, no. 3 (2016): 600–607, https://doi.org/10.1016/j.clnu.2015.05.005.

99. Ettore Cittadini, "The Role of Inositol Supplementation in Patients with Polycystic Ovary Syndrome, with Insulin Resistance, Undergoing the Low-Dose Gonadotropin Ovulation Induction Regimen," *Fertility and Sterility* 95, no. 8 (2011): e57, https://doi.org/10.1016/j.fertnstert.2011.03.100.

100. Osamu Negishi et al., "Content of Methylated Inositols in Familiar Edible Plants."

101. Nedim Tetik and Esra Yüksel, "Ultrasound-Assisted Extraction of D-Pinitol from Carob Pods Using Response Surface Methodology," *Ultrasonics Sonochemistry* 21, no. 2 (2014): 860–65, https://doi.org/10.1016/j.ultsonch .2013.09.008.

102. J. V. Tarazona and F. Sanz, "Toxicity of Fractions Obtained from the Legume Species *Astragalus lusitanicus* Lam. *Lusitanicus*," *Toxicon: Official Journal of the International Society on Toxinology* 28, no. 2 (1990): 235–37, https://www .ncbi.nlm.nih.gov/pubmed/2339438.

103. Daniela M. Biondi et al., "Dihydrostilbene Derivatives from *Glycyrrhiza glabra* Leaves," *Journal of Natural Products* 68, no. 7 (2005): 1099–1102, https://doi.org/10.1021/np050034q.

104. Ji-Yi Zhang et al., "Global Reprogramming of Transcription and Metabolism in *Medicago truncatula* during Progressive Drought and after Rewatering," *Plant, Cell & Environment* 37, no. 11 (2014): 2553–76, https://doi.org/10.1111/pce.12328.

105. David D. Rahn et al., "Vaginal Estrogen for Genitourinary Syndrome of Menopause: A Systematic Review," *Obstetrics and Gynecology* 124, no. 6 (2014): 1147–56, https://doi.org /10.1097/AOG.0000000000000526.

106. Yanping Zhang et al., "Effects of Puerarin on Cholinergic Enzymes in the Brain of Ovariectomized Guinea Pigs," *The International Journal of Neuroscience* 123, no. 11 (2013): 783–91, https://doi.org/10.3109/00207454.2013.803103.

107. Verapol Chandeying and Surachai Lamlertkittikul, "Challenges in the Conduct of Thai Herbal Scientific Study: Efficacy and Safety of Phytoestrogen, *Pueraria mirifica* (Kwao Keur Kao), Phase I, in the Alleviation of Climacteric Symptoms in Perimenopausal Women," *Chotmaihet Thangphaet (Journal of the Medical Association of Thailand)* 90, no. 7 (2007): 1274–80, https://www.ncbi.nlm.nih.gov /pubmed/17710964; Surachai Lamlertkittikul and Verapol Chandeying, "Efficacy and Safety of *Pueraria mirifica* (Kwao Kruea Khao) for the Treatment of Vasomotor Symptoms in Perimenopausal Women: Phase II Study," *Chotmaihet Thangphaet (Journal of the Medical Association of Thailand)* 87, no. 1 (2004): 33–40, https://www.ncbi.nlm .nih.gov/pubmed/14971532.

108. Wacharaporn Tiyasatkulkovit et al., "Upregulation of Osteoblastic Differentiation Marker MRNA Expression in Osteoblast-like UMR106 Cells by Puerarin and Phytoestrogens from *Pueraria mirifica*," *Phytomedicine: International Journal of Phytotherapy and Phytopharmacology* 19, no. 13 (2012): 1147–55, https://doi .org/10.1016/j.phymed.2012.07.010.

109. Jittima Manonai et al., "Effects and Safety of *Pueraria mirifica* on Lipid Profiles and Biochemical Markers of Bone Turnover Rates in Healthy Postmenopausal Women," *Menopause* 15, no. 3 (2008): 530–35, https://doi.org /10.1097/gme.0b013e31815c5fd8.

110. Manonai et al., "Effects and Safety of *Pueraria mirifica*."

111. Abdul Malik Tyagi et al., "Formononetin Reverses Established Osteopenia in Adult Ovariectomized Rats," *Menopause* 19, no. 8 (2012): 856–63, https://doi.org/10.1097/ gme.0b013e31824f9306; Ilona Kaczmarczyk-Sedlak et al., "Effect of Formononetin on Mechanical Properties and Chemical Composition of Bones in Rats with Ovariectomy-Induced Osteoporosis," *Evidence-Based Complementary and Alternative Medicine: ECAM* 2013 (2013): 457052, https://doi.org/10.1155/2013/457052.

112. Dong Wook Lim et al., "Effects of Dietary Isoflavones from *Puerariae radix* on Lipid and Bone Metabolism in Ovariectomized Rats," *Nutrients* 5, no. 7 (2013): 2734–46, https://doi.org/10.3390/nu5072734.

113. Elise F. Saunier et al., "Estrogenic Plant Extracts Reverse Weight Gain and Fat Accumulation without Causing Mammary Gland or Uterine Proliferation," *PLoS One* 6, no. 12 (2011): e28333, https://doi.org/10.1371/journal. pone.0028333; Ji-Feng Wang et al., "Effects of *Radix puerariae* Flavones on Liver Lipid Metabolism in Ovariectomized Rats," *World Journal of Gastroenterology* 10, no. 13 (2004): 1967–70, https://www.ncbi.nlm.nih.gov/pubmed/15222048.

114. Shinichi Okamura et al., "*Pueraria mirifica* Phytoestrogens Improve Dyslipidemia in Postmenopausal Women

Probably by Activating Estrogen Receptor Subtypes," *The Tohoku Journal of Experimental Medicine* 216, no. 4 (2008): 341–51, https://www.ncbi.nlm.nih.gov/pubmed/19060449.

115. Verapol Chandeying and Malinee Sangthawan, "Efficacy Comparison of *Pueraria mirifica* (PM) against Conjugated Equine Estrogen (CEE) with/without Medroxyprogesterone Acetate (MPA) in the Treatment of Climacteric Symptoms in Perimenopausal Women: Phase III Study," *Chotmaihet Thangphaet (Journal of the Medical Association of Thailand)* 90, no. 9 (2007): 1720–26, https://www.ncbi.nlm.nih.gov/pubmed/17957910.

116. Guohui Li et al., "Chemical Constituents from Roots of *Pueraria lobata*," *Zhongguo Zhong Yao Za Zhi = Zhongguo Zhongyao Zazhi (China Journal of Chinese Materia Medica)* 35, no. 23 (2010): 3156–60, https://www.ncbi.nlm.nih.gov/pubmed/21355238; Dewu Zhang et al., "Isoflavones from Vines of *Pueraria lobata*," *Zhongguo Zhong Yao Za Zhi = Zhongguo Zhongyao Zazhi (China Journal of Chinese Materia Medica)* 34, no. 24 (2009): 3217–20, https://www.ncbi.nlm.nih.gov/pubmed/20353004; Jing Lu et al., "Simultaneous Determination of Isoflavones, Saponins and Flavones in *Flos puerariae* by Ultra Performance Liquid Chromatography Coupled with Quadrupole Time-of-Flight Mass Spectrometry," *Chemical & Pharmaceutical Bulletin* 61, no. 9 (2013): 941–51, https://www.ncbi.nlm.nih.gov/pubmed/23759517; Cuijuan Xue et al., "Determination of Five Components in *Pueraria labta* decoction with Reference Extraction Method," *Zhongguo Zhong Yao Za Zhi = Zhongguo Zhongyao Zazhi (China Journal of Chinese Materia Medica)* 37, no. 16 (2012): 2388–91, https://www.ncbi.nlm.nih.gov/pubmed/23234135.

117. Hajime Sugiyama et al., "Insight into Estrogenicity of Phytoestrogens Using in Silico Simulation," *Biochemical and Biophysical Research Communications* 379, no. 1 (2009): 139–44, https://doi.org/10.1016/j.bbrc.2008.12.046; Shinichi Okamura et al., "*Pueraria mirifica* Phytoestrogens Improve Dyslipidemia in Postmenopausal Women Probably by Activating Estrogen Receptor Subtypes," *The Tohoku Journal of Experimental Medicine* 216, no. 4 (2008): 341–51, https://www.ncbi.nlm.nih.gov/pubmed/19060449.

118. Dian-Xia Xing et al., "The Estrogenic Effect of Formononetin and Its Effect on the Expression of Rats' Atrium Estrogen Receptors," *Zhong Yao Cai = Zhongyaocai (Journal of Chinese Medicinal Materials)* 33, no. 9 (2010): 1445–49, https://www.ncbi.nlm.nih.gov/pubmed/21243777.

119. Jeong-Eun Huh et al., "Biphasic Positive Effect of Formononetin on Metabolic Activity of Human Normal and Osteoarthritic Subchondral Osteoblasts," *International Immunopharmacology* 10, no. 4 (2010): 500–507, https://doi.org/10.1016/j.intimp.2010.01.012.

120. Abdul Malik Tyagi et al., "Formononetin Reverses Established Osteopenia in Adult Ovariectomized Rats," *Menopause* 19, no. 8 (2012): 856–63, https://doi.org/10.1097/gme.0b013e31824f9306.

121. Yue Wang et al., "Puerarin Stimulates Proliferation and Differentiation and Protects against Cell Death in Human Osteoblastic MG-63 Cells via ER-Dependent MEK/ERK and PI3K/Akt Activation," *Phytomedicine: International Journal of Phytotherapy and Phytopharmacology* 20, no. 10 (2013): 787–96, https://doi.org/10.1016/j.phymed.2013.03.005.

122. Yue Wang et al., "Puerarin Concurrently Stimulates Osteoprotegerin and Inhibits Receptor Activator of NF-KB Ligand (RANKL) and Interleukin-6 Production in Human Osteoblastic MG-63 Cells," *Phytomedicine: International Journal of Phytotherapy and Phytopharmacology* 21, no. 8–9 (2014): 1032–36, https://doi.org/10.1016/j.phymed.2014.04.012; Wang et al., "Puerarin Stimulates Proliferation and Differentiation."

123. Wacharaporn Tiyasatkulkovit et al., "Upregulation of Osteoblastic Differentiation Marker MRNA Expression in Osteoblast-like UMR106 Cells by Puerarin and Phytoestrogens from *Pueraria mirifica*," *Phytomedicine: International Journal of Phytotherapy and Phytopharmacology* 19, no. 13 (2012): 1147–55, https://doi.org/10.1016/j.phymed.2012.07.010.

124. Jichan Nie and Xishi Liu, "Leonurine Attenuates Hyperalgesia in Mice with Induced Adenomyosis," *Medical Science Monitor: International Medical Journal of Experimental and Clinical Research* 23 (2017): 1701–6, https://www.ncbi.nlm.nih.gov/pubmed/28389633.

125. Miaomiao Jia et al., "Leonurine Exerts Antidepressant-Like Effects in the Chronic Mild Stress-Induced Depression Model in Mice by Inhibiting Neuroinflammation," *The International Journal of Neuropsychopharmacology* 20, no. 11 (2017): 886–95, https://doi.org/10.1093/ijnp/pyx062.

126. Mingsan Miao et al., "Effect of Motherwort Total Alkaloids on the Prostate Hyperplasia Mice Model of Pathological Changes of Related Tissue Morphology Induced by the Fetal Urogenital Sinus Implants," *Saudi Pharmaceutical Journal: SPJ: The Official Publication of the Saudi Pharmaceutical Society* 25, no. 4 (2017): 601–6, https://doi.org/10.1016/j.jsps.2017.04.030.

127. Mingsan Miao et al., "The Effect of Leonuri Herba Alkaloids on Senile BPH (Male and Female Hormone Induced) Model Rats," *Saudi Journal of Biological Sciences* 24, no. 3 (2017): 630–33, https://doi.org/10.1016/j.sjbs.2017.01.035.

128. Friedemann Gaube et al., "Gene Expression Profiling Reveals Effects of *Cimicifuga racemosa* (L.) NUTT. (Black Cohosh) on the Estrogen Receptor Positive Human Breast Cancer Cell Line MCF-7," *BMC Pharmacology* 7 (2007): 11, https://doi.org/10.1186/1471-2210-7-11.

129. Linda Saxe Einbond et al., "Growth Inhibitory Activity of Extracts and Compounds from Cimicifuga Species on Human Breast Cancer Cells," *Phytomedicine: International Journal of Phytotherapy and Phytopharmacology* 15, no. 6–7 (2008): 504–11, https://doi.org/10.1016/j.phymed.2007.09.017.

130. Mee-Ra Rhyu et al., "Black Cohosh (*Actaea Racemosa, Cimicifuga Racemosa*) Behaves as a Mixed Competitive Ligand and Partial Agonist at the Human Mu Opiate Receptor," *Journal of Agricultural and Food Chemistry* 54, no. 26 (2006): 9852–57, https://doi.org/10.1021/jf062808u.

131. Mahnaz Shahnazi et al., "Effect of Black Cohosh (*Cimicifuga Racemosa*) on Vasomotor Symptoms in Postmenopausal Women: A Randomized Clinical Trial," *Journal of Caring Sciences* 2, no. 2 (2013): 105–13, https://doi.org/10.5681/jcs.2013.013.

132. Sakineh Mohammad-Alizadeh-Charandabi et al., "Efficacy of Black Cohosh (*Cimicifuga Racemosa* L.) in Treating Early Symptoms of Menopause: A Randomized Clinical Trial," *Chinese Medicine* 8, no. 1 (2013): 20, https://doi.org/10.1186/1749-8546-8-20.

133. F. Dehghan et al., "The Effect of Relaxin on the Musculoskeletal System," *Scandinavian Journal of Medicine & Science in Sports* 24, no. 4 (2014): e220–229, https://doi.org/10.1111/sms.12149.

134. Lori C. Sakoda et al., "Glutathione S-Transferase M1 and P1 Polymorphisms and Risk of Breast Cancer and Fibrocystic Breast Conditions in Chinese Women," *Breast Cancer Research and Treatment* 109, no. 1 (2008): 143–55, https://doi.org/10.1007/s10549-007-9633-5.

135. Wenjuan Shen et al., "Effects of Tanshinone on Hyperandrogenism and the Quality of Life in Women with Polycystic Ovary Syndrome: Protocol of a Double-Blind, Placebo-Controlled, Randomised Trial," *BMJ Open* 3, no. 10 (2013): e003646, https://doi.org/10.1136/bmjopen-2013-003646.

136. Qi Ye et al., "Casticin, a Flavonoid Isolated from *Vitex rotundifolia*, Inhibits Prolactin Release In Vivo and In Vitro," *Acta Pharmacologica Sinica* 31, no. 12 (2010): 1564–68, https://doi.org/10.1038/aps.2010.178.

137. Sarah M. Nielsen et al., "The Breast-Thyroid Cancer Link: A Systematic Review and Meta-Analysis," *Cancer Epidemiology, Biomarkers & Prevention: A Publication of the American Association for Cancer Research, Cosponsored by the American Society of Preventive Oncology* 25, no. 2 (2016): 231–38, https://doi.org/10.1158/1055-9965.EPI-15-0833.

138. Peter P. A. Smyth, "The Thyroid, Iodine and Breast Cancer," *Breast Cancer Research: BCR* 5, no. 5 (2003): 235–38, https://doi.org/10.1186/bcr638.

139. Frederick R. Stoddard et al., "Iodine Alters Gene Expression in the MCF7 Breast Cancer Cell Line: Evidence for an Anti-Estrogen Effect of Iodine," *International Journal of Medical Sciences* 5, no. 4 (2008): 189–96, https://www.ncbi.nlm.nih.gov/pubmed/18645607.

140. A. R. Carmichael, "Can *Vitex agnus-castus* Be Used for the Treatment of Mastalgia? What Is the Current Evidence?" *Evidence-Based Complementary and Alternative Medicine: ECAM* 5, no. 3 (2008): 247–50, https://doi.org/10.1093/ecam/nem074; Tolga Dinç and Faruk Coşkun, "Comparison of *Fructus agni casti* and Flurbiprofen in the Treatment of Cyclic Mastalgia in Premenopausal Women," *Ulusal Cerrahi Dergisi* 30, no. 1 (2014): 34–38, https://doi.org/10.5152/UCD.2014.2409.

141. Qi Ye et al., "Casticin, a Flavonoid Isolated from *Vitex rotundifolia*, Inhibits Prolactin Release in Vivo and in Vitro," *Acta Pharmacologica Sinica* 31, no. 12 (2010): 1564–68, https://doi.org/10.1038/aps.2010.178.

142. Athar Rasekhjahromi et al., "Herbal Medicines and Ovarian Hyperstimulation Syndrome: A Retrospective Cohort Study," *Obstetrics and Gynecology International* 2016 (2016): 7635185, https://doi.org/10.1155/2016/7635185.

143. Rasekhjahromi et al., "Herbal Medicines and Ovarian Hyperstimulation Syndrome."

144. Wenjuan Shen et al., "Effects of Tanshinone on Hyperandrogenism and the Quality of Life in Women with Polycystic Ovary Syndrome: Protocol of a Double-Blind, Placebo-Controlled, Randomised Trial," *BMJ Open* 3, no. 10 (2013): e003646, https://doi.org/10.1136/bmjopen-2013-003646.

145. Shen et al., "Effects of Tanshinone on Hyperandrogenism."

146. Yan Li et al., "Effect of Berberine on Insulin Resistance in Women with Polycystic Ovary Syndrome: Study Protocol for a Randomized Multicenter Controlled Trial," *Trials* 14 (2013): 226, https://doi.org/10.1186/1745-6215-14-226.

147. Byoung-Seob Ko et al., "Changes in Components, Glycyrrhizin and Glycyrrhetinic Acid, in Raw Glycyrrhiza Uralensis Fisch, Modify Insulin Sensitizing and Insulinotropic Actions," *Bioscience, Biotechnology, and Biochemistry* 71, no. 6 (2007): 1452–61, https://doi.org/10.1271/bbb.60533.

148. Chia Hui Apphia Eu et al., "Glycyrrhizic Acid Improved Lipoprotein Lipase Expression, Insulin Sensitivity, Serum Lipid and Lipid Deposition in High-Fat Diet-Induced Obese Rats," *Lipids in Health and Disease* 9 (2010): 81, https://doi.org/10.1186/1476-511X-9-81.

149. Apphia Eu et al., "Glycyrrhizic Acid Improved Lipoprotein Lipase Expression."

150. Kaku Nakagawa et al., "Licorice Flavonoids Suppress Abdominal Fat Accumulation and Increase in Blood Glucose Level in Obese Diabetic KK-A(y) Mice," *Biological & Pharmaceutical Bulletin* 27, no. 11 (2004): 1775–78, https://www.ncbi.nlm.nih.gov/pubmed/15516721.

151. T. Tanaka, "A Novel Anti-Dysmenorrhea Therapy with Cyclic Administration of Two Japanese Herbal Medicines," *Clinical and Experimental Obstetrics & Gynecology* 30, no. 2–3 (2003): 95–98, https://www.ncbi.nlm.nih.gov/pubmed/12854851.

152. K. Takahashi et al., "Effect of a Traditional Herbal Medicine (Shakuyaku-Kanzo-to) on Testosterone Secretion in Patients with Polycystic Ovary Syndrome Detected by Ultrasound," *Nihon Sanka Fujinka Gakkai Zasshi* 40, no. 6 (1988): 789–92, https://www.ncbi.nlm.nih.gov/pubmed/3292675.

153. Farzaneh Zamansoltani et al., "Antiandrogenic Activities of *Glycyrrhiza glabra* in Male Rats," *International Journal of Andrology* 32, no. 4 (2009): 417–22, https://doi.org/10.1111/j.1365-2605.2009.00944.x; Decio Armanini et al., "Licorice

Reduces Serum Testosterone in Healthy Women," *Steroids* 69, no. 11–12 (2004): 763–66, https://doi.org/10.1016/j .steroids.2004.09.005.

154. T. Takeuchi et al., "Effect of Traditional Herbal Medicine, Shakuyaku-Kanzo-to on Total and Free Serum Testosterone Levels," *The American Journal of Chinese Medicine* 17, no. 1–2 (1989): 35–44, https://doi.org/10.1142 /S0192415X89000073.

155. K. Takahashi et al., "Effect of a Traditional Herbal Medicine (Shakuyaku-Kanzo-to) on Testosterone Secretion."

156. Decio Armanini et al., "Licorice Reduces Serum Testosterone in Healthy Women."

157. E. A. S. Al-Dujaili et al., "Liquorice and Glycyrrhetinic Acid Increase DHEA and Deoxycorticosterone Levels in Vivo and in Vitro by Inhibiting Adrenal SULT2A1 Activity," *Molecular and Cellular Endocrinology* 336, no. 1–2 (2011): 102–9, https://doi.org/10.1016/j.mce.2010.12.011; D. J. Morris et al., "Endogenous Inhibitors (GALFs) of 11beta-Hydroxysteroid Dehydrogenase Isoforms 1 and 2: Derivatives of Adrenally Produced Corticosterone and Cortisol," *The Journal of Steroid Biochemistry and Molecular Biology* 104, no. 3–5 (2007): 161–68, https://doi .org/10.1016/j.jsbmb.2007.03.020.

158. R. A. Josephs et al., "Liquorice Consumption and Salivary Testosterone Concentrations," *Lancet* 358, no. 9293 (2001): 1613–14, https://doi.org/10.1016/S0140-6736(01)06664-8; Xue Ma et al., "Environmental Inhibitors of 11β-Hydroxy-steroid Dehydrogenase Type 2," *Toxicology* 285, no. 3 (2011): 83–89, https://doi.org/10.1016/j.tox.2011.04.007.

159. Paul Grant and Shamin Ramasamy, "An Update on Plant Derived Anti-Androgens," *International Journal of Endocrinology and Metabolism* 10, no. 2 (2012): 497–502, https://doi.org/10.5812/ijem.3644.

160. Farzaneh Zamansoltani et al., "Antiandrogenic Activities of *Glycyrrhiza glabra* in Male Rats," *International Journal of Andrology* 32, no. 4 (2009): 417–22, https://doi.org/10.1111 /j.1365-2605.2009.00944.x.

161. Decio Armanini et al., "Treatment of Polycystic Ovary Syndrome with Spironolactone plus Licorice," *European Journal of Obstetrics, Gynecology, and Reproductive Biology* 131, no. 1 (2007): 61–67, https://doi.org/10.1016/j.ejogrb .2006.10.013.

162. Hai-Ning Yuan et al., "A Randomized, Crossover Comparison of Herbal Medicine and Bromocriptine against Risperidone-Induced Hyperprolactinemia in Patients with Schizophrenia," *Journal of Clinical Psychopharmacology* 28, no. 3 (2008): 264–370, https://doi.org/10.1097/JCP .0b013e318172473c; Deanna L. Kelly et al., "Treating Symptomatic Hyperprolactinemia in Women with Schizophrenia: Presentation of the Ongoing DAAMSEL Clinical Trial (Dopamine Partial Agonist, Aripiprazole, for the Management of Symptomatic Elevated Prolactin)," *BMC Psychiatry* 13 (2013): 214, https://doi.org/10.1186 /1471-244X-13-214.

163. Timo E. Strandberg et al., "Preterm Birth and Licorice Consumption during Pregnancy," *American Journal of Epidemiology* 156, no. 9 (2002): 803–5, https://www.ncbi .nlm.nih.gov/pubmed/12396997.

164. T. E. Strandberg et al., "Birth Outcome in Relation to Licorice Consumption during Pregnancy," *American Journal of Epidemiology* 153, no. 11 (2001): 1085–88, https:// www.ncbi.nlm.nih.gov/pubmed/11390327.

165. June-Seek Choi et al., "Fetal and Neonatal Outcomes in Women Reporting Ingestion of Licorice (*Glycyrrhiza uralensis*) during Pregnancy," *Planta Medica* 79, no. 2 (2013): 97–101, https://doi.org/10.1055/s-0032-1328102.

166. Katri Räikkönen et al., "Maternal Prenatal Licorice Consumption Alters Hypothalamic-Pituitary-Adrenocortical Axis Function in Children," *Psychoneuroendocrinology* 35, no. 10 (2010): 1587–93, https://doi.org/10.1016/j.psyneuen .2010.04.010.

167. Paul M. Stewart, "Tissue-Specific Cushing's Syndrome, 11beta-Hydroxysteroid Dehydrogenases and the Redefinition of Corticosteroid Hormone Action," *European Journal of Endocrinology* 149, no. 3 (2003): 163–68, https://www.ncbi.nlm.nih.gov/pubmed/12943516; E. N. Levtchenko et al., "From Gene to Disease; 'Apparent Mineralocorticoid Excess' Syndrome, a Syndrome with an Apparent Excess of Mineral Corticoids," *Nederlands Tijdschrift Voor Geneeskunde* 151, no. 12 (2007): 692–94, https://www.ncbi.nlm.nih.gov/pubmed/17447595; Helena E. Miettinen et al., "Licorice-Induced Hypertension and Common Variants of Genes Regulating Renal Sodium Reabsorption," *Annals of Medicine* 42, no. 6 (2010): 465–74, https://doi.org/10.3109/07853890.2010 .499133; Mitsuo Inada, "A 87-year-old Woman with Mineralocorticoid Excess Due to 11 beta-HSD2 Deficiency," *Nihon Ronen Igakkai Zasshi (Japanese Journal of Geriatrics)* 44, no. 4 (2007): 513–16, https://www.ncbi.nlm.nih.gov /pubmed/17827812; Stan H. M. van Uum et al., "Cortisol, 11beta-Hydroxysteroid Dehydrogenases, and Hypertension," *Seminars in Vascular Medicine* 4, no. 2 (2004): 121–28, https://doi.org/10.1055/s-2004-835369.

168. Tatsumasa Mae et al., "A Licorice Ethanolic Extract with Peroxisome Proliferator-Activated Receptor-Gamma Ligand-Binding Activity Affects Diabetes in KK-Ay Mice, Abdominal Obesity in Diet-Induced Obese C57BL Mice and Hypertension in Spontaneously Hypertensive Rats," *The Journal of Nutrition* 133, no. 11 (2003): 3369–77, https:// doi.org/10.1093/jn/133.11.3369.

169. B. Marinov et al., "Our Experience with Management of Inherited Thrombophilia during Pregnancy: Preliminary Report," *Akusherstvo I Ginekologiia* 50 Suppl 2 (2011): 28–31, https://www.ncbi.nlm.nih.gov/pubmed/22524136; Ruediger Schellenberg et al., "Dose-Dependent Efficacy of the *Vitex agnus-castus* Extract Ze 440 in Patients Suffering from Premenstrual Syndrome," *Phytomedicine: International Journal of Phytotherapy and*

Phytopharmacology 19, no. 14 (2012): 1325–31, https://doi .org/10.1016/j.phymed.2012.08.006; A. Milewicz et al., "*Vitex agnus-castus* Extract in the Treatment of Luteal Phase Defects Due to Latent Hyperprolactinemia. Results of a Randomized Placebo-Controlled Double-Blind Study," *Arzneimittel-Forschung* 43, no. 7 (1993): 752–56, https:// www.ncbi.nlm.nih.gov/pubmed/8369008.

170. M. Halaska et al., "Treatment of Cyclical Mastodynia Using an Extract of *Vitex agnus-castus*: Results of a Double-Blind Comparison with a Placebo," *Ceska Gynekologie* 63, no. 5 (1998): 388–92, https://www.ncbi.nlm.nih.gov /pubmed/9818496.

171. Margaret Diana van Die et al., "*Vitex agnus-castus* (Chaste-Tree/Berry) in the Treatment of Menopause-Related Complaints," *Journal of Alternative and Complementary Medicine* 15, no. 8 (2009): 853–62, https://doi.org/10.1089 /acm.2008.0447.

172. Donna E. Webster et al., "Opioidergic Mechanisms Underlying the Actions of *Vitex agnus-castus* L," *Biochemical Pharmacology* 81, no. 1 (2011): 170–77, https://doi.org/10.1016/j.bcp.2010.09.013; B. Meier et al., "Pharmacological Activities of *Vitex agnus-castus* Extracts In Vitro," *Phytomedicine: International Journal of Phytotherapy and Phytopharmacology* 7, no. 5 (2000): 373–81, https://doi.org/10.1016/S0944-7113(00)80058-6; D. E. Webster et al., "Activation of the Mu-Opiate Receptor by *Vitex agnus-castus* Methanol Extracts: Implication for Its Use in PMS," *Journal of Ethnopharmacology* 106, no. 2 (2006): 216–21, https://doi.org/10.1016/j.jep.2005.12.025.

173. N. A. Ibrahim et al., "Gynecological Efficacy and Chemical Investigation of *Vitex agnus-castus* L. Fruits Growing in Egypt," *Natural Product Research* 22, no. 6 (2008): 537–46, https://doi.org/10.1080/14786410701592612.

174. Rattan L. Sharma et al., "A New Iridoid Glycoside from *Vitex negundo Linn* (Verbenacea)," *Natural Product Research* 23, no. 13 (2009): 1201–9, https://doi.org/10.1080/14786410802696494; Parvinder Pal Singh et al., "Synthesis of Novel Lipidated Iridoid Glycosides as Vaccine Adjuvants: 6-O-Palmitoyl Agnuside Elicit Strong Th1 and Th2 Response to Ovalbumin in Mice," *International Immunopharmacology* 17, no. 3 (2013): 593–600, https://doi.org/10.1016/j.intimp .2013.07.018; Ayşe Kuruüzüm-Uz et al., "Glucosides from *Vitex agnus-castus*," *Phytochemistry* 63, no. 8 (2003): 959–64, https://www.ncbi.nlm.nih.gov/pubmed/12895546.

175. Shenghong Li et al., "Compounds from the Fruits of the Popular European Medicinal Plant *Vitex agnus-castus* in Chemoprevention via NADP(H):Quinone Oxidoreductase Type 1 Induction," *Evidence-Based Complementary and Alternative Medicine: ECAM* 2013 (2013): 432829, https:// doi.org/10.1155/2013/432829; W. Wuttke et al., "Chaste Tree (*Vitex agnus-castus*)—Pharmacology and Clinical Indications," *Phytomedicine: International Journal of Phytotherapy and Phytopharmacology* 10, no. 4 (2003): 348–57, https://doi.org/10.1078/094471103322004866.

176. Hubertus Jarry et al., "Evidence for Estrogen Receptor Beta-Selective Activity of *Vitex agnus-castus* and Isolated Flavones," *Planta Medica* 69, no. 10 (2003): 945–47, https:// doi.org/10.1055/s-2003-45105.

177. J. Liu et al., "Isolation of Linoleic Acid as an Estrogenic Compound from the Fruits of *Vitex agnus-castus* L. (Chaste-Berry)," *Phytomedicine: International Journal of Phytotherapy and Phytopharmacology* 11, no. 1 (2004): 18–23, https://doi.org/10.1078/0944-7113-00331.

178. Yuan Hu et al., "Evaluation of the Estrogenic Activity of the Constituents in the Fruits of *Vitex rotundifolia* L. for the Potential Treatment of Premenstrual Syndrome," *The Journal of Pharmacy and Pharmacology* 59, no. 9 (2007): 1307–12, https://doi.org/10.1211/jpp.59.9.0016; J. Liu et al., "Isolation of Linoleic Acid as an Estrogenic Compound from the Fruits of *Vitex agnus-castus* L. (Chaste-Berry)," *Phytomedicine: International Journal of Phytotherapy and Phytopharmacology* 11, no. 1 (2004): 18–23, https://doi.org /10.1078/0944-7113-00331.

179. J. Liu et al., "Evaluation of Estrogenic Activity of Plant Extracts for the Potential Treatment of Menopausal Symptoms," *Journal of Agricultural and Food Chemistry* 49, no. 5 (2001): 2472–79, https://www.ncbi.nlm.nih.gov/pubmed/11368622.

180. Suwagmani Das et al., "Reproduction in Male Rats Is Vulnerable to Treatment with the Flavonoid-Rich Seed Extracts of *Vitex negundo*," *Phytotherapy Research: PTR* 18, no. 1 (2004): 8–13, https://doi.org/10.1002/ptr.1352.

181. S. K. Bhargava, "Antiandrogenic Effects of a Flavonoid-Rich Fraction of *Vitex negundo* Seeds: A Histological and Biochemical Study in Dogs," *Journal of Ethnopharmacology* 27, no. 3 (1989): 327–39, https://www.ncbi.nlm.nih.gov /pubmed/2615438.

182. Jean-Jacques Dugoua et al., "Safety and Efficacy of Chastetree (*Vitex agnus-castus*) during Pregnancy and Lactation," *Journal Canadien de Pharmacologie (The Canadian Journal of Clinical Pharmacology)* 15, no. 1 (2008): e74–79, https://www.ncbi.nlm.nih.gov/pubmed/18204102.

183. Pamela Stratton and Karen J. Berkley, "Chronic Pelvic Pain and Endometriosis: Translational Evidence of the Relationship and Implications," *Human Reproduction Update* 17, no. 3 (2011): 327–46, https://doi.org/10.1093 /humupd/dmq050.

184. Nick Dalton-Brewer, "The Role of Complementary and Alternative Medicine for the Management of Fibroids and Associated Symptomatology," *Current Obstetrics and Gynecology Reports* 5 (2016): 110–18, https://doi.org/10.1007 /s13669-016-0156-0.

185. Soumia Brakta et al., "Role of Vitamin D in Uterine Fibroid Biology," *Fertility and Sterility* 104, no. 3 (2015): 698–706, https://doi.org/10.1016/j.fertnstert.2015.05.031.

186. Darlene K. Taylor and Phyllis C. Leppert, "Treatment for Uterine Fibroids: Searching for Effective Drug Therapies," *Drug Discovery Today. Therapeutic Strategies* 9, no. 1 (2012): e41–49, https://doi.org/10.1016/j.ddstr.2012.06.001.

187. H. Nagasawa et al., "Further Study on the Effects of Motherwort (*Leonurus sibiricus* L.) on Preneoplastic and Neoplastic Mammary Gland Growth in Multiparous GR/A Mice," *Anticancer Research* 12, no. 1 (1992): 141–43, https://www.ncbi.nlm.nih.gov/pubmed/1567160.

188. Nagasawa et al., "Further Study on the Effects of Motherwort (*Leonurus sibiricus* L.)."

189. Carlos Rb Gama et al., "Clinical Assessment of *Tribulus terrestris* Extract in the Treatment of Female Sexual Dysfunction," *Clinical Medicine Insights. Women's Health* 7 (2014): 45–50, https://doi.org/10.4137/CMWH.S17853; Nagendra Singh Chauhan et al., "A Review on Plants Used for Improvement of Sexual Performance and Virility," *BioMed Research International* 2014 (2014): 868062, https://doi.org/10.1155/2014/868062.

190. Kar Wah Leung and Alice St Wong, "Ginseng and Male Reproductive Function," *Spermatogenesis* 3, no. 3 (2013): e26391, https://doi.org/10.4161/spmg.26391.

191. Arif Adimoelja and Ali Fuchih Siauw, "AB69. Phyto-Androgenic Androgens in Men's Health, Sex and Aging FX," *Translational Andrology and Urology* 3, no. Suppl 1 (2014), https://doi.org/10.3978/j.issn.2223-4683.2014.s069.

192. Desislava Abadjieva and Elena Kistanova, "*Tribulus terrestris* Alters the Expression of Growth Differentiation Factor 9 and Bone Morphogenetic Protein 15 in Rabbit Ovaries of Mothers and F1 Female Offspring," *PLoS One* 11, no. 2 (2016): e0150400, https://doi.org/10.1371/journal.pone.0150400.

193. Sang-Won Park et al., "Effect of SA1, a Herbal Formulation, on Sexual Behavior and Penile Erection," *Biological & Pharmaceutical Bulletin* 29, no. 7 (2006): 1383–86, https://www.ncbi.nlm.nih.gov/pubmed/16819173.

194. R. Santti et al., "Phytoestrogens: Potential Endocrine Disruptors in Males," *Toxicology and Industrial Health* 14, no. 1–2 (1998): 223–37, https://doi.org/10.1177/074823379801400114.

195. Li Li et al., "Non-Monotonic Dose-Response Relationship in Steroid Hormone Receptor-Mediated Gene Expression," *Journal of Molecular Endocrinology* 38, no. 5 (2007): 569–85, https://doi.org/10.1677/JME-07-0003; L. Payen et al., "The Drug Efflux Pump MRP2: Regulation of Expression in Physiopathological Situations and by Endogenous and Exogenous Compounds," *Cell Biology and Toxicology* 18, no. 4 (2002): 221–33, https://www.ncbi.nlm.nih.gov/pubmed/12206135.

196. Tze-Chen Hsieh and Joseph M. Wu, "Targeting CWR22Rv1 Prostate Cancer Cell Proliferation and Gene Expression by Combinations of the Phytochemicals EGCG, Genistein and Quercetin," *Anticancer Research* 29, no. 10 (2009): 4025–32, https://www.ncbi.nlm.nih.gov/pubmed/19846946; Anja Bliedtner et al., "Effects of Genistein and Estrogen Receptor Subtype-Specific Agonists in ArKO Mice Following Different Administration Routes," *Molecular and Cellular Endocrinology* 314, no. 1 (2010): 41–52, https://doi.org/10.1016/j.mce.2009.07.032.

197. Andrea Dueregger et al., "The Use of Dietary Supplements to Alleviate Androgen Deprivation Therapy Side Effects during Prostate Cancer Treatment," *Nutrients* 6, no. 10 (2014): 4491–4519, https://doi.org/10.3390/nu6104491.

198. Hironobu Sasano et al., "New Developments in Intracrinology of Human Breast Cancer: Estrogen Sulfatase and Sulfotransferase," *Annals of the New York Academy of Sciences* 1155 (2009): 76–79, https://doi.org/10.1111/j.1749-6632.2008.03683.x.

199. Gilles Frenette et al., "Estrogen Sulfotransferase Is Highly Expressed along the Bovine Epididymis and Is Secreted into the Intraluminal Environment," *Journal of Andrology* 30, no. 5 (2009): 580–89, https://doi.org/10.2164/jandrol.108.006668.

200. Yue Chen et al., "Genistein Induction of Human Sulfotransferases in HepG2 and Caco-2 Cells," *Basic & Clinical Pharmacology & Toxicology* 103, no. 6 (2008): 553–59, https://doi.org/10.1111/j.1742-7843.2008.00316.x.

201. Jorge Raul Pasqualini, "Estrogen Sulfotransferases in Breast and Endometrial Cancers," *Annals of the New York Academy of Sciences* 1155 (2009): 88–98, https://doi.org/10.1111/j.1749-6632.2009.04113.x.

202. Laurie P. Volak et al., "Curcuminoids Inhibit Multiple Human Cytochromes P450, UDP-Glucuronosyltransferase, and Sulfotransferase Enzymes, Whereas Piperine Is a Relatively Selective CYP3A4 Inhibitor," *Drug Metabolism and Disposition: The Biological Fate of Chemicals* 36, no. 8 (2008): 1594–1605, https://doi.org/10.1124/dmd.108.020552.

203. Joanna M. Day et al., "Design and Validation of Specific Inhibitors of 17beta-Hydroxysteroid Dehydrogenases for Therapeutic Application in Breast and Prostate Cancer, and in Endometriosis," *Endocrine-Related Cancer* 15, no. 3 (2008): 665–92, https://doi.org/10.1677/ERC-08-0042.

204. A. Krazeisen et al., "Phytoestrogens Inhibit Human 17beta-Hydroxysteroid Dehydrogenase Type 5," *Molecular and Cellular Endocrinology* 171, no. 1–2 (2001): 151–62, https://www.ncbi.nlm.nih.gov/pubmed/11165023; J. C. Le Bail et al., "Effects of Phytoestrogens on Aromatase, 3beta and 17beta-Hydroxysteroid Dehydrogenase Activities and Human Breast Cancer Cells," *Life Sciences* 66, no. 14 (2000): 1281–91, https://www.ncbi.nlm.nih.gov/pubmed/10755463.

205. Thomas E. Spires et al., "Identification of Novel Functional Inhibitors of 17beta-Hydroxysteroid Dehydrogenase Type III (17beta-HSD3)," *The Prostate* 65, no. 2 (2005): 159–70, https://doi.org/10.1002/pros.20279.

206. A. Albini et al., "Exogenous Hormonal Regulation in Breast Cancer Cells by Phytoestrogens and Endocrine Disruptors," *Current Medicinal Chemistry* 21, no. 9 (2014): 1129–45, https://www.ncbi.nlm.nih.gov/pubmed/24304271.

207. T. Nesselhut et al., "Untersuchungen zur proliferativen Potenz von Phytopharmaka mit Ostrogenahnlicher Wirkung bei Mammakarzinom-Zellen," *Archives of Gynecology and Obstetrics*. 254 (1993): 817–18.

208. *Remifemin: Active Substance: Cimicifuga Fluid Extract Herbal Gynecologic Remedy* (Salzgitter, Germany: Schaper & Brümmer, 2002), http://www.medref.se/rem_sem/4_scienbroch.pdf; T. Nesselhut, S. Borth, and W. Kuhn, "Influence of *Cimicifuga racemosa* Extracts on the In Vitro Proliferation of Mammalian Carcinoma Cells," submitted for publication, 1998; Rishma Walji et al., "Black Cohosh (*Cimicifuga racemosa* [L.] *Nutt.*): Safety and Efficacy for Cancer Patients," *Supportive Care in Cancer: Official Journal of the Multinational Association of Supportive Care in Cancer* 15, no. 8 (2007): 913–21, https://doi.org/10.1007/s00520-007-0286-z; Linda Saxe Einbond et al., "Actein and a Fraction of Black Cohosh Potentiate Antiproliferative Effects of Chemotherapy Agents on Human Breast Cancer Cells," *Planta Medica* 72, no. 13 (2006): 1200–1206, https://doi.org/10.1055/s-2006-947225.

209. H. Nagasawa et al., "Effects of Motherwort (*Leonurus sibiricus* L.) on Preneoplastic and Neoplastic Mammary Gland Growth in Multiparous GR/A Mice," *Anticancer Research* 10, no. 4 (1990): 1019–23, https://www.ncbi.nlm.nih.gov/pubmed/2382973; H. Nagasawa et al., "Further Study on the Effects of Motherwort (*Leonurus sibiricus* L.) on Preneoplastic and Neoplastic Mammary Gland Growth in Multiparous GR/A Mice," *Anticancer Research* 12, no. 1 (1992): 141–43, https://www.ncbi.nlm.nih.gov/pubmed/1567160.

210. Min Ji Bak et al., "Role of Dietary Bioactive Natural Products in Estrogen Receptor-Positive Breast Cancer," *Seminars in Cancer Biology* 40–41 (2016): 170–91, https://doi.org/10.1016/j.semcancer.2016.03.001.

211. Heidi Fritz et al., "Soy, Red Clover, and Isoflavones and Breast Cancer: A Systematic Review," *PLoS One* 8, no. 11 (2013): e81968, https://doi.org/10.1371/journal.pone.0081968.

212. Tareisha L. Dunlap et al., "Red Clover Aryl Hydrocarbon Receptor (AhR) and Estrogen Receptor (ER) Agonists Enhance Genotoxic Estrogen Metabolism," *Chemical Research in Toxicology* 30, no. 11 (2017): 2084–92, https://doi.org/10.1021/acs.chemrestox.7b00237.

213. Jittima Manonai et al., "Effects and Safety of *Pueraria mirifica* on Lipid Profiles and Biochemical Markers of Bone Turnover Rates in Healthy Postmenopausal Women," *Menopause* 15, no. 3 (2008): 530–35, https://doi.org/10.1097/gme.0b013e31815c5fd8.

214. Woo Kyoung Kim et al., "Radish (*Raphanus sativus* L. Leaf) Ethanol Extract Inhibits Protein and MRNA Expression of ErbB(2) and ErbB(3) in MDA-MB-231 Human Breast Cancer Cells," *Nutrition Research and Practice* 5, no. 4 (2011): 288–93, https://doi.org/10.4162/nrp.2011.5.4.288.

215. Alhaji U. N'jai et al., "Spanish Black Radish (*Raphanus sativus* L. Var. *niger*) Diet Enhances Clearance of DMBA and Diminishes Toxic Effects on Bone Marrow Progenitor Cells," *Nutrition and Cancer* 64, no. 7 (2012): 1038–48, https://doi.org/10.1080/01635581.2012.714831.

216. Ashok Subramanian et al., "Oestrogen Producing Enzymes and Mammary Carcinogenesis: A Review," *Breast Cancer Research and Treatment* 111, no. 2 (2008): 191–202, https://doi.org/10.1007/s10549-007-9788-0.

217. Ferdinando Auricchio et al., "Sex-Steroid Hormones and EGF Signalling in Breast and Prostate Cancer Cells: Targeting the Association of Src with Steroid Receptors," *Steroids* 73, no. 9–10 (2008): 880–84, https://doi.org/10.1016/j.steroids.2008.01.023.

218. Li Li et al., "Non-Monotonic Dose-Response Relationship in Steroid Hormone Receptor-Mediated Gene Expression," *Journal of Molecular Endocrinology* 38, no. 5 (2007): 569–85, https://doi.org/10.1677/JME-07-0003; L. Payen et al., "The Drug Efflux Pump MRP2: Regulation of Expression in Physiopathological Situations and by Endogenous and Exogenous Compounds," *Cell Biology and Toxicology* 18, no. 4 (2002): 221–33, https://www.ncbi.nlm.nih.gov/pubmed/12206135.

219. Tze-Chen Hsieh and Joseph M. Wu, "Targeting CWR22Rv1 Prostate Cancer Cell Proliferation and Gene Expression by Combinations of the Phytochemicals EGCG, Genistein and Quercetin," *Anticancer Research* 29, no. 10 (2009): 4025–32, https://www.ncbi.nlm.nih.gov/pubmed/19846946; Anja Bliedtner et al., "Effects of Genistein and Estrogen Receptor Subtype-Specific Agonists in ArKO Mice Following Different Administration Routes," *Molecular and Cellular Endocrinology* 314, no. 1 (2010): 41–52, https://doi.org/10.1016/j.mce.2009.07.032.

220. A. Morani et al., "Biological Functions and Clinical Implications of Oestrogen Receptors Alfa and Beta in Epithelial Tissues," *Journal of Internal Medicine* 264, no. 2 (2008): 128–42, https://doi.org/10.1111/j.1365-2796.2008.01976.x.

221. R. Santti et al., "Phytoestrogens: Potential Endocrine Disruptors in Males," *Toxicology and Industrial Health* 14, no. 1–2 (1998): 223–37, https://doi.org/10.1177/074823379801400114.

222. Enrico de Andrade et al., "Study of the Efficacy of Korean Red Ginseng in the Treatment of Erectile Dysfunction," *Asian Journal of Andrology* 9, no. 2 (2007): 241–44, https://doi.org/10.1111/j.1745-7262.2007.00210.x.

223. Aedín Cassidy et al., "Dietary Flavonoid Intake and Incidence of Erectile Dysfunction," *The American Journal of Clinical Nutrition* 103, no. 2 (2016): 534–41, https://doi.org/10.3945/ajcn.115.122010.

224. Kazim Sahin et al., "Comparative Evaluation of the Sexual Functions and NF-KB and Nrf2 Pathways of Some Aphrodisiac Herbal Extracts in Male Rats," *BMC Complementary and Alternative Medicine* 16, no. 1 (2016): 318, https://doi.org/10.1186/s12906-016-1303-x; Hyun-Myung Choi et al., "Cinnamomi Cortex (*Cinnamomum verum*) Suppresses Testosterone-Induced Benign Prostatic Hyperplasia by Regulating 5α-Reductase," *Scientific Reports* 6 (2016): 31906, https://doi.org/10.1038/srep31906; Carla Gonzales-Arimborgo et al., "Acceptability, Safety, and Efficacy of Oral Administration of Extracts of Black or Red Maca

(*Lepidium meyenii*) in Adult Human Subjects: A Randomized, Double-Blind, Placebo-Controlled Study," *Pharmaceuticals* 9, no. 3 (2016), https://doi.org/10.3390/ph9030049.

225. Monica G. Ferrini et al., "Treatment with a Combination of Ginger, L-Citrulline, Muira Puama and *Paullinia cupana* Can Reverse the Progression of Corporal Smooth Muscle Loss, Fibrosis and Veno-Occlusive Dysfunction in the Aging Rat," *Andrology: Open Access* 4, no. 1 (2015), https://doi.org/10.4172/2167-0250.1000132.

226. W. F. Chiou et al., "Vasorelaxing Effect of Coumarins from *Cnidium monnieri* on Rabbit Corpus Cavernosum," *Planta Medica* 67, no. 3 (2001): 282–84, https://doi.org/10.1055/s-2001-12013.

227. J. Chen et al., "Effect of the Plant-Extract Osthole on the Relaxation of Rabbit *Corpus cavernosum* Tissue In Vitro," *The Journal of Urology* 163, no. 6 (2000): 1975–80, https://www.ncbi.nlm.nih.gov/pubmed/10799242; L. P. Qin et al., "Effects of Osthol and Total-Coumarins from *Cnidium monnieri* on Immunological Function in Kidney Yang Deficiency Mice," *Zhongguo Zhong Xi Yi Jie He Za Zhi Zhongguo Zhongxiyi Jiehe Zazhi (Chinese Journal of Integrated Traditional and Western Medicine)* 15, no. 9 (1995): 547–49, https://www.ncbi.nlm.nih.gov/pubmed/8704443.

228. Juanli Yuan et al., "Effects of Osthol on Androgen Level and Nitric Oxide Synthase Activity in Castrate Rats," *Zhong Yao Cai = Zhongyaocai (Journal of Chinese Medicinal Materials)* 27, no. 7 (2004): 504–6, https://www.ncbi.nlm.nih.gov/pubmed/15551966.

229. Yuan-Shan Zhu and Guang-Huan Sun, "5α-Reductase Isozymes in the Prostate," *Journal of Medical Sciences* 25, no. 1 (2005): 1–12, https://www.ncbi.nlm.nih.gov/pubmed/18483578.

230. J. Drsata, "Enzyme Inhibition in the Drug Therapy of Benign Prostatic Hyperplasia," *Casopis Lekaru Ceskych* 141, no. 20 (2002): 630–35, https://www.ncbi.nlm.nih.gov/pubmed/12515038.

231. Teri L. Wadsworth et al., "Effects of Dietary Saw Palmetto on the Prostate of Transgenic Adenocarcinoma of the Mouse Prostate Model (TRAMP)," *The Prostate* 67, no. 6 (2007): 661–73, https://doi.org/10.1002/pros.20552.

232. Fouad K. Habib et al., "*Serenoa repens* (Permixon) Inhibits the 5alpha-Reductase Activity of Human Prostate Cancer Cell Lines without Interfering with PSA Expression," *International Journal of Cancer* 114, no. 2 (2005): 190–94, https://doi.org/10.1002/ijc.20701.

233. Drsata, "Enzyme Inhibition in the Drug Therapy of Benign Prostatic Hyperplasia."

234. Fouad K. Habib et al., "*Serenoa repens* (Permixon) Inhibits the 5alpha-Reductase Activity."

235. Wadsworth et al., "Effects of Dietary Saw Palmetto on the Prostate."

236. Carla Gonzales-Arimborgo et al., "Acceptability, Safety, and Efficacy of Oral Administration of Extracts of Black or Red Maca (*Lepidium meyenii*) in Adult Human Subjects: A Randomized, Double-Blind, Placebo-Controlled Study," *Pharmaceuticals* 9, no. 3 (2016), https://doi.org/10.3390/ph9030049.

237. Ana C. Ruiz-Luna et al., "*Lepidium meyenii* (Maca) Increases Litter Size in Normal Adult Female Mice," *Reproductive Biology and Endocrinology: RB&E* 3 (2005): 16, https://doi.org/10.1186/1477-7827-3-16.

238. Ingrid Melnikovova et al., "Effect of *Lepidium meyenii* Walp. on Semen Parameters and Serum Hormone Levels in Healthy Adult Men: A Double-Blind, Randomized, Placebo-Controlled Pilot Study," *Evidence-Based Complementary and Alternative Medicine: ECAM* 2015 (2015): 324369, https://doi.org/10.1155/2015/324369.

239. Gustavo F. Gonzales et al., "Red Maca (*Lepidium meyenii*) Reduced Prostate Size in Rats," *Reproductive Biology and Endocrinology: RB&E* 3 (2005): 5, https://doi.org/10.1186/1477-7827-3-5.

240. Yongzhong Zhang et al., "Effect of Ethanolic Extract of *Lepidium meyenii Walp* on Serum Hormone Levels in Ovariectomized Rats," *Indian Journal of Pharmacology* 46, no. 4 (2014): 416–19, https://doi.org/10.4103/0253-7613.135955.

241. Mie Nishimura et al., "Pumpkin Seed Oil Extracted from *Cucurbita maxima* Improves Urinary Disorder in Human Overactive Bladder," *Journal of Traditional and Complementary Medicine* 4, no. 1 (2014): 72–74, https://doi.org/10.4103/2225-4110.124355; Heeok Hong et al., "Effects of Pumpkin Seed Oil and Saw Palmetto Oil in Korean Men with Symptomatic Benign Prostatic Hyperplasia," *Nutrition Research and Practice* 3, no. 4 (2009): 323–27, https://doi.org/10.4162/nrp.2009.3.4.323.

242. Drsata, "Enzyme Inhibition in the Drug Therapy of Benign Prostatic Hyperplasia."

243. G. Salvati et al., "Effects of *Panax ginseng* C.A. Meyer Saponins on Male Fertility," *Panminerva Medica* 38, no. 4 (1996): 249–54, https://www.ncbi.nlm.nih.gov/pubmed/9063034.

244. Ameneh Hasanzadeh Khosh et al., "Ameliorative Effects of *Achillea millefolium* Inflorescences Alcoholic Extract on Nicotine-Induced Reproductive Toxicity in Male Rat: Apoptotic and Biochemical Evidences," *Veterinary Research Forum: An International Quarterly Journal* 8, no. 2 (2017): 97–104, https://www.ncbi.nlm.nih.gov/pubmed/28785383.

245. Claudia Montani et al., "Genistein Is an Efficient Estrogen in the Whole-Body throughout Mouse Development," *Toxicological Sciences: An Official Journal of the Society of Toxicology* 103, no. 1 (2008): 57–67, https://doi.org/10.1093/toxsci/kfn021.

246. Kentaro Matsumura et al., "Involvement of the Estrogen Receptor Beta in Genistein-Induced Expression of P21(Waf1/Cip1) in PC-3 Prostate Cancer Cells," *Anticancer Research* 28, no. 2A (2008): 709–14, https://www.ncbi.nlm.nih.gov/pubmed/18507011; Nobuyuki Kikuno et al., "Genistein Mediated Histone Acetylation and Demethylation Activates Tumor Suppressor Genes in

Prostate Cancer Cells," *International Journal of Cancer* 123, no. 3 (2008): 552–60, https://doi.org/10.1002/ijc.23590.

247. Robert Michael Hermann et al., "In Vitro Studies on the Modification of Low-Dose Hyper-Radiosensitivity in Prostate Cancer Cells by Incubation with Genistein and Estradiol," *Radiation Oncology* 3 (2008): 19, https://doi.org/10.1186/1748-717X-3-19.

248. Nagi B. Kumar et al., "Safety of Purified Isoflavones in Men with Clinically Localized Prostate Cancer," *Nutrition and Cancer* 59, no. 2 (2007): 169–75, https://doi.org/10.1080/01635580701432660.

249. John M. Pendleton et al., "Phase II Trial of Isoflavone in Prostate-Specific Antigen Recurrent Prostate Cancer after Previous Local Therapy," *BMC Cancer* 8 (2008): 132, https://doi.org/10.1186/1471-2407-8-132.

250. S. Saeidnia et al., "A Review on Phytochemistry and Medicinal Properties of the Genus *Achillea*," *Daru: Journal of Faculty of Pharmacy, Tehran University of Medical Sciences* 19, no. 3 (2011): 173–86, https://www.ncbi.nlm.nih.gov/pubmed/22615655.

251. S. Qureshi et al., "Toxicity Studies on *Alpinia galanga* and *Curcuma longa*," *Planta Medica* 58, no. 2 (1992): 124–27, https://doi.org/10.1055/s-2006-961412; Mahta Mazaheri et al., "Molecullar and Biochemical Effect of Alcohlic Extract of *Alpinia galanga* on Rat Spermatogenesis Process," *Iranian Journal of Reproductive Medicine* 12, no. 11 (2014): 765–70, https://www.ncbi.nlm.nih.gov/pubmed/25709632.

252. Saeed Samarghandian et al., "Antiproliferative Activity and Induction of Apoptotic by Ethanolic Extract of *Alpinia galanga* Rhizhome in Human Breast Carcinoma Cell Line," *BMC Complementary and Alternative Medicine* 14 (2014): 192, https://doi.org/10.1186/1472-6882-14-192.

253. S. B. Jaju et al., "Isolation of β-Sitosterol Diglucosyl Caprate from *Alpinia galanga*," *Pharmacognosy Research* 2, no. 4 (2010): 264–66, https://doi.org/10.4103/0974-8490.69129.

254. Dong Wook Lim and Yun Tai Kim, "Anti-Osteoporotic Effects of *Angelica sinensis* (Oliv.) Diels Extract on Ovariectomized Rats and Its Oral Toxicity in Rats," *Nutrients* 6, no. 10 (2014): 4362–72, https://doi.org/10.3390/nu6104362.

255. Jun Shen et al., "The Antidepressant Effect of *Angelica sinensis* Extracts on Chronic Unpredictable Mild Stress-Induced Depression Is Mediated via the Upregulation of the BDNF Signaling Pathway in Rats," *Evidence-Based Complementary and Alternative Medicine: ECAM* 2016 (2016): 7434692, https://doi.org/10.1155/2016/7434692.

256. Marco K. C. Hui et al., "Polysaccharides from the Root of *Angelica sinensis* Protect Bone Marrow and Gastrointestinal Tissues against the Cytotoxicity of Cyclophosphamide in Mice," *International Journal of Medical Sciences* 3, no. 1 (2006): 1–6, https://www.ncbi.nlm.nih.gov/pubmed/16421623; Hye Won Lee et al., "Dangguijihwang-Tang and Dangguijakyak-San Prevent Menopausal Symptoms and Dangguijihwang-Tang Prevents Articular Cartilage Deterioration in Ovariectomized Obese Rats with

Monoiodoacetate-Induced Osteoarthritis," *Evidence-Based Complementary and Alternative Medicine: ECAM* 2017 (2017): 5658681, https://doi.org/10.1155/2017/5658681.

257. Akram Ahangarpour et al., "Effects of Hydro-Alcoholic Extract from *Arctium lappa* L. (Burdock) Root on Gonadotropins, Testosterone, and Sperm Count and Viability in Male Mice with Nicotinamide/ Streptozotocin-Induced Type 2 Diabetes," *The Malaysian Journal of Medical Sciences: MJMS* 22, no. 2 (2015): 25–32, https://www.ncbi.nlm.nih.gov/pubmed/26023292.

258. Tingting Feng et al., "Arctigenin Inhibits STAT3 and Exhibits Anticancer Potential in Human Triple-Negative Breast Cancer Therapy," *Oncotarget* 8, no. 1 (2017): 329–44, https://doi.org/10.18632/oncotarget.13393.

259. Piwen Wang et al., "Arctigenin in Combination with Quercetin Synergistically Enhances the Antiproliferative Effect in Prostate Cancer Cells," *Molecular Nutrition & Food Research* 59, no. 2 (2015): 250–61, https://doi.org/10.1002/mnfr.201400558.

260. Mashitha Vinod Pise et al., "Immunomodulatory Potential of Shatavarins Produced from *Asparagus racemosus* Tissue Cultures," *Journal of Natural Science, Biology, and Medicine* 6, no. 2 (2015): 415–20, https://doi.org/10.4103/0976-9668.160025.

261. Ping Liu et al., "Anti-Aging Implications of *Astragalus membranaceus* (Huangqi): A Well-Known Chinese Tonic," *Aging and Disease* 8, no. 6 (2017): 868–86, https://doi.org/10.14336/AD.2017.0816.

262. Xiaolong Jiang et al., "Effects of Treatment with *Astragalus membranaceus* on Function of Rat Leydig Cells," *BMC Complementary and Alternative Medicine* 15 (2015): 261, https://doi.org/10.1186/s12906-015-0776-3.

263. Wonnam Kim et al., "*Astragalus membranaceus* Augment Sperm Parameters in Male Mice Associated with CAMP-Responsive Element Modulator and Activator of CREM in Testis," *Journal of Traditional and Complementary Medicine* 6, no. 3 (2016): 294–98, https://doi.org/10.1016/j.jtcme.2015.10.002.

264. V. R. Ravikumar et al., "In-Vitro Anti- Inflammatory Activity of Aqueous Extract of Leaves of *Plectranthus amboinicus* (Lour.) Spreng," *Ancient Science of Life* 28, no. 4 (2009): 7–9, https://www.ncbi.nlm.nih.gov/pubmed/22557324; Yung-Jia Chiu et al., "Analgesic and Antiinflammatory Activities of the Aqueous Extract from *Plectranthus amboinicus* (Lour.) Spreng. Both In Vitro and In Vivo," *Evidence-Based Complementary and Alternative Medicine: ECAM* 2012 (2012): 508137, https://doi.org/10.1155/2012/508137.

265. Virginie Follin-Arbelet et al., "The Natural Compound Forskolin Synergizes with Dexamethasone to Induce Cell Death in Myeloma Cells via BIM," *Scientific Reports* 5 (2015): 13001, https://doi.org/10.1038/srep13001.

266. Jingjing Yan et al., "Up-Regulation on Cytochromes P450 in Rat Mediated by Total Alkaloid Extract from *Corydalis yanhusuo*," *BMC Complementary and Alternative Medicine* 14 (2014): 306, https://doi.org/10.1186/1472-6882-14-306.

267. Jiyun Ahn et al., "Eleutheroside E, an Active Component of *Eleutherococcus senticosus*, Ameliorates Insulin Resistance in Type 2 Diabetic Db/Db Mice," *Evidence-Based Complementary and Alternative Medicine: ECAM* 2013 (2013): 934183, https://doi.org/10.1155/2013/934183.

268. Hyun Jun Park et al., "Restoration of Spermatogenesis Using a New Combined Herbal Formula of *Epimedium koreanum Nakai* and *Angelica gigas Nakai* in an Luteinizing Hormone-Releasing Hormone Agonist-Induced Rat Model of Male Infertility," *The World Journal of Men's Health* 35, no. 3 (2017): 170–77, https://doi.org/10.5534/wjmh.17031.

269. Fu Li et al., "Flavonoid Glycosides Isolated from *Epimedium brevicornum* and Their Estrogen Biosynthesis-Promoting Effects," *Scientific Reports* 7, no. 1 (2017): 7760, https://doi.org/10.1038/s41598-017-08203-7.

270. Liming Xue et al., "Comparative Effects of Er-Xian Decoction, Epimedium Herbs, and Icariin with Estrogen on Bone and Reproductive Tissue in Ovariectomized Rats," *Evidence-Based Complementary and Alternative Medicine: ECAM* 2012 (2012): 241416, https://doi.org/10.1155/2012/241416.

271. Hui-Li Tan et al., "Anti-Cancer Properties of the Naturally Occurring Aphrodisiacs: Icariin and Its Derivatives," *Frontiers in Pharmacology* 7 (2016): 191, https://doi.org/10.3389/fphar.2016.00191.

272. Jinsheng Hong et al., "Icaritin Synergistically Enhances the Radiosensitivity of 4T1 Breast Cancer Cells," *PLoS ONE* 8, no. 8 (2013), https://doi.org/10.1371/journal.pone.0071347.

273. Jing-Shan Tong et al., "Icaritin Causes Sustained ERK1/2 Activation and Induces Apoptosis in Human Endometrial Cancer Cells," *PLoS One* 6, no. 3 (2011): e16781, https://doi.org/10.1371/journal.pone.0016781.

274. Norhazlina Abdul Wahab et al., "The Effect of *Eurycoma longifolia Jack* on Spermatogenesis in Estrogen-Treated Rats," *Clinics* 65, no. 1 (2010): 93–98, https://doi.org/10.1590/S1807-59322010000100014.

275. Mohd Ismail Bin Mohd Tambi and M. Kamarul Imran, "*Eurycoma longifolia Jack* in Managing Idiopathic Male Infertility," *Asian Journal of Andrology* 12, no. 3 (2010): 376–80, https://doi.org/10.1038/aja.2010.7.

276. Ahmad Nazrun Shuid et al., "*Eurycoma longifolia* Upregulates Osteoprotegerin Gene Expression in Androgen-Deficient Osteoporosis Rat Model," *BMC Complementary and Alternative Medicine* 12 (2012): 152, https://doi.org/10.1186/1472-6882-12-152.

277. S. N. Ostad et al., "The Effect of Fennel Essential Oil on Uterine Contraction as a Model for Dysmenorrhea, Pharmacology and Toxicology Study," *Journal of Ethnopharmacology* 76, no. 3 (2001): 299–304, https://www.ncbi.nlm.nih.gov/pubmed/11448553.

278. Mozafar Khazaei et al., "Study of *Foeniculum vulgare* Effect on Folliculogenesis in Female Mice," *International Journal of Fertility & Sterility* 5, no. 3 (2011): 122–27, https://www.ncbi.nlm.nih.gov/pmc/articles/PMC4122825/.

279. Zahra Mahmoudi et al., "Effects of *Foeniculum vulgare* Ethanol Extract on Osteogenesis in Human Mecenchymal Stem Cells," *Avicenna Journal of Phytomedicine* 3, no. 2 (2013): 135–42, https://www.ncbi.nlm.nih.gov/pubmed/25050267.

280. Christine F. Skibola, "The Effect of *Fucus vesiculosus*, an Edible Brown Seaweed, upon Menstrual Cycle Length and Hormonal Status in Three Pre-Menopausal Women: A Case Report," *BMC Complementary and Alternative Medicine* 4 (2004): 10, https://doi.org/10.1186/1472-6882-4-10.

281. Sergey N. Fedorov et al., "Anticancer and Cancer Preventive Properties of Marine Polysaccharides: Some Results and Prospects," *Marine Drugs* 11, no. 12 (2013): 4876–901, https://doi.org/10.3390/md11124876.

282. E. A. S. Al-Dujaili et al., "Liquorice and Glycyrrhetinic Acid Increase DHEA and Deoxycorticosterone Levels In Vivo and In Vitro by Inhibiting Adrenal SULT2A1 Activity," *Molecular and Cellular Endocrinology* 336, no. 1–2 (2011): 102–9, https://doi.org/10.1016/j.mce.2010.12.011.

283. Chia Hui Apphia Eu et al., "Glycyrrhizic Acid Improved Lipoprotein Lipase Expression, Insulin Sensitivity, Serum Lipid and Lipid Deposition in High-Fat Diet-Induced Obese Rats," *Lipids in Health and Disease* 9 (2010): 81, https://doi.org/10.1186/1476-511X-9-81.

284. Hazem Hasan Kataya et al., "Effect of Licorice Extract on the Complications of Diabetes Nephropathy in Rats," *Drug and Chemical Toxicology* 34, no. 2 (2011): 101–8, https://doi.org/10.3109/01480545.2010.510524.

285. Chia Hui Apphia Eu et al., "Glycyrrhizic Acid Improved Lipoprotein Lipase Expression, Insulin Sensitivity, Serum Lipid and Lipid Deposition in High-Fat Diet-Induced Obese Rats," *Lipids in Health and Disease* 9 (2010): 81, https://doi.org/10.1186/1476-511X-9-81.

286. Kaku Nakagawa et al., "Licorice Flavonoids Suppress Abdominal Fat Accumulation and Increase in Blood Glucose Level in Obese Diabetic KK-A(y) Mice," *Biological & Pharmaceutical Bulletin* 27, no. 11 (2004): 1775–78, https://www.ncbi.nlm.nih.gov/pubmed/15516721.

287. Shoko Tsuji et al., "Shakuyaku-Kanzo-to Inhibits Smooth Muscle Contractions of Human Pregnant Uterine Tissue In Vitro," *The Journal of Obstetrics and Gynaecology Research* 38, no. 7 (2012): 1004–10, https://doi.org/10.1111/j.1447-0756.2011.01827.x.

288. Jianwei Jia et al., "Relaxative Effect of Core Licorice Aqueous Extract on Mouse Isolated Uterine Horns," *Pharmaceutical Biology* 51, no. 6 (2013): 744–48, https://doi.org/10.3109/13880209.2013.764536.

289. Yulu Shi et al., "Analgesic and Uterine Relaxant Effects of Isoliquiritigenin, a Flavone from *Glycyrrhiza glabra*," *Phytotherapy Research: PTR* 26, no. 9 (2012): 1410–17, https://doi.org/10.1002/ptr.3715.

290. L. R. Chadwick et al., "The Pharmacognosy of *Humulus lupulus* L. (Hops) with an Emphasis on Estrogenic Properties," *Phytomedicine: International Journal of*

Phytotherapy and Phytopharmacology 13, no. 1–2 (2006): 119–31, https://doi.org/10.1016/j.phymed.2004.07.006.

291. Fatemeh Abdi et al., "Protocol for Systematic Review and Meta-Analysis: Hop (*Humulus lupulus* L.) for Menopausal Vasomotor Symptoms," *BMJ Open* 6, no. 4 (2016): e010734, https://doi.org/10.1136/bmjopen-2015-010734.

292. Chadwick et al., "The Pharmacognosy of *Humulus lupulus* L. (Hops)."

293. Gustavo F. Gonzales et al., "Role of Maca (*Lepidium meyenii*) Consumption on Serum Interleukin-6 Levels and Health Status in Populations Living in the Peruvian Central Andes over 4000 m of Altitude," *Plant Foods for Human Nutrition* 68, no. 4 (2013): 347–51, https://doi.org/10.1007/s11130-013-0378-5.

294. Hao Zhen Cui et al., "Ethanol Extract of *Lycopus lucidus* Elicits Positive Inotropic Effect via Activation of Ca2+ Entry and Ca2+ Release in Beating Rabbit Atria," *Journal of Medicinal Food* 16, no. 7 (2013): 633–40, https://doi.org/10.1089/jmf.2012.2487.

295. Yuanzhang Yao et al., "The Aqueous Extract of *Lycopus lucidus* Turcz Ameliorates Streptozotocin-Induced Diabetic Renal Damage via Inhibiting TGF-B1 Signaling Pathway," *Phytomedicine: International Journal of Phytotherapy and Phytopharmacology* 20, no. 13 (2013): 1160–67, https://doi.org/10.1016/j.phymed.2013.06.004.

296. Christian Vonhoff et al., "Extract of *Lycopus europaeus* L. Reduces Cardiac Signs of Hyperthyroidism in Rats," *Life Sciences* 78, no. 10 (2006): 1063–70, https://doi.org/10.1016/j.lfs.2005.06.014.

297. A.-M. Beer et al., "*Lycopus europaeus* (Gypsywort): Effects on the Thyroidal Parameters and Symptoms Associated with Thyroid Function," *Phytomedicine: International Journal of Phytotherapy and Phytopharmacology* 15, no. 1–2 (2008): 16–22, https://doi.org/10.1016/j.phymed.2007.11.001.

298. Zafari Zangeneh Farideh et al., "Effects of Chamomile Extract on Biochemical and Clinical Parameters in a Rat Model of Polycystic Ovary Syndrome," *Journal of Reproduction & Infertility* 11, no. 3 (2010): 169–74, https://www.ncbi.nlm.nih.gov/pubmed/23926485.

299. Xinghua Long et al., "Apigenin Inhibits Antiestrogen-Resistant Breast Cancer Cell Growth through Estrogen Receptor-Alpha-Dependent and Estrogen Receptor-Alpha-Independent Mechanisms," *Molecular Cancer Therapeutics* 7, no. 7 (2008): 2096–108, https://doi.org/10.1158/1535-7163.MCT-07-2350; Paul Mak et al., "Apigenin Suppresses Cancer Cell Growth through ERbeta," *Neoplasia* 8, no. 11 (2006): 896–904, https://doi.org/10.1593/neo.06538.

300. Kadir Demirci et al., "Does *Melissa officinalis* Cause Withdrawal or Dependence?" *Medical Archives* 69, no. 1 (2015): 60–61, https://doi.org/10.5455/medarh.2015.69.60-61.

301. Marzieh Akbarzadeh et al., "Effect of *Melissa officinalis* Capsule on the Intensity of Premenstrual Syndrome Symptoms in High School Girl Students," *Nursing and Midwifery Studies* 4, no. 2 (2015): e27001, https://doi.org/10.17795/nmsjournal27001.

302. Mohammad-Taghi Moradi et al., "A Review Study on the Effect of Iranian Herbal Medicines against In Vitro Replication of Herpes Simplex Virus," *Avicenna Journal of Phytomedicine* 6, no. 5 (2016): 506–15, https://www.ncbi.nlm.nih.gov/pubmed/27761420.

303. Priyabrata Pattanayak et al., "*Ocimum sanctum* Linn. A Reservoir Plant for Therapeutic Applications: An Overview," *Pharmacognosy Reviews* 4, no. 7 (2010): 95–105, https://doi.org/10.4103/0973-7847.65323; Mukhtar Ahmed et al., "Reversible Anti-Fertility Effect of Benzene Extract of *Ocimum sanctum* Leaves on Sperm Parameters and Fructose Content in Rats," *Journal of Basic and Clinical Physiology and Pharmacology* 13, no. 1 (2002): 51–59, https://www.ncbi.nlm.nih.gov/pubmed/12099405.

304. Mukhtar Ahmed et al., "Effect of Benzene Extract of *Ocimum sanctum* Leaves on Cauda Epididymal Spermatozoa of Rats," *Iranian Journal of Reproductive Medicine* 9, no. 3 (2011): 177–86, https://www.ncbi.nlm.nih.gov/pubmed/26396561.

305. Mukhtar Ahmed et al., "Effect of Benzene Extract of *Ocimum sanctum* Leaves on Cauda Epididymis of Albino Rats," *Journal of Basic and Clinical Physiology and Pharmacology* 20, no. 1 (2009): 29–41, https://www.ncbi.nlm.nih.gov/pubmed/19601393.

306. Jyoti Sethi et al., "Effect of Tulsi (*Ocimum sanctum* Linn.) on Sperm Count and Reproductive Hormones in Male Albino Rabbits," *International Journal of Ayurveda Research* 1, no. 4 (2010): 208–10, https://doi.org/10.4103/0974-7788.76782.

307. Suleman S. Hussain et al., "Extracting the Benefit of Nexrutine for Cancer Prevention," *Current Pharmacology Reports* 1, no. 6 (2015): 365–72, https://www.ncbi.nlm.nih.gov/pubmed/26539341.

308. Gregory P. Swanson et al., "Tolerance of *Phellodendron amurense* Bark Extract (Nexrutine) in Patients with Human Prostate Cancer," *Phytotherapy Research: PTR* 29, no. 1 (2015): 40–42, https://doi.org/10.1002/ptr.5221.

309. Shawn M. Talbott et al., "Effect of *Magnolia officinalis* and *Phellodendron amurense* (Relora) on Cortisol and Psychological Mood State in Moderately Stressed Subjects," *Journal of the International Society of Sports Nutrition* 10, no. 1 (2013): 37, https://doi.org/10.1186/1550-2783-10-37.

310. Isabelle Bekeredjian-Ding et al., "Poke Weed Mitogen Requires Toll-like Receptor Ligands for Proliferative Activity in Human and Murine B Lymphocytes," *PLoS One* 7, no. 1 (2012): e29806, https://doi.org/10.1371/journal.pone.0029806.

311. Zhongbo Liu et al., "Kavalactone Yangonin Induces Autophagy and Sensitizes Bladder Cancer Cells to Flavokawain A and Docetaxel via Inhibition of the MTOR Pathway," *Journal of Biomedical Research* 31, no. 5 (2017): 408–18, https://doi.org/10.7555/JBR.31.20160160.

312. Ying-Qian Liu et al., "Recent Progress on C-4-Modified Podophyllotoxin Analogs as Potent Antitumor Agents,"

Medicinal Research Reviews 35, no. 1 (2015): 1–62, https://doi.org/10.1002/med.21319.

313. C. J. N. Lacey et al., "Randomised Controlled Trial and Economic Evaluation of Podophyllotoxin Solution, Podophyllotoxin Cream, and Podophyllin in the Treatment of Genital Warts," *Sexually Transmitted Infections* 79, no. 4 (2003): 270–75, https://www.ncbi.nlm.nih.gov/pubmed/12902571; G. von Krogh et al., "Self-Treatment Using 0.25%-0.50% Podophyllotoxin-Ethanol Solutions against Penile Condylomata Acuminata: A Placebo-Controlled Comparative Study," *Genitourinary Medicine* 70, no. 2 (1994): 105–9, https://www.ncbi.nlm.nih.gov/pubmed/8206467.

314. A. Edwards et al., "Podophyllotoxin 0.5% v Podophyllin 20% to Treat Penile Warts," *Genitourinary Medicine* 64, no. 4 (1988): 263–65, https://www.ncbi.nlm.nih.gov/pubmed/3169757.

315. Stéphane Larré et al., "Biological Effect of Human Serum Collected before and after Oral Intake of *Pygeum africanum* on Various Benign Prostate Cell Cultures," *Asian Journal of Andrology* 14, no. 3 (2012): 499–504, https://doi.org/10.1038/aja.2011.132.

316. Saleem Ali Banihani, "Radish (*Raphanus sativus*) and Diabetes," *Nutrients* 9, no. 9 (2017), https://doi.org/10.3390/nu9091014.

317. Wei Zhang et al., "Urolithin A Suppresses the Proliferation of Endometrial Cancer Cells by Mediating Estrogen Receptor-α-Dependent Gene Expression," *Molecular Nutrition & Food Research* 60, no. 11 (2016): 2387–95, https://doi.org/10.1002/mnfr.201600048.

318. N. R. Dahiya et al., "A Natural Molecule, Urolithin A, Downregulates Androgen Receptor Activation and Suppresses Growth of Prostate Cancer," *Molecular Carcinogenesis* 57, no. 10 (2018): 1332–41, https://doi.org/10.1002/mc.22848.

319. Andrew Croaker et al., "A Review of Black Salve: Cancer Specificity, Cure, and Cosmesis," *Evidence-Based Complementary and Alternative Medicine: ECAM* 2017 (2017): 9184034, https://doi.org/10.1155/2017/9184034.

320. Andrew Croaker et al., "*Sanguinaria canadensis*: Traditional Medicine, Phytochemical Composition, Biological Activities and Current Uses," *International Journal of Molecular Sciences* 17, no. 9 (2016), https://doi.org/10.3390/ijms17091414.

321. Tiantian She, "Sarsaparilla (Smilax Glabra Rhizome) Extract Inhibits Migration and Invasion of Cancer Cells by Suppressing TGF-B1 Pathway," *PLoS One* 10, no. 3 (2015): e0118287, https://doi.org/10.1371/journal.pone.0118287.

322. Jing Chen et al., "Fraction of Macroporous Resin from *Smilax china* L. Inhibits Testosterone Propionate-Induced Prostatic Hyperplasia in Castrated Rats," *Journal of Medicinal Food* 15, no. 7 (2012): 646–50, https://doi.org/10.1089/jmf.2011.1968.

323. Irma Podolak et al., "Saponins as Cytotoxic Agents: A Review," *Phytochemistry Reviews: Proceedings of the Phytochemical Society of Europe* 9, no. 3 (2010): 425–74, https://doi.org/10.1007/s11101-010-9183-z.

324. Ajay Kesharwani et al., "Anti-HSV-2 Activity of *Terminalia chebula Retz* Extract and Its Constituents, Chebulagic and Chebulinic Acids," *BMC Complementary and Alternative Medicine* 17, no. 1 (2017): 110, https://doi.org/10.1186/s12906-017-1620-8.

325. Alka Dwevedi et al., "Exploration of Phytochemicals Found in *Terminalia* Spp. and Their Antiretroviral Activities," *Pharmacognosy Reviews* 10, no. 20 (2016): 73–83, https://doi.org/10.4103/0973-7847.194048.

326. Belal Naser et al., "*Thuja occidentalis* (Arbor Vitae): A Review of Its Pharmaceutical, Pharmacological and Clinical Properties," *Evidence-Based Complementary and Alternative Medicine: ECAM* 2, no. 1 (2005): 69–78, https://doi.org/10.1093/ecam/neh065.

327. R. Joseph et al., "Successful Treatment of Verruca Vulgaris with *Thuja occidentalis* in a Renal Allograft Recipient," *Indian Journal of Nephrology* 23, no. 5 (2013): 362–64, https://doi.org/10.4103/0971-4065.116316.

328. Raktim Biswas et al., "Thujone-Rich Fraction of *Thuja occidentalis* Demonstrates Major Anti-Cancer Potentials: Evidences from In Vitro Studies on A375 Cells," *Evidence-Based Complementary and Alternative Medicine: ECAM* 2011 (2011): 568148, https://doi.org/10.1093/ecam/neq042.

329. Elham Akhtari et al., "*Tribulus terrestris* for Treatment of Sexual Dysfunction in Women: Randomized Double-Blind Placebo-Controlled Study," *Daru: Journal of Faculty of Pharmacy, Tehran University of Medical Sciences* 22 (2014): 40, https://doi.org/10.1186/2008-2231-22-40.

330. Wenyi Zhu et al., "A Review of Traditional Pharmacological Uses, Phytochemistry, and Pharmacological Activities of *Tribulus terrestris*," *Chemistry Central Journal* 11, no. 1 (2017): 60, https://doi.org/10.1186/s13065-017-0289-x.

331. Nancy L. Booth et al., "Clinical Studies of Red Clover (*Trifolium pratense*) Dietary Supplements in Menopause: A Literature Review," *Menopause* 13, no. 2 (2006): 251–64, https://doi.org/10.1097/01.gme.0000198297.40269.f7.

332. Paolo Mannella et al., "Effects of Red Clover Extracts on Breast Cancer Cell Migration and Invasion," *Gynecological Endocrinology: The Official Journal of the International Society of Gynecological Endocrinology* 28, no. 1 (2012): 29–33, https://doi.org/10.3109/09513590.2011.579660.

333. Pavlína Hloucalová et al., "Determination of Phytoestrogen Content in Fresh-Cut Legume Forage," *Animals: An Open Access Journal from MDPI* 6, no. 7 (2016), https://doi.org/10.3390/ani6070043.

334. Hloucalová et al., "Determination of Phytoestrogen Content in Fresh-Cut Legume Forage."

335. Johanna R. Rochester et al., "Dietary Red Clover (*Trifolium pratense*) Induces Oviduct Growth and Decreases Ovary and Testes Growth in Japanese Quail Chicks," *Reproductive Toxicology* 27, no. 1 (2009): 63–71, https://doi.org/10.1016/j.reprotox.2008.11.056.

336. Tieraona Low Dog, "Menopause: A Review of Botanical Dietary Supplements," *The American Journal of Medicine* 118

Suppl 12B (2005): 98–108, https://doi.org/10.1016/j.amjmed
.2005.09.044; Charlotte Atkinson et al., "Red-Clover-
Derived Isoflavones and Mammographic Breast Density:
A Double-Blind, Randomized, Placebo-Controlled Trial
[ISRCTN42940165]," *Breast Cancer Research: BCR* 6, no. 3
(2004): R170–179, https://doi.org/10.1186/bcr773; Cecília del
Giorno et al., "Effects of *Trifolium pratense* on the Climacteric
and Sexual Symptoms in Postmenopause Women," *Revista da
Associacao Medica Brasileira (1992)* 56, no. 5 (2010): 558–62,
https://www.ncbi.nlm.nih.gov/pubmed/21152828.

337. Pragya Gartoulla and Myo Mint Han, "Red Clover Extract
for Alleviating Hot Flushes in Postmenopausal Women: A
Meta-Analysis," *Maturitas* 79, no. 1 (2014): 58–64, https://doi.
org/10.1016/j.maturitas.2014.06.018; Markus Lipovac et al.,
"The Effect of Red Clover Isoflavone Supplementation over
Vasomotor and Menopausal Symptoms in Postmenopausal
Women," *Gynecological Endocrinology: The Official Journal
of the International Society of Gynecological Endocrinology*
28, no. 3 (2012): 203–7, https://doi.org/10.3109
/09513590.2011.593671; G. Mainini et al., "Nonhormonal
Management of Postmenopausal Women: Effects of a Red
Clover Based Isoflavones Supplementation on Climacteric
Syndrome and Cardiovascular Risk Serum Profile," *Clinical
and Experimental Obstetrics & Gynecology* 40, no. 3 (2013):
337–41, https://www.ncbi.nlm.nih.gov/pubmed/24283160.

338. Annette J. Thomas et al., "Effects of Isoflavones and Amino
Acid Therapies for Hot Flashes and Co-Occurring Symptoms
during the Menopausal Transition and Early Postmenopause:
A Systematic Review," *Maturitas* 78, no. 4 (2014): 263–76,
https://doi.org/10.1016/j.maturitas.2014.05.007.

339. Lipovac et al., "The Effect of Red Clover Isoflavone
Supplementation over Vasomotor and Menopausal
Symptoms."

340. Joanna Thompson Coon et al., "*Trifolium pratense*
Isoflavones in the Treatment of Menopausal Hot
Flushes: A Systematic Review and Meta-Analysis,"
*Phytomedicine: International Journal of Phytotherapy and
Phytopharmacology* 14, no. 2–3 (2007): 153–59, https://doi
.org/10.1016/j.phymed.2006.12.009.

341. Booth et al., "Clinical Studies of Red Clover (*Trifolium
pratense*) Dietary Supplements in Menopause."

342. G. Mainini et al., "Nonhormonal Management of
Postmenopausal Women: Effects of a Red Clover Based
Isoflavones Supplementation on Climacteric Syndrome
and Cardiovascular Risk Serum Profile," *Clinical and
Experimental Obstetrics & Gynecology* 40, no. 3 (2013):
337–41, https://www.ncbi.nlm.nih.gov/pubmed/24283160.

343. Joan Pitkin, "Red Clover Isoflavones in Practice: A
Clinician's View," *The Journal of the British Menopause
Society* 10 Suppl 1 (2004): 7–12, https://www.ncbi.nlm.nih
.gov/pubmed/15107200; Charlotte Atkinson et al., "Red-
Clover-Derived Isoflavones and Mammographic Breast
Density: A Double-Blind, Randomized, Placebo-Controlled
Trial [ISRCTN42940165]," *Breast Cancer Research: BCR* 6,
no. 3 (2004): R170–79, https://doi.org/10.1186/bcr773; Lee
Hooper et al., "Effects of Isoflavones on Breast Density in
Pre- and Post-Menopausal Women: A Systematic Review
and Meta-Analysis of Randomized Controlled Trials,"
Human Reproduction Update 16, no. 6 (2010): 745–60,
https://doi.org/10.1093/humupd/dmq011.

344. J. Chen and L. Sun, "Formononetin-Induced Apoptosis
by Activation of Ras/P38 Mitogen-Activated Protein
Kinase in Estrogen Receptor-Positive Human Breast
Cancer Cells," *Hormones et Metabolisme (Hormone and
Metabolic Research)* 44, no. 13 (2012): 943–48, https://
doi.org/10.1055/s-0032-1321818; Evelyne Reiter et al.,
"Red Clover and Soy Isoflavones—An In Vitro Safety
Assessment," *Gynecological Endocrinology: The Official
Journal of the International Society of Gynecological
Endocrinology* 27, no. 12 (2011): 1037–42, https://doi.org/10
.3109/09513590.2011.588743.

345. Amr Amin et al., "Chemopreventive Activities of *Trigonella
foenum graecum* (Fenugreek) against Breast Cancer," *Cell
Biology International* 29, no. 8 (2005): 687–94, https://doi
.org/10.1016/j.cellbi.2005.04.004.

346. Shivangi Goyal, "Investigating Therapeutic Potential of
Trigonella foenum-graecum L. as Our Defense Mechanism
against Several Human Diseases," *Journal of Toxicology*
2016 (2016): 1250387, https://doi.org/10.1155/2016/1250387.

347. Suresh Kumar and Anupam Sharma, "Anti-Anxiety Activity
Studies on Homoeopathic Formulations of *Turnera
aphrodisiaca Ward*," *Evidence-Based Complementary and
Alternative Medicine: ECAM* 2, no. 1 (2005): 117–19, https://
doi.org/10.1093/ecam/neh069; Anand Swaroop et al.,
"Efficacy of a Novel Fenugreek Seed Extract (*Trigonella
foenum-graecum*, Furocyst) in Polycystic Ovary Syndrome
(PCOS)," *International Journal of Medical Sciences* 12, no. 10
(2015): 825–31, https://doi.org/10.7150/ijms.13024.

348. Anuj Maheshwari et al., "Efficacy of Furosap, a Novel
Trigonella foenum-graecum Seed Extract, in Enhancing
Testosterone Level and Improving Sperm Profile in Male
Volunteers," *International Journal of Medical Sciences* 14,
no. 1 (2017): 58–66, https://doi.org/10.7150/ijms.17256.

349. María del Carmen Avelino-Flores et al., "Cytotoxic Activity
of the Methanolic Extract of *Turnera diffusa Willd* on
Breast Cancer Cells," *Journal of Medicinal Food* 18, no. 3
(2015): 299–305, https://doi.org/10.1089/jmf.2013.0055.

350. Kumar and Sharma, "Anti-Anxiety Activity Studies on
Homoeopathic Formulations of *Turnera aphrodisiaca Ward*."

351. Wafa Siouda and Cherif Abdennour, "Can *Urtica dioica*
Supplementation Attenuate Mercury Intoxication in
Wistar Rats?" *Veterinary World* 8, no. 12 (2015): 1458–65,
https://doi.org/10.14202/vetworld.2015.1458-1465.

352. Mohammad Jafar Golalipour et al., "Protective Effect of *Urtica
dioica* L. (Urticaceae) on Morphometric and Morphologic
Alterations of Seminiferous Tubules in STZ Diabetic Rats,"
Iranian Journal of Basic Medical Sciences 14, no. 5 (2011):
472–77, https://www.ncbi.nlm.nih.gov/pubmed/23493848.

353. Cyrus Jalili et al., "Protective Effect of *Urtica dioica* L. against Nicotine-Induced Damage on Sperm Parameters, Testosterone and Testis Tissue in Mice," *Iranian Journal of Reproductive Medicine* 12, no. 6 (2014): 401–8, https://www.ncbi.nlm.nih.gov/pubmed/25071848.

354. Hamid Reza Moradi et al., "The Histological and Histometrical Effects of *Urtica dioica* Extract on Rat's Prostate Hyperplasia," *Veterinary Research Forum: An International Quarterly Journal* 6, no. 1 (2015): 23–29, https://www.ncbi.nlm.nih.gov/pubmed/25992248.

355. Alireza Ghorbanibirgani et al., "The Efficacy of Stinging Nettle (*Urtica dioica*) in Patients with Benign Prostatic Hyperplasia: A Randomized Double-Blind Study in 100 Patients," *Iranian Red Crescent Medical Journal* 15, no. 1 (2013): 9–10, https://doi.org/10.5812/ircmj.2386.

356. Beatrix Roemheld-Hamm, "Chasteberry," *American Family Physician* 72, no. 5 (2005): 821–24, https://www.ncbi.nlm.nih.gov/pubmed/16156340; M. Diana van Die et al., "*Vitex agnus-castus* Extracts for Female Reproductive Disorders: A Systematic Review of Clinical Trials," *Planta Medica* 79, no. 7 (2013): 562–75, https://doi.org/10.1055/s-0032-1327831; Claudia Daniele et al., "*Vitex agnus-castus*: A Systematic Review of Adverse Events," *Drug Safety* 28, no. 4 (2005): 319–32, https://www.ncbi.nlm.nih.gov/pubmed/15783241.

— INDEX —

Page numbers in *italics* refer to figures and illustrations. Page numbers followed by *t* refer to tables.

— ABOUT THE AUTHOR —

Shelly Fry of Battle Ground

Dr. Jill Stansbury is a naturopathic physician with 30 years of clinical experience. She served as the chair of the Botanical Medicine Department of the National University of Natural Medicine in Portland, Oregon, for more than 20 years. She remains on the faculty, teaching herbal medicine and medicinal plant chemistry and leading ethnobotany field courses in the Amazon. Dr. Stansbury presents numerous original research papers each year and writes for health magazines and professional journals. She serves on scientific advisory boards for several medical organizations. She is the author of *Herbal Formularies for Health Professionals*, Volumes 1 and 2, and is the coauthor of *The PCOS Health and Nutrition Guide and Herbs for Health and Healing*. Dr. Stansbury lives in Battle Ground, Washington, and is the medical director of Battle Ground Healing Arts. She also runs an herbal apothecary, offering the best quality medicines from around the world, featuring many of her own custom tea formulas, blends, powders, and medicinal foods.

www.healingartsapothecary.org

HERBAL FORMULARIES FOR HEALTH PROFESSIONALS

This comprehensive five-volume set by Dr. Jill Stansbury serves as a practical and necessary reference manual for herbalists, physicians, nurses, and allied health professionals everywhere. This set is organized by body system, and each volume includes hundreds of formulas to treat common health conditions, as well as formulas that address specific energetic or symptomatic presentations.

VOLUME 1

DIGESTION AND ELIMINATION

INCLUDING THE GASTROINTESTINAL SYSTEM, LIVER AND GALLBLADDER, URINARY SYSTEM, AND THE SKIN

9781603587075
Hardcover • $59.95

VOLUME 2

CIRCULATION AND RESPIRATION

INCLUDING THE CARDIOVASCULAR, PERIPHERAL VASCULAR, PULMONARY, AND RESPIRATORY SYSTEMS

9781603587983
Hardcover • $44.95

VOLUME 4

NEUROLOGY, PSYCHIATRY, AND PAIN MANAGEMENT

INCLUDING COGNITIVE AND NEUROLOGIC CONDITIONS AND EMOTIONAL CONDITIONS

9781603588560
Hardcover
Available Spring 2020

VOLUME 5

IMMUNOLOGY, ORTHOPEDICS, AND OTOLARYNGOLOGY

INCLUDING ALLERGIES, THE IMMUNE SYSTEM, THE MUSCULOSKELETAL SYSTEM, AND THE EYES, EARS, NOSE, AND THROAT

9781603588577
Hardcover
Available Spring 2021

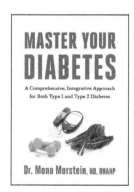